THE LIBRARY
ST. MARY'S COLLEGE OF MARYLAND
ST. MARY'S CITY, MARYLAND 20686

D1565478

Plant Dormancy:
Physiology, Biochemistry and Molecular Biology

Plant Dormancy:
Physiology, Biochemistry and Molecular Biology

Edited by

G.A. Lang
*Irrigated Agriculture Research and Extension Center
Washington State University
Prosser, WA
USA*

CAB INTERNATIONAL

CAB INTERNATIONAL
Wallingford
Oxon OX10 8DE
UK

Tel: +44 (0)1491 832111
Fax: +44 (0)1491 833508
E-mail: cabi@cabi.org

CAB INTERNATIONAL
198 Madison Avenue
New York, NY 10016-4314
USA

Tel: +1 212 726 6490
Fax: +1 212 686 7993
E-mail: cabi-nao@cabi.org

©CAB INTERNATIONAL 1996. All rights reserved. No part of this publication may be reproduced in any form or by any means, electronically, mechanically, by photocopying, recording or otherwise, without the prior permission of the copyright owners.

A catalogue record for this book is available from the British Library, London, UK

ISBN 0 85198 978 0

First printed 1996
Reprinted 1997

Printed and bound in the UK by Biddles Ltd., Guildford and King's Lynn

Contents

Contributors		ix
Preface		xix

Part I Seed Dormancy Systems and Concepts

1	Natural History of Seed Dormancy A. CARL LEOPOLD	3
2	Is Failure of Seeds to Germinate during Development a Dormancy-related Phenomenon? J. DEREK BEWLEY AND BRUCE DOWNIE	17
3	Control and Manipulation of Seed Dormancy ANWAR A. KHAN	29
4	A Physiological Comparison of Seed and Bud Dormancy FRANK G. DENNIS, JR	47

Part II Bud Dormancy Systems And Concepts

5	Dormancy and Symplasmic Networking at the Shoot Apical Meristem CHRIS VAN DER SCHOOT	59
6	A New Conceptual Approach to Bud Dormancy in Woody Plants JACQUES CRABBÉ AND PAUL BARNOLA	83
7	Development of Dormancy in Tissue-Cultured Lily Bulblets and Apple Shoots GEERT-JAN M. DE KLERK AND MEREL M. GERRITS	115
8	Dormancy in Tuberous Organs: Problems and Perspectives JEFFREY C. SUTTLE	133

Part III Physiology/Temperature, Light, Stress

9 A Physiological Comparison of Vernalization and
Dormancy Chilling Requirement   147
JAMES D. METZGER

10 Dormancy Breakage by Chilling: Phytochrome, Calcium
and Calmodulin   157
JAMES D. ROSS

11 Conifer Bud Dormancy and Stress Resistance: A
Forestry Perspective   171
FRANCINE J. BIGRAS

12 Early Development of Bud Dormancy in Conifer Seedlings   193
JOANNE E. MACDONALD

13 Near-Lethal Stress and Bud Dormancy in Woody Plants   201
MICHAEL WISNIEWSKI, LESLIE H. FUCHIGAMI, JÖRG J. SAUTER, ABBAS SHIRAZI AND LIPING ZHEN

Part IV Biochemistry

14 Structural Requirements of the ABA Molecule for
Maintenance of Dormancy in Excised Wheat Embryos   213
SUZANNE R. ABRAMS, PATRICIA A. ROSE AND M.K. WALKER-SIMMONS

15 Changes in Hormone Sensitivity in Relation to Onset
and Breaking of Sunflower Embryo Dormancy   221
MARIE-THÉRÈSE LE PAGE-DEGIVRY, JACQUELINE BIANCO, PHILIPPE BARTHE AND GINETTE GARELLO

16 Processes at the Plasma Membrane and Plasmalemma
ATPase during Dormancy   233
GILLES PÉTEL AND MICHEL GENDRAUD

17 Carbohydrate Metabolism as a Physiological Regulator
of Seed Dormancy   245
MICHAEL E. FOLEY

18 Chemical Mechanisms of Breaking Seed Dormancy   257
MARC ALAN COHN

Part V Molecular Biology

19 Molecular Analysis of Turion Formation in *Spirodela
polyrrhiza*: a Model System for Dormant Bud Induction   269
CHERYL C. SMART

20 Characterization of Genes Expressed When Dormant
Seeds of Cereals and Wild Grasses are Hydrated and
Remain Growth-Arrested   283
M.K. WALKER-SIMMONS AND PETER J. GOLDMARK

21	Analysis of cDNA Clones for Differentially Expressed Genes in Dormant and Non-dormant Wild Oat (*Avena fatua* L.) Embryos RUSSELL R. JOHNSON, HARWOOD J. CRANSTON, MARTHA E. CHAVERRA AND WILLIAM E. DYER	293
22	Photoperiod-Associated Gene Expression during Dormancy in Woody Perennials GARY D. COLEMAN AND TONY H.H. CHEN	301

Part VI Dormancy Modelling

23	Population-Based Models Describing Seed Dormancy Behaviour: Implications for Experimental Design and Interpretation KENT J. BRADFORD	313
24	An Integrating Model for Seed Dormancy Cycling: Characterization of Reversible Sensitivity HENK W.M. HILHORST, MARIA P.M. DERKX AND CEES M. KARSSEN	341
25	Modelling Climatic Regulation of Bud Dormancy SCHUYLER D. SEELEY	361

Index   377

Contributors

Name	Address	Chapter number	Chapter title
Abrams, Suzanne R.	Plant Biotechnology Institute National Research Council of Canada Saskatoon Saskatchewan S7N 0W9 Canada	14	Structural Requirements of the ABA Molecule for Maintenance of Dormancy in Excised Wheat Embryos
Barnola, Paul	University of Nancy I Biology of Woody Plants BP 239 F-54506 Vandoeuvre-lès-Nancy France	6	A New Conceptual Approach to Bud Dormancy in Woody Plants
Barthe, Philippe	Laboratoire de Physiologie Végétale Université de Nice – Sophia Antipolis 06108 Nice Cedex 2 France	15	Changes in Hormone Sensitivity in Relation to Onset and Breaking of Sunflower Embryo Dormancy
Bewley, J. Derek	Department of Botany University of Guelph Guelph Ontario N1G 2W1 Canada	2	Is Failure of Seeds to Germinate during Development a Dormancy-Related Phenomenon?

Name	Address	Chapter number	Chapter title
Bianco, Jacqueline	Laboratoire de Physiologie Végétale Université de Nice – Sophia Antipolis 06108 Nice Cedex 2 France	15	Changes in Hormone Sensitivity in Relation to Onset and Breaking of Sunflower Embryo Dormancy
Bigras, Francine J.	Natural Resources Canada Canadian Forest Service – Quebec Region 1055 rue du PEPS PO Box 3800 Sainte-Foy Quebec G1V 4C7 Canada	11	Conifer Bud Dormancy and Stress Resistance: A Forestry Perspective
Bradford, Kent J.	Department of Vegetable Crops University of California Davis CA 95616-8631 USA	23	Population-Based Models Describing Seed Dormancy Behaviour: Implications for Experimental Design and Interpretation
Chaverra, Martha E.	Department of Plant, Soil and Environmental Sciences Montana State University Bozeman MT 59717 USA	21	Analysis of cDNA Clones for Differentially Expressed Genes in Dormant and Non-dormant Wild Oat (*Avena fatua* L.) Embryos
Chen, Tony H. H.	Department of Horticulture Oregon State University Corvallis OR 97331 USA	22	Photoperiod-Associated Gene Expression during Dormancy in Woody Perennials
Cohn, Marc Alan	Department of Plant Pathology and Crop Physiology Louisiana State University Agricultural Center Baton Rouge LA 70803 USA	18	Chemical Mechanisms of Breaking Seed Dormancy

Name	Address	Chapter number	Chapter title
Coleman, Gary D.	Department of Horticulture University of Maryland College Park MD 20742-5611 USA	22	Photoperiod-Associated Gene Expression during Dormancy in Woody Perennials
Crabbé, Jacques	Faculty of Agronomical Sciences Applied Plant Morphogenesis B-5030 Gembloux Belgium	6	A New Conceptual Approach to Bud Dormancy in Woody Plants
Cranston, Harwood J.	Department of Plant, Soil and Environmental Sciences Montana State University Bozeman MT 59717 USA	21	Analysis of cDNA Clones for Differentially Expressed Genes in Dormant and Non-dormant Wild Oat (*Avena fatua* L.) Embryos
De Klerk, Geert-Jan M.	Centre for Plant Tissue Culture Research PO Box 85 2160 AB Lisse The Netherlands	7	Development of Dormancy in Tissue-Cultured Lily Bulblets and Apple Shoots
Dennis, Frank G., Jr	Department of Horticulture Michigan State University East Lansing MI 48824-1325 USA	4	A Physiological Comparison of Seed and Bud Dormancy
Derkx, Maria P. M.	Research Institute for Nursery Stock Rijneveld 153 2770 AC Boskoop The Netherlands	24	An Integrating Model for Seed Dormancy Cycling: Characterization of Reversible Sensitivity
Downie, Bruce	Department of Botany University of Guelph Guelph Ontario N1G 2W1 Canada	2	Is Failure of Seeds to Germinate during Development a Dormancy-Related Phenomenon?

Name	Address	Chapter number	Chapter title
Dyer, William E.	Department of Plant, Soil and Environmental Sciences Montana State University Bozeman MT 59717 USA	21	Analysis of cDNA Clones for Differentially Expressed Genes in Dormant and Non-dormant Wild Oat (*Avena fatua* L.) Embryos
Foley, Michael E.	Purdue University Department of Botany and Plant Pathology West Lafayette IN 47907-1155 USA	17	Carbohydrate Metabolism as a Physiological Regulator of Seed Dormancy
Fuchigami, Leslie H.	Department of Horticulture Oregon State University Corvallis OR 97331 USA	13	Near-Lethal Stress and Bud Dormancy in Woody Plants
Garello, Ginette	Laboratoire de Physiologie Végétale Université de Nice – Sophia Antipolis 06108 Nice Cedex 2 France	15	Changes in Hormone Sensitivity in Relation to Onset and Breaking of Sunflower Embryo Dormancy
Gendraud, Michel	Unité Associée Bioclimatologie – PIAF (INRA – Université Blaise Pascal) 4 rue Ledru 63038 Clermont-Ferrand Cedex 01 France	16	Processes at the Plasma Membrane and Plasmalemma ATPase during Dormancy
Gerrits, Merel M.	Centre for Plant Tissue Culture Research PO Box 85 2160 AB Lisse The Netherlands	7	Development of Dormancy in Tissue-Cultured Lily Bulblets and Apple Shoots

Name	Address	Chapter number	Chapter title
Goldmark, Peter J.	DJR Research Star Route 69 Okanogan WA 98840 USA	20	Characterization of Genes Expressed When Dormant Seeds of Cereals and Wild Grasses are Hydrated and Remain Growth-Arrested
Hilhorst, Henk W. M.	Department of Plant Physiology Wageningen Agricultural University Arboretumlaan 4 NL-6703 BD Wageningen The Netherlands	24	An Integrating Model for Seed Dormancy Cycling: Characterization of Reversible Sensitivity
Johnson, Russell R.	Department of Plant, Soil and Environmental Sciences Montana State University Bozeman MT 59717 USA	21	Analysis of cDNA Clones for Differentially Expressed Genes in Dormant and Non-dormant Wild Oat (*Avena fatua* L.) Embryos
Karssen, Cees M.	Department of Plant Physiology Wageningen Agricultural University Arboretumlaan 4 NL-6703 BD Wageningen The Netherlands	24	An Integrating Model for Seed Dormancy Cycling: Characterization of Reversible Sensitivity
Khan, Anwar A.	Department of Horticultural Sciences New York State Agricultural Experiment Station Cornell University Geneva NY 14456 USA	3	Control and Manipulation of Seed Dormancy

Name	Address	Chapter number	Chapter title
Lang, Gregory A.	Irrigated Agriculture Research and Extension Center Washington State University 24106 N. Bunn Rd Prosser WA 99350-9687 USA	(Editor)	
Leopold, A. Carl	Boyce Thompson Institute for Plant Research Ithaca NY 14853 USA	1	Natural History of Seed Dormancy
Le Page-Degivry, Marie-Thérèse	Laboratoire de Physiologie Végétale Université de Nice – Sophia Antipolis 06108 Nice Cedex 2 France	15	Changes in Hormone Sensitivity in Relation to Onset and Breaking of Sunflower Embryo Dormancy
MacDonald, Joanne E.	Natural Resources Canada Canadian Forest Service PO Box 6028 St John's Newfoundland A1C 5X8 Canada	12	Early Development of Bud Dormancy in Conifer Seedlings
Metzger, James D.	Department of Horticulture and Crop Science 2001 Fyffe Court Ohio State University Columbus OH 43210-1096 USA	9	A Physiological Comparison of Vernalization and Dormancy Chilling Requirement

Name	Address	Chapter number	Chapter title
Pétel, Gilles	Unité Associée Bioclimatologie - PIAF (INRA - Université Blaise Pascal) 4 rue Ledru 63038 Clermont-Ferrand Cedex 01 France	16	Processes at the Plasma Membrane and Plasmalemma ATPase during Dormancy
Rose, Patricia A.	Plant Biotechnology Institute National Research Council of Canada Saskatoon Saskatchewan S7N 0W9 Canada	14	Structural Requirements of the ABA Molecule for Maintenance of Dormancy in Excised Wheat Embryos
Ross, James D.	Department of Botany University of Reading Whiteknights Reading Berkshire RG6 2AS United Kingdom	10	Dormancy Breakage by Chilling: Phytochrome, Calcium and Calmodulin
Sauter, Jörg J.	Botanical Institut Christian Albrecht Universität Kiel Germany	13	Near-Lethal Stress and Bud Dormancy in Woody Plants
Seeley, Schuyler D.	Plants, Soils and Biometeorology Utah State University Logan UT 84322-4820 USA	25	Modelling Climatic Regulation of Bud Dormancy
Shirazi, Abbas	Department of Horticulture Oregon State University Corvallis OR 97331 USA	13	Near-Lethal Stress and Bud Dormancy in Woody Plants

Name	Address	Chapter number	Chapter title
Smart, Cheryl C.	Institute of Plant Sciences Swiss Federal Institute of Technology (ETH) Zürich Universitätstrasse 2 CH-8092 Zürich Switzerland	19	Molecular Analysis of Turion Formation in *Spirodela polyrrhiza* – a Model System for Dormant Bud Induction
Suttle, Jeffrey C.	US Department of Agriculture Northern Crop Science Laboratory PO Box 5677 State University Station Fargo ND 58105-5677 USA	8	Dormancy in Tuberous Organs: Problems and Perspectives
van der Schoot, Chris	Agrotechnological Research Institute (ATO-DLO) Bornsesteeg 59 PO Box 17 6700 AA Wageningen The Netherlands	5	Dormancy and Symplasmic Networking at the Shoot Apical Meristem
Walker-Simmons, M. K.	United States Department of Agriculture Agricultural Research Service Washington State University Pullman WA 99164-6420 USA	14	Structural Requirements of the ABA Molecule for Maintenance of Dormancy in Excised Wheat Embryos
		20	Characterization of Genes Expressed When Dormant Seeds of Cereals and Wild Grasses are Hydrated and Remain Growth-Arrested

Name	Address	Chapter number	Chapter title
Wisniewski, Michael	USDA-ARS 45 Wiltshire Road Kearneysville WV 25430 USA	13	Near-Lethal Stress and Bud Dormancy in Woody Plants
Zhen, Liping	Department of Horticulture Oregon State University Corvallis OR 97331 USA	13	Near-Lethal Stress and Bud Dormancy in Woody Plants

Preface

Dormancy is the term used generically to encompass the processes that constitute a programmed inability for growth in various types of plant meristematic apices, often in spite of suitable environmental conditions. Dormancy phenomena are manifested in seeds, tubers, bulbs, and aerial vegetative and reproductive buds, and they occur during significant portions of the annual life cycle. These processes are highly evolved to confer various advantages to the plant, from developmental synchronization to orchestration of plant architecture to survival of environmental stresses. The mechanisms that regulate these processes are diverse and specific, yet their expression can be dynamic and, quite possibly, interrelated. The quest for a better understanding of plant dormancy is not merely academic, but rather of great consequence to sustainable and efficient procurement of food and fibre for nutritional and economic benefits.

There is much to be discovered regarding the genetic regulation of dormancy, as well as its physiological and biochemical manifestations and interactions. Comprehensive comparisons across the plant kingdom of types of dormancy, experimental systems for studying dormancy, mechanistic theories for dormancy regulation and various approaches to dormancy research have been scarce or non-existent. The *1st International Symposium on Plant Dormancy* (Corvallis, Oregon, USA) was conceived to *begin* addressing these issues and to initiate a dynamic forum for idea exchange, paradigm revision, and the advancement of knowledge. The *Symposium* was designed to both concentrate critical thought on specific problems in dormancy and foster the extension of individual scientists' research creativity through interdisciplinary interactions with specialists from other areas of dormancy. This book comprises selected topics from the more than 100 presentations at the *Symposium*, which was attended by scientists from 22 countries.

The authors of this book's chapters have attempted to probe the cutting edges of current plant dormancy research, as well as current limitations and what the future of dormancy research may hold. A diversity of dormancy-exhibiting plant systems are discussed from evolutionary, theoretical and mechanistic viewpoints in Parts I and II. Physiological processes associated with the apparent regulation of dormancy by temperature, photoperiod and/or stress are addressed in Part III. Experiments to study the biochemistry and molecular biology associated with dormancy are often difficult to design and interpret, since dormancy is a complex and dynamic state that may involve important biophysical changes in concert with those chemical processes that are tied to gene expression; several different biochemical and molecular approaches are presented in Parts IV and V. Finally, when partial, but insufficient, knowledge of a phenomenon exists, models are often valuable tools for attempting to explain gaps in our understanding and to provide frameworks for further experimentation; several authors present such models for consideration in Part VI. The book should be considered by no means comprehensive, as the inclusion of numerous other stimulating papers from the *Symposium* (let alone all divergent viewpoints and biological systems in current dormancy research) would have required several volumes. However, it is hoped that this volume will provide a glimpse of the technical achievements, challenging interpretations, formidable research barriers and optimistic insights that characterized the *1st International Symposium*, presenting a multidisciplinary foundation against which our future progress in plant dormancy research can be measured.

I would like to thank both the individual chapter authors and those *Symposium* participants whose works could not be accommodated herein, for their dedication to the experimental pursuit and eventual elucidation of a difficult and challenging subject area. I also wish to express my gratitude to my official co-organizers of the *Symposium*, Dr Anwar Khan, Dr Tony Chen, and Dr Rebecca Darnell for their suggestions, reviews and various organizational tasks, as well as to my unofficial co-organizer, Dr Suzanne Lang, who, in addition to contributing the above, also provided longstanding patience with this entire process. Finally, this book could not be complete without my appreciation to CAB INTERNATIONAL's Tim Hardwick for his initial consultations and Amanda Horsfall for her patience, persistence and good cheer in bringing this beast to see the light of day.

Gregory A. Lang
Prosser, Washington, USA

I Seed Dormancy Systems and Concepts

1 Natural History of Seed Dormancy

A. Carl Leopold
Boyce Thompson Institute for Plant Research, Ithaca, NY 14853, USA

Introduction

On entering into a discussion of dormancy, two aspects of contemporary plant science are particularly worrisome. First, the focus on reductionism has the potential to cause science to drift inexorably away from the real biology of living organisms, increasingly separating us from the natural history of our science. This drift has been abetted by the sequence of research fads which tend to explain all plant functions on the basis of individual phenomena. Such fads have included hormones, then nucleic acids and now molecular biology. While the importance of these reductionist sectors in understanding plant biology cannot be denied, skills gained in biochemical and molecular analysis should not be allowed to gradually insulate scientists from the beauty and elegance of living plants and their splendid adaptations to environmental events.

Second, the present focus on the regulation of dormancy by perception mechanisms does not, in any case, seem to be providing an understanding of the dramatic manner in which growth can be abruptly turned on or turned off. Even if the precise chemistry or physics of perception may be revealed, whether it be through abscisic acid, ethylene, phytochrome or another regulatory system, an understanding of the mechanism of the dormant state in which growth can be summarily and completely arrested and then activated, does not necessarily follow.

Consequently, this chapter will eschew any discussion of perception mechanisms, focusing instead on some aspects of the natural history of dormancy. Examples of seed dormancy will be used to underscore this approach and illuminate the remarkable, and sometimes astonishing, orchestrations of developmental processes that plants have evolved in adapting to their unique environments.

Why should dormancy be an important implement for adaptive regulation? There are four general conditions under which dormancy is an important regulator utilized by plants.

1. Seasonal synchrony. Dormancy benefits the survival of a species or taxon by synchronizing growth to favourable seasons. It is analogous to diapause in insects.

2. Widening the range for germination. Dormancy provides options that widen the range of circumstances or locations for germination, thus improving the potential for plant survival.

3. Utilizing erratic opportunities. Dormancy adapts the germination response to specific erratic or occasional signals.

4. Exploiting other organisms. Dormancy facilitates seed dispersal through the assistance of animals.

It is quite possible that, through evolution, dormancy has become not only the most dramatic, but also the most adaptable, regulatory function in plant development and survival. The ability to suspend all growth, even when water and other media for growth are present, is not only a spectacular type of control, but also an impressive mode of adaptation.

Seasonal Synchrony

By restricting the onset of germination or growth to a given time, dormancy can nicely synchronize flowering. This is often the case for bud dormancy in flowering perennials. Synchrony of flowering is also often obtained for the seeds of annual species in which flowering is not regulated by photo- or thermoperiodism. Synchrony of germination can produce synchrony of flowering.

One of the most direct linkages to seasonal change that a seed or bud may have is through the perception of temperature. To illustrate, the dormancy of rice is relieved by elevated temperatures, and the higher the temperature, the more rapidly the seeds emerge from dormancy (Fig. 1.1) (Roberts, 1965). In this way, seed germination can be adjusted to the onset of the growing season.

A remarkable variation of the temperature device is the requirement for alternating temperatures to break dormancy. For example, constant temperatures as high as 35°C are ineffective in bringing about *Bermuda* grass seed germination, whereas even a 10°C alternation between day and night results in high levels of germination (Fig. 1.2) (Harrington, 1923). We can infer that temperature alternation would be a more dependable indicator of the onset of the growing season than would be any single temperature.

The more complex ability of some seeds to cycle dormancy off and on makes it possible to synchronize germination with appropriate seasons and to avoid inappropriate seasons. An example is the removal and reinduction of

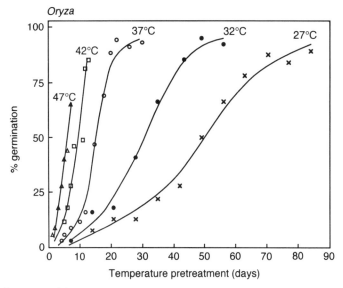

Fig. 1.1. Dormancy of rice seeds can be broken by elevated temperatures. Note that the treatments are additive, and the response reflects the extent of warming – as would be expected to relate to the proximity of the onset of summer. (From Roberts, 1965.)

dormancy in seeds of some annuals, such as *Arabidopsis* and *Ambrosia* (Baskin and Baskin, 1985). In these cases, temperatures associated with the favourable season can stimulate the breaking of dormancy, and unfavourable temperatures can reinduce dormancy (Fig. 1.3). As further ornamentation of the temperature/seasonal effect, a species such as *Ambrosia*, which favours germination in the spring, has an on/off pattern to seasonal periodicity that contrasts sharply with *Arabidopsis*, which favours germination in the late summer.

One may presume this sort of seasonal synchrony of dormancy to be due to a sensing of seasonal temperatures (Kummerov, 1965). There are instances, however, of other species that go through annual cycles even when stored in the laboratory at relatively uniform temperatures, as in the case of a desert *Mesembryanthemum* species (Gutterman, 1980). These seeds germinate in winter months, with this seasonal cycling continuing in stored seeds for several years. Temperature treatments did not alter the seasonal pattern.

Even more complex mechanisms must be involved in species for which seed dormancy is regulated by photoperiodic phenomena (Evenari, 1965). For example, *Begonia* seeds will germinate only at photoperiods longer than about 12 h; this would synchronize germination with summer month day lengths (Fig. 1.4) (Nagao *et al.*, 1959). To illustrate that *Begonia* seeds are actually under the classic photoperiodic control, experimental application of

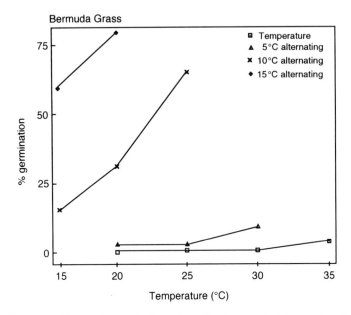

Fig. 1.2. Dormancy of Bermuda grass is almost insensitive to a constant temperature (lowest curve) but recognizes alternating temperatures such as would reflect the onset of the growing season. (From Harrington, 1923.)

a brief light interruption in the middle of a 14 h night has been shown to induce germination just as if it were a long photoperiod (Fig. 1.4). For some species, then, the entire phytochrome system must be present in the seed, including the clock mechanism, for tracking day and night reactions.

There are other systems for seasonal synchronization, but the ones illustrated here can at least outline the truly remarkable range of physiological cues that seeds utilize to adapt plant growth to seasonal changes. Some of these cues lead to the inference that some seeds have evolved the ability to serve as quantitative sensors of temperature, and others the ability to identify the more complicated phenomena of temperature cycling. It is astonishing that simple seeds can have the complete capability for photoperiodic timekeeping of circadian cycles and, on a greater scale, the capability even for time-keeping of annual calendar cycles.

Widening the Range for Survival

In many species of annual weeds, a dispersion of the time of germination can be beneficial. Spreading germination over time can often improve the chances for success of establishment. There are several means by which dormancy is employed to spread out the time of germination. In some weed species, dimorphic seeds are produced – some being non-dormant and others being

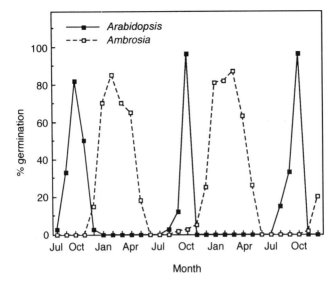

Fig. 1.3. Some seeds shift into and out of dormancy by the seasons. Here *Arabidopsis* seeds become non-dormant in the summer, whereas ragweed becomes non-dormant in the winter. Dormancy is reinstated after those events, thus constituting an annual cycling of dormancy. (From Baskin and Baskin, 1985.)

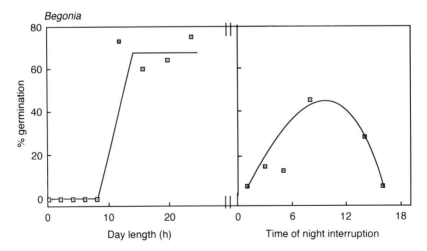

Fig. 1.4. Seeds of *Begonia* respond to photoperiod. Dormancy is relieved by long days (over 12 h), but in a short-day regime (7 h) a 30 min night interruption will produce the long-day response in the classical manner of photoperiodism. (From Nagao *et al.*, 1959.)

dormant. This dormancy dimorphism is especially common among *Compositae* and *Chenopodiaceae*. A grass (*Avena fatua*) and a composite (*Astericus pygmaeus*) illustrate a sequential level of dormancy. Individual seeds along a given fruiting axis develop with progressively increasing dormancy. The first seed in the panicle is the first seed to lose dormancy; after it is shed, then the next proximal seed becomes non-dormant, and so on (Koller, 1969), providing a linear array of germination times.

An enormous spread of germination times is obtained by some seeds through the intervention of hard seed-coats. Hard seeds can defer germination for very long periods of time. The longest proved survival of germinability for any seed at moderate temperatures has been reported for the hard-seeded *Lotus nelumbo*, which has retained germinability for over 400 years (Priestley and Posthumus, 1982). Seeds frozen in the Arctic permafrost may survive for millennia (Porsild *et al.*, 1967).

A simple device for spreading germination time is the gradual loss of dormancy over a period of months or years (Fig. 1.5) (Olatoye and Hall, 1973). The gradual erosion of dormancy may be considered a fail-safe

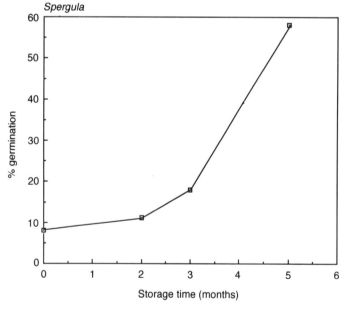

Fig. 1.5. Dormancy of pigweed seeds dissipates gradually over a period of months. (From Olatoye and Hall, 1973.)

mechanism, which serves to guarantee ultimate germination, even if none of the seasonal or other dormancy-breaking signals is received.

Each of these devices serves to increase the potential for survival by spreading the time vector for germination.

Utilizing Erratic Opportunities

The soil in a typical cultivated field is permeated with truly large numbers of dormant seeds. This 'seed bank' may involve numbers running between 20,000 and 100,000 seeds per square metre of soil, each waiting for some signal to initiate germination (Harper, 1977). The very act of tilling the soil, usually done to destroy weed seedlings, will stimulate many of the buried weed seeds to germinate. There are at least two components leading to this stimulation: in some instances, the seeds contacting the cultivator or plough experience damage to the structure of the hilum or strophiole of the hard seed-coat, such that hydration can proceed and dormancy is broken (Baskin and Baskin, 1989). In other cases, the seeds may be stimulated to germinate by exposure to light as the soil is turned. The involvement of light is illustrated by the data in Fig. 1.6, where disc harrowing in the light led to a substantially greater germination of seedlings than harrowing in the dark (Scopel et al., 1994).

The keen sensitivity of some seeds to light is illustrated in Fig. 1.7, where dormant *Arabidopsis* seeds are shown to achieve 50% germination upon exposure to only 2.5 μmol m^{-2} of light (Cone and Spruit, 1983), which is in the intensity range of bright moonlight. The requirement for light is responsible in part for the suspended germination of seeds which are situated too deep in the soil. Experiments by Wesson and Wareing (1969) showed that sunlight on a field soil was sufficient to stimulate the germination of weed seeds several centimetres deep in the soil. Exclusion of light in control plots entirely prevented the depth-sensitive regulation of germination (Fig. 1.8). A part of this germination gradient into the ground was also attributed to limitation of gas exchange at the deeper locations.

Many aggressive (weedy) species are adapted to invade terrain following a fire. And, of course, fire has been adopted as a dormancy-breaking signal in some of these species. This type of dormancy-breaking signal is common to chaparral shrubs, *Eucalyptus* trees and, especially, the serotinous pines. In the latter case, the serotinous cones may remain sealed for many years until there is an occurrence of fire. The heat of the fire melts the resin which seals the cone, and the seeds are released. In *Pinus banksiana*, the sealing resin is melted at 50°C, and the seeds can survive extraordinary temperatures, as high as 150°C – a remarkable feat for any living system (Beaufait, 1960). In *Rhus laurina* and several other shrubs, seed dormancy is broken by the heat of fire (Wright, 1931). Quite separate from the dormancy regulation by the heat of

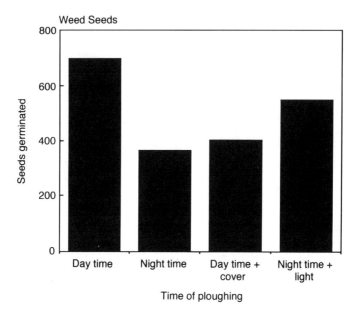

Fig. 1.6. Disc harrowing a field in the daytime causes more weed seed germination than does harrowing in the dark. A light above the harrow at night can result in germination similar to that after harrowing in the daytime. (From Scopel *et al.*, 1994.)

fire, seeds of some species can break dormancy upon contact with smoke, or even with charred wood or extracts of charred lignin (Keeley and Pizzorno, 1986).

The seed-coat may regulate germination in many ways. First, dormancy may be imposed by a hard seed-coat, most particularly among the *Leguminoseae* (Justice and Bass, 1978). The barrier is most often a limitation of water entry. In other cases, it regulates dormancy as a barrier to gas exchange (Shull, 1913). In many instances, it contains inhibitors which can keep the seed in the dormant condition. In any case, the coat-imposed dormancy is generally relieved by physical breakage. In addition to the example of the dormancy-breaking in response to tillage, seed dormancy may be broken in prairie soils as a response to animal tramping. In arid situations it may respond to erosion by wind-blown sand or abrasion by moving gravel (Harper, 1977). The relief of dormancy by disruption of the seed-coat barrier probably reflects damage to the hilum or strophiole (Baskin and Baskin, 1989) and is most common among weed species that thrive on disturbed soils.

Another function of the seed-coat involves the presence of inhibitors that may serve as precise measures of rainfall (Went, 1949). For example, inhibitors in the seeds of some desert species, such as *Euphorbia* and *Pectis* suspend them in the dormant state. Occurrence of sufficient rainfall to elute

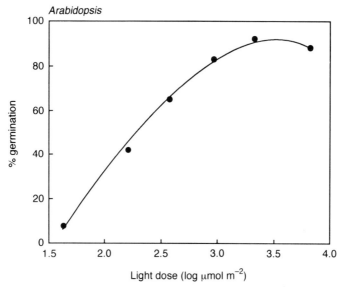

Fig. 1.7. The remarkable sensitivity of some seeds to light is illustrated by *Arabidopsis*, in which dormancy release is evident even at 2 μmol m^{-2}. (From Cone and Spruit, 1983.)

the inhibitors from the seed-coat relieves the dormancy. The result is the well-known flush of desert annuals flowering after rains. A parallel situation occurs in another desert annual, *Baeria*, which is held in dormancy by the high salt concentrations in its seed-coat and the surrounding soil (Koller, 1969); when rainfall elutes the salt, lowering the osmotic concentration, germination of the seeds is permitted. These inhibitor-mediated dormancy systems can each serve to bring on germination just at the start of a rainy spell.

Many species of orchids utilize commensal fungi as regulators of their germination. The seeds are extremely small, with undifferentiated embryos. They are thus limited to very small amounts of substrates. Certain fungi growing near the seed will provide the orchid seed with sugars and nitrogenous substrates, which then permit germination of the seed to proceed (Lang, 1965). In this case, germination is allowed only when there is an appropriate commensal fungus present.

Among parasitic angiosperms, it is common for the seeds to lie dormant in the soil until a root of an appropriate host grows in the vicinity. *Striga*, a parasite of sorghum and maize, and *Orobanche*, a parasite of legumes and composites, are prominent members of this class. The dormancy of their seeds is relieved in response to the diffusion of some solutes from the roots of an appropriate host plant. To promote germination, the optimal distance for diffusion of the solutes from the host plant is within 6 mm; in some instances the dormancy may be broken at a distance as far as 16 mm (Brown, 1965). The

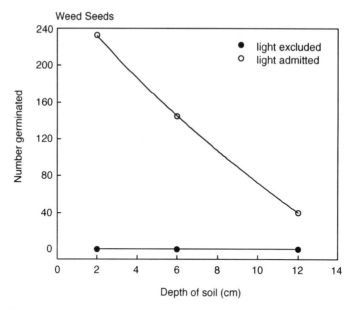

Fig. 1.8. Weed seeds are sensitive to the depth of their location in the soil, and they will remain dormant at greater depths. This sensing of soil depth is partly a function of the light sensitivity of the seeds. (From Wesson and Wareing, 1967.)

solutes responsible for the seed dormancy response probably include a cytokinin, thiourea and a keto-sugar. This dormancy control serves as a search-and-find mechanism by which a parasite locates its host plant; it is another elegant adaptation for efficient recognition of optimal opportunities for germination.

In these various means by which seeds take advantage of erratic events to coordinate germination with appropriate occasions, we can see a truly diverse range of seed capabilities – including capabilities for extremely fine light sensitivity, to withstand (and respond to) fire, to sense disturbances to the soil, to measure fine differences in rainfall or to detect tiny amounts of chemicals distinctive to a host plant species.

Exploiting Other Organisms

The dispersal of seeds can be a major contribution to the spread of a species. Seed dispersal is often linked to dormancy as a component of an exploitation of animals. The seeds of spermatophytes are characteristically endowed with a supply of food reserves. Not only do these accumulated substrates serve to sustain the germinating seedling, but also they often appeal to various animals as food, thereby making possible an exploitation of animals in achieving seed mobility. In addition to the foodstuffs in the seeds, many angiosperms appeal to animals through foodstuffs in the fruit. Hard seeds embedded in the edible fruit are commonly adapted to passage through the animal gut without damage.

Of course, there may be a substantial mortality of seeds in passing through the animal gut. For example, seeds of the common grains pass through the guts of cows with greater frequency of viability than through horses (Murray, 1986). Seeds in which the seed-coat has been penetrated by an insect are usually digested. And, most relevant for our discussion, seeds which attempt to germinate while within the gut are of course victims of digestion. But in many cases some seeds pass through the digestive tract, retaining their germinability and shedding their dormancy.

Carriers of seeds include the songbirds. The rows of cherry seedlings or cedar seedlings along fence lines are common consequences of the proclivity of songbirds to perch on fences. Cattle are frequent seed carriers; in semiarid regions, cattle browse mesquite plants, and transmit at least some germinable seeds through their faeces. Reptiles such as lizards consume cactus fruits, and then pass the seeds in a non-dormant condition. Elephants eat the fruits of *Acacia*, and one can often see *Acacia* seedlings germinating in elephant dung (Murray, 1986). Monkeys are great carriers of dormant seeds; it is estimated that the larger monkeys may distribute as many as 45 species of seeds through their faeces (Kuroda, 1993).

The *New York Times* (July 1994) reported evidence that certain tropical solanaceous fruits contain a laxative. Observations of birds that ate the fruits indicated that the presence of the laxative caused seed passage through the bird's gut to occur in half the time required in the absence of the laxative. We can presume an advantage to the seed in terms of a greater proportion of survivors associated with a briefer time period in the bird's gut.

An impressive level of specific seed/animal adaptation is the case of a wild *Lycopersicon* which grows in the Galapagos Islands and is eaten by the Galapagos tortoise. The tortoise deposits non-dormant seeds in its faeces (Rick and Bowman, 1961). All efforts to break dormancy by chemical treatment or by passage through other animals were found to be ineffective. Only the gut of the Galapagos tortoise was effective in breaking this seed dormancy.

A dramatic case of specific adaptation of dormancy to animal consumption is the aardvark. This fructivore excavates the underground fruits of a *Cucumis* and eats the fruit, including the seeds. The animal subsequently digs a hole for defecation, serving thus not only to provide mobility but even to plant the now non-dormant seed (van der Pijl, 1969).

And, finally, the tiny spores of *Isoetes* are eaten by earthworms. Whereas the earthworms provide for only a modest mobility for the spores that they have eaten, the mobility is greatly amplified when the worms are eaten by thrushes, which subsequently pass the seeds in their faeces (van der Pijl, 1969).

The nature of the changes in the seed that lead to the loss of dormancy as it passes through the gut of a bird or animal are not known - whether dormancy is broken by the pH, by regulator chemicals or by enzymic or abrasive actions.

Conclusion

Seeds are usually characterized in the biological literature as little bundles of poorly differentiated cells, surrounded by a scaly seed-coat - a sort of inert bag of deoxyribonucleic acid (DNA) awaiting the chance to become something else. Knowing of the capability for regulated dormancy, however, we can perceive seeds as having quite extraordinary, dynamic abilities. Some can switch either into or out of dormancy, and sometimes do that through several cycles. Some can show impressive thermosensing, as well as photosensing and hydrosensing. Not only can these seeds detect and respond to these thermal, photic and light signals, they can integrate their responses into cyclic events, through a memory effect. Such elaborate timing schemes range from diurnal to even annual cycles. Some seeds have the ability to carry out all of the complex integrations of photoperiodism, including the complex time-measuring capabilities. Other seeds can measure rainfall; some can recognize the opportunities that will occur following cultivation, disturbance of the soil or the advent of fire. In the case of some parasitic plants, chemosensing is used to establish whether their host species is growing nearby. Other seeds are adapted to exploit animals to provide mobility, utilizing their passage through the intestinal tract as a mode of transport and dormancy release.

Do we dare to think of seeds as little bags of DNA in poorly differentiated cells? Or is the structural simplicity of seeds misleading, tending to conceal their remarkable capabilities as sophisticated sensors of the environment, as clever exploiters of other organisms, as precise measurers of timing for events in the world above them, producing predictive information on probable changes in the environment and facilitating adaptation to these changes through the dramatic ability for regulation of dormancy?

References

Baskin, J.M. and Baskin, C.C. (1985) The annual dormancy cycle in buried weed seeds. *BioScience* 35, 492-498.

Baskin, J.M. and Baskin, C.C. (1989) Physiology of dormancy and germination in relation to seed bank ecology. In: Leck, M.A., Parker, V.T. and Simpson, R.L. (eds) *Ecology of Seed Banks*. Academic Press, New York, pp. 53-69.

Beaufait, W.R. (1960) Some effects of high temperature on the cones and seeds of jack pine. *Forest Science* 6, 194-199.
Brown, R. (1965) The germination of angiospermous parasite-seeds. *Encyclopedia of Plant Physiology* 15, 925-932.
Cone, J.W. and Spruit, C.J.P. (1983) Imbibition conditions and seed dormancy of *Arabidopsis thaliana*. *Physiologia Plantarum* 59, 416-420.
Evenari, M. (1965) Light and seed dormancy. *Encyclopedia of Plant Physiology* 15, 804-847.
Gutterman, Y. (1980) Annual rhythm and position effect in the germinability of *Mesembryanthemum nodiflorum*. *Israel Journal of Botany* 29, 93-97.
Harper, J.L. (1977) *Population Biology of Plants*. Academic Press, New York.
Harrington, G.T. (1923) Use of alternating temperatures in the germination of seeds. *Journal of Agricultural Research* 23, 295-332.
Justice, O.L. and Bass, L.N. (1978) *Principles and Practices of Seed Storage*. USDA Agriculture Handbook No. 506, Washington.
Keeley, S.C. and Pizzorno, M. (1986) Charred wood stimulated germination of California chaparral. *American Journal of Botany* 73, 1289-1297.
Koller, D. (1969) The physiology of dormancy and survival of desert plants. In: Woolhouse, H.W. (ed.) *Dormancy and Survival*. Society for Experimental Biologists Symposium No. 23, Company of Biologists, Cambridge, pp. 449-469.
Kummerov, J. (1965) Endogen-rhytmische Schwanken der Keimfahigkeit von Samen. *Encyclopedia of Plant Physiology* 15, 721-726.
Kuroda, S. (1993) Creation of reserves in tropical forests and researchers. In: Kawanabe, H., Ohgushi, T. and Higashi, M. (eds) *Symbiosphere: Ecological Complexity*. International Union of Biological Scientists Special Issue No. 29, p. 53-58.
Lang, A. (1965) Effects of some internal and external conditions on seed germination. *Encyclopedia of Plant Physiology* 15, 848-893.
Murray, D.R. (1986) *Seed Dispersal*. Academic Press, New York.
Nagao, M., Esashi, Y., Tanaka, T., Kumagai, T. and Fukumoto, S. (1959) Effects of photoperiod and gibberellin on germination of seeds of *Begonia*. *Plant and Cell Physiology* 1, 39-48.
Olatoye, S.T. and Hall, M.A. (1973) Interactions of ethylene and light on dormant weed seeds. In: Heydecker, M.A. (ed.) *Seed Ecology*. Pennsylvania State University, University Park, pp. 223-249.
Porsild, A.E., Harrington, C.R. and Mulligan, G.A. (1967) *Lupinus arcticus* grown from seeds of Pleistocene age. *Science* 158, 113-114.
Priestley, D.A. and Posthumus, M.A. (1982) Extreme longevity of lotus seeds from Pulatien. *Nature* 299, 148-149.
Rick, C.M. and Bowman, R.I. (1961) Galapagos tomatoes and tortoises. *Evolution* 15, 407-417.
Roberts, E.H. (1965) Dormancy in rice seed: IV, varietal responses to temperatures. *Journal of Experimental Botany* 16, 341-349.
Scopel, A.L., Ballare, C.L. and Radosevich, S.R. (1994) Photostimulation of seed germination during tillage. *New Phytologist* 126, 145-152.
Shull, C.A. (1913) Semipermeability of seed coats. *Botanical Gazette* 56, 169-199.
van der Pijl, L. (1969) *Principles of Seed Dispersal*. Springer Verlag, Berlin.
Went, F.W. (1949) Ecology of desert plants II. *Ecology* 30, 1-13.

Wesson, G. and Wareing, P.F. (1969) The induction of light sensitivity in weed seeds by burial. *Journal of Experimental Botany* 20, 413-425.

Wright, E. (1931) The effect of high temperatures on seed germination. *Journal of Forestry* 29, 679-687.

2 Is Failure of Seeds to Germinate during Development a Dormancy-related Phenomenon?

J. DEREK BEWLEY AND BRUCE DOWNIE
Department of Botany, University of Guelph, Guelph, Ontario N1G 2W1, Canada

Introduction

Germination of developing seeds on the parent plant (i.e. viviparous or precocious germination) is a relatively rare phenomenon, and its incidence is frequently associated with deficiences in abscisic acid (ABA) synthesis or sensitivity (Fong *et al.*, 1983; McCarty and Carson, 1991). Sometimes, removal of the developing seed from its surrounding fruit tissues is sufficient to permit germination, e.g. seeds within fleshy fruits (Welbaum and Bradford, 1988; Berry and Bewley, 1992) or non-fleshy fruits (Fischer *et al.*, 1988), but more commonly germination of the seed cannot occur until near-maturity (i.e. after the onset of maturation drying). This raises the obvious questions: how, during development, is germination of the seed prevented? How are the constraints to germination eventually overcome to permit its occurrence?

There are analogies between the failure of developing seeds to germinate and the phenomenon of dormancy. For example, when an isolated developing seed is placed in conditions of adequate temperature and moisture, it fails to germinate. This is in accord with an accepted definition of dormancy (Bewley and Black, 1994). Likewise, in order to induce germination, the seed must perceive a condition that is necessary to induce germination, but not to sustain it. This 'discontinuity' of conditions, i.e. one to break dormancy and one to permit germination, is a phenomenon associated with, for example, red light-induced germination (Borthwick *et al.*, 1952) and the breaking of dormancy by chilling (Bradbeer, 1968). Here light and cold are required by the imbibed seed, for periods as short as a few seconds in the former case, to as many as several months in the latter, but for germination itself darkness and warmth are then sufficient.

Desiccation – a Developmental Switch?

For the developing seed, the discontinuous condition that is most frequently required is desiccation (Adams and Rinne, 1980), as illustrated below for the castor bean seed. In some cases, desiccation seems not to be necessary as a perturbation for relieving dormancy, e.g. horse chestnut (*Aesculus hippocastanum*) (Tompsett and Pritchard, 1993). However, slight desiccation might be necessary before treatments which alleviate dormancy can be effective. Moreover, subjecting *Acer pseudoplatanus* seeds to chilling to remove dormancy is effective only after the seeds have undergone some moisture loss (Thomas *et al.*, 1973; Hong and Ellis, 1990).

Castor bean seeds removed from the capsule up to 45 days after pollination (DAP) fail to germinate when placed on water at 25°C (Fig. 2.1A). Only those at 50 DAP and more mature stages germinate well (at least to the seedling stage), but subsequent development is poor compared with that following germination of fully mature seeds. If the seeds are dried following removal from the capsule and then placed on water, germination is completed by most seeds, especially from 30 DAP onwards. This shows that developing seeds have the potential to germinate, but do not unless first given a treatment (desiccation) to release the embryos from the constraints of the surrounding structures, either in the seed, the capsule or both. The requirement for drying can be circumvented if the developing castor bean embryos

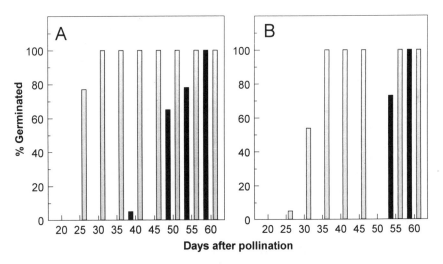

Fig. 2.1. (A) Germination of intact developing castor bean seeds isolated from the capsule at various days after pollination and placed directly on water (dark bars) or desiccated prior to incubation on water (grey bars). (B) Germination of isolated castor bean embryos placed on water following their dissection from intact developing seeds (grey bars). Intact seed germination (dark bars) as in (A). (Based on data in Kermode and Bewley, 1985a, 1988.)

are dissected and placed directly on water. Within days, the isolated embryos germinate fully if dissected after about 35 DAP; hence removal from the surrounding endosperm and/or seed-coat is sufficient. This situation has obvious analogies with seed-coat dormancy, where detachment from the structures surrounding the embryo obviates the requirement for a dormancy-breaking treatment (Bewley and Black, 1994).

Desiccation of the developing seeds, leading to their germination upon subsequent rehydration, has a profound effect upon their metabolism. This is largely, of course, because the seeds have not completed their maturation at the time of desiccation, and hence their synthesis is still geared towards the production of storage reserves. Upon rehydration following premature drying of castor and *Phaseolus* beans, as examples, synthesis of storage proteins ceases, and proteins associated with germination and subsequent postgerminative events are synthesized (Kermode and Bewley, 1985b; Kermode *et al.*, 1985; Misra and Bewley, 1985; Rosenberg and Rinne, 1986). This 'switch' in metabolism occurs at the level of the genome, for transcription of storage protein and other embryogenic messages is suppressed permanently, and messages for proteins associated with postgerminative seedling establishment are transcribed (Kermode *et al.*, 1989; Bewley and Marcus, 1990). Thus, as summarized in Fig. 2.2, drying imposed prematurely during development or occurring naturally at the completion of seed maturation results in the off-regulation of developmental genes and the on-regulation of those involved in germination and postgermination. Treatments thereafter to prevent germination, e.g. osmotica or high concentrations of ABA, do not cause the embryo to revert to a developmental mode, and hence the off-regulation caused by drying is permanent. This will be examined later in relation to studies on alfalfa.

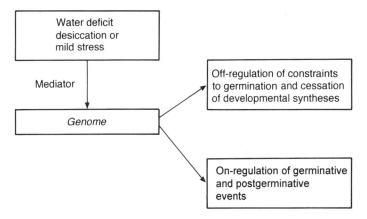

Fig. 2.2. Schematic representation of the effect of desiccation or partial drying on the switch from development to germination.

Of interest, though, is the observation that relatively small decreases in water content will achieve the same effect. Maintaining seeds isolated during development in a fully hydrated condition does not lead to germination, nor do developmental events cease; storage protein synthesis can continue for several days after isolation (Kermode and Bewley, 1985b). This demonstrates, again, that prevention of germination and the switch to a germinative mode of metabolism operate within the seed itself and, in the absence of the 'discontinuity' in conditions, metabolism is unaffected. When a decline in water content of some 5-10% is achieved during development, upon subsequent return to full hydration storage protein mobilization occurs, there is an increase in postgerminative enzymes associated with reserve mobilization, including those for gluconeogenesis from lipids, and the genes for developmental storage proteins are permanently repressed (Fig. 2.3) (Kermode and Bewley, 1989). Since the so-called 'recalcitrant' seeds also experience decreases (greater than 5-10%) in moisture content as they approach maturity (Thomas *et al.*, 1973; Dodd *et al.*, 1989), partial desiccation has been accepted as the developmental switch for such seeds as well (Hong and Ellis, 1990).

Desiccation Perception/Signalling

As alluded to in Fig. 2.2, a critical question involves identification of the moderator, or signal, that perceives the desiccation or water stress signal and transmits this to the genome such that its activity is on- or off-regulated. One possibility that has been suggested is ABA (Kermode, 1990), a regulator that also maintains seeds in a developmental mode and prevents some mature seeds from germinating, thus simulating dormancy (LePage-Degivry *et al.*, 1990; LePage-Degivry and Garello, 1992). To document the effects of ABA on germinability during development, studies on alfalfa (*Medicago sativa*) seeds will be used as an illustration. When removed from the pod and placed in water, the developing intact seeds do not germinate until at least late stage VII, and stage VIII when drying of the seed has commenced during the final stages of maturation (Fig. 2.4). When the embryo is removed from the seed and placed on water, germination commences a little earlier, but it is promoted most noticeably when the embryo is placed on a nutrient medium (Murashige and Skoog, 1962). Even embryos isolated as early in development as stage III are capable of considerable germination, although, interestingly, germinability declines during midmaturation, and then increases again as more mature stages are reached (Fig. 2.4). Thus, there appear to be two superimposed phenomena in operation: the inability of the developing intact seed to germinate (as in castor bean, Fig. 2.1A) and the inability of the isolated embryo to germinate during mid-development (unlike castor bean, Fig. 2.1B).

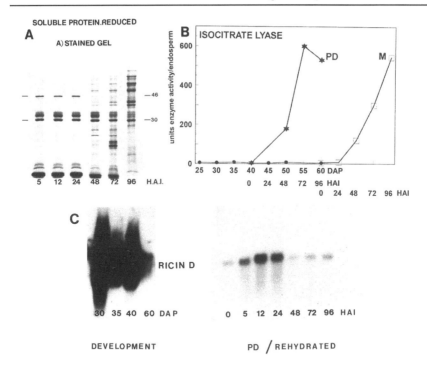

Fig. 2.3. The response of developing castor bean seeds (40 days after pollination (DAP)) to partial desiccation. (A) Stained gel of changes in soluble endosperm proteins from 40 DAP castor beans during the first 96 h after imbibition. Note the decline after 48 h in the major storage proteins present at 5 h and the increase in other proteins associated with postgerminative development. (B) Increase in isocitrate lyase activity (a postgerminative enzyme) in response to partial desiccation (PD). Enzyme activity following normal maturation drying is shown in curve (M), and the amount present during development in curve (●). (C) The fate of the messenger ribonucleic acid (mRNA) for ricin D (a developmentally regulated storage protein) in response to partial desiccation. Northern blot analysis using a complementary deoxyribonucleic acid (cDNA) probe to ricin D shows its presence during development (30–40 DAP). Following partial drying at 35 DAP, ricin D mRNA expression is suppressed on return to full hydration (PD/REHY), with its appearance occurring only transiently at 12 to 24 h after imbibition. (Based on data in Kermode and Bewley, 1989, and unpublished data.)

If dried at midmaturation, isolated whole seeds will germinate upon subsequent rehydration (Xu, 1993) in a similar way to castor bean, confirming the phenomenon of whole-seed 'dormancy' during development. What, then, is the nature of inhibition of isolated embryo germination, and is this analogous to embryo dormancy?

That there are inhibitors of germination contained in the seed-coat and endosperm tissue surrounding the developing embryo was confirmed by incubating mature, highly germinable embryos in their presence (Fig. 2.5A).

Inhibitory substances leached from dissected seed-coats, and especially endosperms at stages V–VII, prevented germination of otherwise highly germinable embryos. Since developing seeds are known to contain ABA (Bewley and Black, 1994), and both the concentration of this growth regulator and varying embryo sensitivity to it have been shown to control sprouting in wheat embryos (Walker-Simmons, 1987), an analysis of the seed parts for ABA was carried out. During development, there is a substantial increase in the amount of ABA in both the endosperm and testa, especially in the former, between stages V and VII (Fig. 2.5B). This may account for the inhibition of germination of mature embryos when incubated with these structures. Interestingly, there is a peak of ABA in the embryos at stage VII also, which explains why isolated embryos at this stage fail to germinate – their endogenous ABA content is too high. Confirmation of an inhibitory role for the ABA

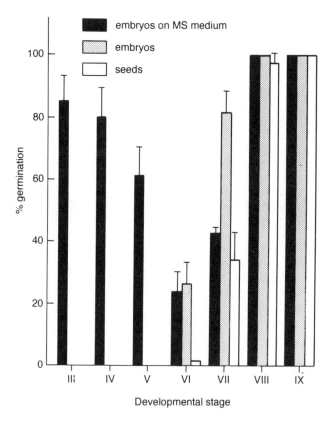

Fig. 2.4. Germination of developing alfalfa embryos and seeds during various stages (III–IX) of development. Seeds or embryos were incubated on water or Murashige and Skoog (MS) (1962) medium, plus 3% sucrose, and germination recorded 5 days after removal from the pod (seeds) or dissection (embryos). (After Xu *et al.*, 1990.)

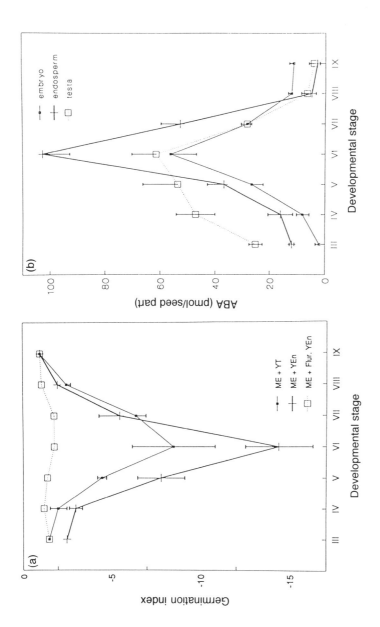

Fig. 2.5 (a) Germination of mature embryos (ME) of alfalfa incubated with endosperms (YEn) or seed coats (YT) from different stages of development of the seeds. Mature embryos were also incubated with endosperms of developing seeds from fluridone-sprayed pods (Flur. YEn). Germination index is a rate of germination, the percentage of embryos germinated on any given day after the start of incubation. (B) Abscisic acid content of the embryo and surrounding tissues during alfalfa seed development. (Based on data in Xu et al., 1990.)

located in the surrounding structures comes from experiments using the ABA synthesis inhibitor, fluridone (Fong *et al.*, 1983). When applied to developing pods, fluridone prevents an increase in ABA within the seed (Fig. 2.5B). Also, when endosperms are dissected at all stages of development from fluridone-treated seeds and are incubated with mature embryos, there is no inhibition of germination; thus the inhibitory component, ABA, has been removed. One surprising result emanating from these experiments was the observation that intact, developing, fluridone-treated seeds, with no appreciable ABA, did not germinate, especially in light of the many reports (e.g. Robichaud *et al.*, 1980; Karssen *et al.*, 1983) of precocious germination and/or reduced dormancy in ABA-insensitive or deficient mutants. This suggests that factors other than ABA alone account for inhibition of whole-seed germination during development. There is evidence that the internal osmotic environment could play a role, also (Xu *et al.*, 1990; Xu and Bewley, 1991, 1993).

In relation to the inhibition of germination during development, and its release by premature drying, it is notable that, as in castor bean, upon rehydration the metabolism of desiccated alfalfa seeds switches from a developmental to a germinative mode (Xu, 1993; Xu and Bewley, 1993). Following desiccation, neither osmoticum nor ABA can elicit the synthesis of stored reserves; the genes are off-regulated. In contrast, when undesiccated stage VII seeds are incubated in water, many will eventually complete germination, and mobilization of stored reserves, including proteins, occurs. Even in this postgerminative state, however, both ABA and osmoticum will induce the synthesis of storage proteins and on-regulate their genes. Hence some residual capacity to conduct developmental events is retained if developing alfalfa embryos are not desiccated, and this is expressed in the presence of ABA and osmoticum, even after germination is completed. Drying reduces or eliminates the sensitivity of seeds to ABA and osmoticum, and loss of developmental-type 'dormancy' may be the result of a loss of sensitivity to such 'dormancy'-inducing agents. In alfalfa, sensitivity to ABA and osmoticum changes during development and is at its lowest in the seed following maturation drying (Xu and Bewley, 1991).

In conclusion, a number of questions have been raised, and some interesting similarities between the germination capacity of developing seeds/embryos and mature seeds have been discussed, in relation to both whole-seed dormancy and embryo dormancy. Is the continuation of dormancy in 'classically' dormant mature dry seeds because ABA content or sensitivity to ABA (or a combination of both) is retained past the stage of maturation drying? Are there other effects of ABA, or other endogenous constraints that are imposed during development, which are retained after maturation? Also pertinent is the question of when, during development, the more permanent dormancy (e.g. that which cannot be broken by desiccation alone) is really acquired. There is the possibility, of course, that postmaturation-expressed dormancy and the temporary developmental dormancy described here are unrelated, and any similarities are merely coincidental. Certainly, this is one of the several areas of dormancy research that warrants further investigation.

References

Adams, C.A. and Rinne, R.W. (1980) Moisture content as a controlling factor in seed development and germination. *International Review of Cytology* 68, 1-8.

Berry, T. and Bewley, J.D. (1992) A role for the surrounding fruit tissues in preventing the germination of tomato (*Lycopersicon esculentum*) seeds: a consideration of the osmotic environment and abscisic acid. *Plant Physiology* 100, 951-957.

Bewley, J.D. and Black, M. (1994) *Seeds: Physiology of Development and Germination*, 2nd edn. Plenum, New York.

Bewley, J.D. and Marcus, A. (1990) Gene expression in seed development and germination. In: Cohn, W.E. and Moldave, K. (eds) *Progress in Nucleic Acid Research and Molecular Biology*. Academic Press, Orlando, pp. 165-193.

Borthwick, H.A., Hendricks, S.B., Parker, M., Toole, E.H. and Toole, V.H. (1952) A reversible photoreaction controlling seed germination. *Proceedings of the National Academy of Sciences, USA* 38, 662-666.

Bradbeer, J.W. (1968) Studies on seed dormancy. 4. The role of endogenous inhibitors and gibberellins on the dormancy and germination of *Corylus avellana* L. seeds. *Planta* 73, 899-901.

Dodd, M.C., Van Staden, J. and Smith, M.T. (1989) Seed development in *Podocarpus henkelii*: an ultrastructural and biochemical study. *Annals of Botany* 64, 297-310.

Fischer, W., Bergfeld, R., Plachy, C., Schäfer, R. and Schopfer, P. (1988) Accumulation of storage materials, precocious germination and development of desiccation tolerance during seed maturation in mustard (*Sinapis alba* L.). *Botanica Acta* 1, 344-354.

Fong, F., Smith, J.D. and Koehler, D.E. (1983) Early events in maize seed development: 1-methyl-3-phenyl-5-(3-[trifluoromethyl]phenyl-4-(1H)-pyridinone induction of vivipary. *Plant Physiology* 73, 899-901.

Hong, T.D. and Ellis, R.H. (1990) A comparison of maturation drying, germination, and desiccation tolerance between developing seeds of *Acer pseudoplatanus* L. and *Acer platanoides* L. *New Phytologist* 116, 589-596.

Karssen, C.M., Brinkhorst-Van der Swan, D.L.C., Breekland, A.C. and Koornneef, M. (1983) Induction of dormancy during seed development by endogenous abscisic acid: studies on abscisic acid deficient genotypes of *Arabidopsis thaliana* (L.) Heynh. *Planta* 157, 158-165.

Kermode, A.R. (1990) Regulatory mechanisms involved in the transition from seed development to germination. *Critical Reviews in Plant Sciences* 9, 155-195.

Kermode, A.R. and Bewley, J.D. (1985a) The role of maturation drying in the transition from seed development to germination. I. Acquisition of desiccation-tolerance and germinability during development of *Ricinus communis* L. seeds. *Journal of Experimental Botany* 36, 1906-1915.

Kermode, A.R. and Bewley, J.D. (1985b) The role of maturation drying in the transition from seed development to germination. II. Post-germinative enzyme production and soluble protein synthetic pattern changes within the endosperm of *Ricinus communis* L. seeds. *Journal of Experimental Botany* 36, 1916-1927.

Kermode, A.R. and Bewley, J.D. (1988) The role of maturation drying in the transition from seed development to germination. V. Responses of the immature castor bean embryo to isolation from the whole seed: a comparison with premature desiccation. *Journal of Experimental Botany* 39, 487-497.

Kermode, A.R. and Bewley, J.D. (1989) Developing seeds of *Ricinus communis* L. when detached and maintained in an atmosphere of high relative humidity, switch to a germinative mode without requirement for complete desiccation. *Plant Physiology* 90, 702-707.

Kermode, A.R., Gifford, D.J. and Bewley, J.D. (1985) The role of maturation drying in the transition of seed development to germination. III. Insoluble protein synthetic pattern changes within the endosperm of *Ricinus communis* L. seeds. *Journal of Experimental Botany* 36, 1928-1936.

Kermode, A.R., Pramanik, S.K. and Bewley, J.D. (1989) The role of maturation drying in the transition from seed development to germination. VI. Desiccation-induced changes in the messenger RNA populations within the endosperm of *Ricinus communis* L. seeds. *Journal of Experimental Botany* 40, 33-41.

LePage-Degivry, M.T., Barthe, M.T. and Garello, G. (1990) Involvement of endogenous abscisic acid in the onset and release of *Helianthus annuus* embryo dormancy. *Plant Physiology* 92, 1164-1168.

LePage-Degivry, M.T. and Garello, G. (1992) *In situ* abscisic acid synthesis: a requirement for induction of embryo dormancy in *Helianthus annuus*. *Plant Physiology* 98, 1386-1390.

McCarty, D.R. and Carson, C.B. (1991) The molecular genetics of seed maturation in maize. *Physiologia Plantarum* 81, 267-272.

Misra, S. and Bewley, J.D. (1985) Reprogramming of protein synthesis from a developmental to a germinative mode induced by desiccation of the axes of *Phaseolus vulgaris*. *Plant Physiology* 78, 876-882.

Murashige, T. and Skoog, F. (1962) A revised medium for rapid growth and bioassays with tobacco tissue cultures. *Physiologia Plantarum* 15, 473-497.

Robichaud, C.S., Wong, J. and Sussex, I.M. (1980) Control of *in vitro* growth of viviparous embryo mutants of maize by abscisic acid. *Developmental Genetics* 1, 325-330.

Rosenberg, L.A. and Rinne, R.W. (1986) Moisture loss as a prerequisite for seedling growth in soybean seeds (*Glycine max* L. Merr.). *Journal of Experimental Botany* 37, 1663-1674.

Thomas, H., Webb, D.P. and Wareing, P.F. (1973) Seed dormancy in *Acer*: maturation in relation to dormancy in *Acer pseudoplatanus*. *Journal of Experimental Botany* 24, 958-967.

Tompsett, P.B. and Pritchard, H.W. (1993) Water status changes during development in relation to the germination and desiccation tolerance of *Aesculus hippocastanum* L. seeds. *Annals of Botany* 71, 107-116.

Walker-Simmons, M. (1987) ABA levels and sensitivity in developing wheat embryos of sprouting resistant and susceptible cultivars. *Plant Physiology* 84, 61-66.

Welbaum, G.E. and Bradford, K.E. (1988) Water relations of seed development and germination in muskmelon (*Cucumis melo* L.): I. Water relations of seed and fruit development. *Plant Physiology* 86, 406-411.

Xu, N. (1993) Regulation of storage protein synthesis in alfalfa (*Medicago sativa* L.) by osmotic potential and abscisic acid. PhD thesis, University of Guelph.

Xu, N. and Bewley, J.D. (1991) Sensitivity to abscisic acid and osmoticum changes during embryogenesis of alfalfa (*Medicago sativa*). *Journal of Experimental Botany* 42, 821-826.

Xu, N. and Bewley, J.D. (1993) Effects of inhibition of ABA synthesis by fluridone on seed germination, storage protein synthesis and desiccation tolerance in alfalfa (*Medicago sativa* L.). In: Côme, D. and Corbineau, F. (eds) *Basic and Applied Aspects of Seed Biology*. ASFIS, Paris, pp. 109-114.

Xu, N., Coulter, K.M. and Bewley, J.D. (1990) Abscisic acid and osmoticum prevent germination of developing alfalfa (*Medicago sativa*) embryos, but only osmoticum maintains the synthesis of developmental proteins. *Planta* 182, 382-390.

3 Control and Manipulation of Seed Dormancy

ANWAR A. KHAN
Department of Horticultural Sciences, New York State Agricultural Experiment Station, Cornell University, Geneva, NY 14456, USA

Introduction

Dormancy induction/release and germination are two dynamic, energy-requiring seed processes (Khan and Zeng, 1985). While a dormant state can be changed to a non-dormant state and vice versa, germination is irreversible (Fig. 3.1). This seems particularly true for seeds in which phytochrome, moist-chilling and gibberellin (GA) exert a control. In many other seed types, including grass and cereal seeds, dormancy can be released by GA or moist-chilling, even though a clear photocontrol mechanism may be lacking. The

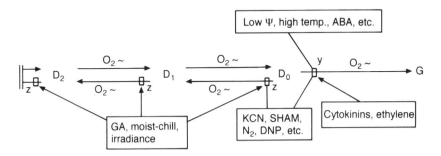

Fig. 3.1. A generalized scheme showing two dynamic energy related processes, a reversible dormancy induction/release phase and a non-reversible germination phase. ψ = Water potential; ~ = requirement of respiratory energy; D_0 = no dormancy; D_1 and D_2 = different stages of dormancy; G = germination; GA = gibberellic acid; O_2 = oxygen; z = block to dormancy induction, action site for preventing dormancy induction or initiating dormancy release; y = block to germination, action site for stress imposition and alleviation (from Khan and Andreoli, 1993, with some modification).

seeds of many cultivated species have little or no dormancy and are barely responsive to GA, light or moist-chilling. It was proposed several years ago that GA, more than any other hormone, is implicated in seed dormancy release and germination of mature seeds, and absence of GA resulted in dormancy induction (Khan, 1971). Recent findings are consistent with the above but show, additionally, that induction/release of dormancy is a reversible process and operates independently of germination (Khan, 1994a,b). Contrary to popular belief, germination is incidental to the release or prevention of dormancy; the presence of GA (as well as other factors, such as light and moist-chilling, which mimic GA effects) is not necessary once dormancy release has been accomplished (Khan and Andreoli, 1993). Other hormones, such as cytokinins and ethylene, are not involved in dormancy release but are needed if stressful factors limit germination (visible growth).

The above view of dormancy and germination differs from that embraced by some workers, who consider dormancy and germination to be a continuum. Rather, it advocates that seed dormancy is determined by abscisic acid (ABA) produced during seed development and its effect manifests itself in germination inhibition in mature or immature seeds (or embryos) during soaking. According to this view, ABA and GA modulate dormancy and germination (Karssen *et al.*, 1989; Le Page-Degivry and Garello, 1992).

Physiological separation of dormancy induction and release from germination requires a reevaluation and better definition of the differences between germination inhibition and dormancy induction, and between germination and dormancy release. Other fundamental questions need to be addressed, such as whether GA, light and moist-chilling are needed for dormancy release or for germination. Is separation of primary and secondary dormancy based on physiological grounds or on mere convenience? Is ABA an inhibitor of germination or a dormancy inducer? This chapter reviews briefly the studies from this and other laboratories on fundamental aspects of dormancy and factors controlling dormancy induction and release. An attempt is made to address questions posed by separation of dormancy induction from germination inhibition and dormancy release from germination. The ability to reversibly induce and release (or prevent) dormancy, independently of germination, has opened up new possibilities for seed management and crop production, some of which are outlined here.

Separation of Dormancy Release and Induction from Germination and its Inhibition

Reversibility of induction and release

Several examples of reversibility of dormancy induction and release, in the absence of germination, can be found in the literature. With 'Mesa 659' lettuce (*Lactuca sativa* L.) seeds, a 35°C soak induced dormancy; this dormancy was prevented or released by GA or irradiation (Khan, 1980/81). Similarly, at 25°C dormancy was induced by treatment with tetcyclacis during a 2-day osmotic soak in -1.2 MPa polyethylene glycol (PEG); this dormancy was prevented upon addition of GA (Khan, 1994a). The primary dormancy in perennial goosefoot (*Chenopodium bonus-henricus* L.) seeds was released by moist-chilling; transfer to -0.86 MPa PEG solution reinduced dormancy, which was again released by a transfer to GA (Khan and Karssen, 1980). The dormancy of common ragweed (*Ambrosia artimisiifolia* L.) seeds was released by winter chilling in the field; dormancy was reinduced in the spring and the induced dormancy was released by application of a hormone mixture containing GA (Samimy and Khan, 1983b). Studies on cyclic changes in seed dormancy patterns of various species (Taylorson, 1972; Karssen, 1980/81; Baskin and Baskin, 1985) further indicate that a reversible dormancy induction/release mechanism, independent of germination, might operate in seed.

Embryo growth potential

A seed becomes dormant when growth potential (expansive force) of the embryo falls below the restraining force of its covering structures (e.g. endosperm, pericarp, palea, lemma) (Khan and Samimy, 1982; Conner and Conner, 1988). In a non-dormant or quiescent state, a seed is in a unique position to either revert to a dormant state (under stressful conditions) or to germinate (under non-stressful conditions). A unique aspect of the embryo, surrounded by covering structures, is its ability (to varying degrees) to reversibly increase or decrease its potential in response to environmental and chemical stimuli. In photosensitive seeds (e.g. lettuce, curly dock (*Rumex crispus* L.)), a reduction in embryo growth potential occurs by soaking seeds in darkness under germination prohibitive conditions, e.g. at supraoptimal temperatures or in low water potential media (Khan and Samimy, 1982; Samimy and Khan, 1983a). GA, irradiation, or moist-chilling enhances embryo growth potential or prevents it from declining, thus ensuring that a seed remains in a non-dormant state (Khan and Samimy, 1982; Powell *et al.*, 1984; Geneve, 1991). Thermal shock, anaesthetics and other membrane active agents also initiate changes that enhance embryo growth potential, thereby releasing dormancy (Taylorson and Hendricks, 1980/81; Hemmat *et al.*, 1985).

In seeds having postharvest dormancy, such as grasses and cereals, dry after-ripening during prolonged storage or a few days of moist-chilling renders the embryo non-dormant, presumably by increasing the embryo growth potential to a level sufficient to penetrate the covering structures. An induction of dormancy, however, has been reported in oats (*Avena fatua* L.), indicating that embryo growth potential could decrease in seeds of the grass family as well (Fregeau and Burrows, 1989).

Participation of endosperm in dormancy control is indicated in several cases. In seeds of a GA-deficient mutant tomato (*Lycopersicon esculentum* Mill.), digestion of the endosperm (which serves as a barrier to germination) was shown to be under the control of GA-mediated enzymes (Groot *et al.*, 1988). The same may be true in other seeds, including pepper and carrot (Watkins *et al.*, 1985; Davidowicz-Grzegorzewska and Maguire, 1993). Lowering of water potential in GA-deficient mutant tomato seeds was not related to an increase in embryo growth potential; endosperm weakening opposite the radicle tip determined germinability (Ni and Bradford, 1993). Changes in covering structures other than endosperm, e.g. pericarp, also may affect the dormant state (Khan and Samimy, 1982; Hemmat *et al.*, 1985; West and Marousky, 1989; Williams *et al.*, 1989). In other cases, both embryo growth potential and changes in covering structures may influence the dormant state (Khan and Samimy, 1982; Hemmat *et al.*, 1985; Hoff, 1987; Cairns and De Villiers, 1989; Geneve, 1991; Weston *et al.*, 1992).

Acquisition of dormancy generally does not result in complete suspension of growth in excised or isolated embryo, it only reduces the capacity of the embryo to generate a force that is high enough to penetrate the covering structures. It is well known that embryos excised from dormant seeds, including those with deep dormancy (e.g. *Malus* spp., *Acer* spp., *Panax ginseng*), are capable of some growth (Nikolaeva, 1977). The embryo covering structures, along with factors enhancing or decreasing embryo growth potential, thus play an important role in ensuring that seed dormancy is retained, intensified, or released. The terms 'primary' vs. 'secondary' dormancy have been used in the context of dormancy that is induced during seed development vs. that induced by stressful factors (e.g. water stress) following harvest. There is no physiological basis for separating the two dormancies. Both primary and secondary dormancy in lettuce, for example, can be released by irradiation, moist-chilling and GA treatments. In a study with 'Grand Rapids' lettuce seeds, secondary dormancy was found to be merely an extension of the dormancy already existing at the time of harvest (Khan and Zeng, 1985).

Both primary and secondary dormancy may be a manifestation of reduced embryo growth potential brought about by stressful environmental and chemical factors. The imposition and alleviation of stress can serve to bring about changes in embryo growth potential. With an increase in temperature, the growth potential of non-dormant lettuce embryos decreased and at 35°C the potential was less than that needed to remove the restraint offered by the covering structures (Takeba and Matsubara, 1979). Intact curly dock seeds germinated well at 20 to 25°C in the light, but poorly at 30°C (Samimy and Khan, 1983a). Scarification of the seed coat with sulphuric acid caused complete germination at 30°C, indicating that a reduction in seed coat restraint was necessary to compensate for the decreased growth potential of the embryo and to bring about germination at the higher temperature.

Several growth regulators, notably ethylene, cytokinin and fusicoccin (a fungal toxin), have been shown to prevent a decrease in embryo growth or germination potential under stressful conditions (see Khan *et al.*, 1993). Stresses on germination caused by osmotic forces (Braun and Khan, 1976; Negm and Smith, 1978; Abeles, 1986), high temperature (Sharples, 1973; Braun and Khan, 1976) or seed coat restraint, including restraint imposed by artificial seed coat or pellet (Khan *et al.*, 1976; Khan and Prusinski, 1989), were alleviated by the addition of kinetin, ethylene (or ethephon), fusicoccin, or their combination. These chemicals did not influence dormancy release, however, but rather influenced germination by alleviating stress (Khan and Andreoli, 1993).

To summarize, embryos have the inherent capacity for reversibility of growth. Imposition of stress over a long period can decrease embryo growth potential or intensify dormancy. Unlike dormancy releasing factors (e.g. light, GA), which can enhance growth potential of a dormant embryo, stress alleviating factors (e.g. cytokinin, ethylene) work only under stressful conditions (e.g. high temperature, low ψ, etc.) on non-dormant embryos; they are incapable of acting once the growth potential is decreased or dormancy is induced (Khan and Samimy, 1982; Khan and Prusinski, 1989).

GA synthesis and synthesis inhibitors

The advent of highly active GA synthesis inhibitors such as ancymidol, tetcyclacis, uniconazol and paclobutrazol (Rademacher, 1991), with high specificity for kaurene oxidation in the GA biosynthetic pathway, renewed research on GA participation in seed germination (Gardner, 1983; Karssen *et al.*, 1989; Nambara *et al.*, 1991). These inhibitors induced dormancy in non-dormant seeds of lettuce, tomato, pepper, carrot, celery and other species, and the induced dormancy was prevented or reversed upon addition of GA (Khan, 1994a,b). Dormancy was induced more readily in darkness than in light in lettuce seeds, while tomato, pepper and carrot were rendered dormant equally well in light or darkness. These studies implicate GA in the reversible dormancy induction/release process described earlier (Khan and Zeng, 1985).

Dormancy induction by GA synthesis inhibitors, and its prevention and release by GA, occurred in lettuce seeds under conditions both favourable (e.g. 25°C) and unfavourable (low water potential, 35°C, 5°C) for germination (Khan *et al.*, 1992; Khan, 1994a), indicating that germination was not a requirement for the dormancy induction/release process. Like GA, other factors that mimic the GA effect, such as light or moist-chilling, also released dormancy induced by GA synthesis inhibitors (Khan, 1994a; Fig. 3.2). In tomato and pepper seeds, dormancy induced by tetcyclacis was not released by irradiation or moist-chilling. Unlike tetcyclacis, ABA did not induce dormancy nor did it interfere with the dormancy-preventing effects of GA or light in lettuce or pepper seeds (Khan *et al.*, 1992; Ilyas, 1993; Khan, 1994a). The fact that dormancy can be induced in ABA-deficient mutant seeds of *Arabidopsis thaliana* (Hilhorst and Karssen, 1992) indicates further that ABA is not a dormancy factor.

Mature dormant or non-dormant seeds contain very little active GA (see Khan, 1982). There is no evidence that GAs produced during seed development are utilized for dormancy release in mature seeds. Studies with 'Emperor' lettuce seeds showed that tetcyclacis was effective in inducing dormancy when present during the initial hours of soaking in water. When the presoaking period was extended beyond 6 to 8 h, it was ineffective in inducing dormancy (Khan, 1994a) (Fig. 3.3). Abscisic acid was ineffective when applied similarly. When tetcyclacis was applied in combination with GA during initial hours, dormancy induction was prevented. Similarly, a complete escape from dormancy induction occurred if tetcyclacis was applied to pepper seeds hydrated for 2 days at 15°C in a low water potential medium (Ilyas, 1993). These studies indicate that GA is synthesized during the early hydration period, enabling the seeds to escape from the dormancy-sensitive phase of imbibition. Separation of dormancy induction/release

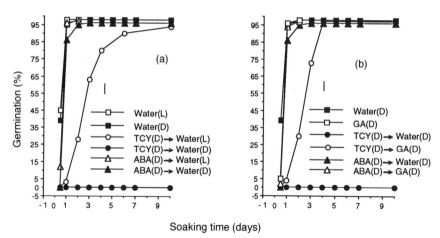

Fig. 3.2. Induction of dormancy in non-dormant 'Montello' lettuce seeds by tetcyclacis (TCY) and its release by (a) light and (b) gibberellins A_4 and A_7 (GA_{4+7}). After a 24 h presoak in 50 μM TCY or 50 μM ABA in darkness, the seeds were washed and dried in darkness and germinated (a) in water in light or darkness and (b) in water or 1 mM GA_{4+7} in darkness at 25°C. Water(L) = seeds germinated in water in light; Water(D) = germinated in water in dark; GA(D) = germinated in GA solution in dark; TCY(D)→Water(L) = dormancy induced by TCY in dark, germination in water in light (arrow indicates transfer from one medium to another); TCY(D)→Water(D) = same as above except germinated in dark; TCY(D)→GA(D) = same as above except germinated in GA solution in dark; ABA(D)→Water(L) = ABA treatment in dark, germinated in water in light; ABA(D)→Water(D) = same as above except germinated in dark; ABA(D)→GA(D) = same as above except germinated in GA solution in dark (data from Khan, 1994a).

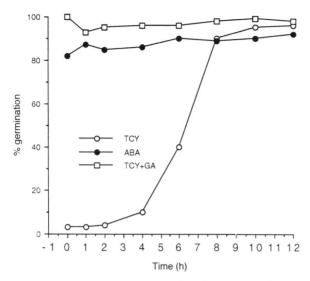

Fig. 3.3. An escape from the dormancy-inducing effect of tetcyclacis (TCY) applied to 'Emperor' lettuce seeds at various times of soaking. Seeds were soaked in water for up to 12 h in darkness at 25°C, transferred to 50 mM TCY, 50 μM ABA or 50 mM TCY + 1 mM GA_{4+7} solution for 24 h in darkness at 25°C, and then washed, and germinated in water at 25°C in darkness for 10 days. Bar indicates $LSD_{0.05}$ (part of the data from Khan, 1994a).

processes from germination provides a unique opportunity to manipulate dormancy, or events associated with it, in developing novel seed management and crop production strategies.

Manipulation of Dormancy in Agriculture

In a field setting, the interrelationships of dormancy release and induction with germination and its prevention are of particular importance. The release and induction of dormancy are brought about by factors (Table 3.1) that are distinctly different from those inducing and preventing germination (Fig. 3.4). A clear understanding of the interrelationships and factors controlling them is needed in order to effectively control and manipulate seed dormancy.

Preventing dormancy induction and improving growth potential

Prolonged exposure of photosensitive seeds, such as lettuce, celery and endive (*Chicorium endivia* L.), to GA or irradiation during suspension of germination by a low water potential medium, such as -1.2 MPa PEG solution (osmoconditioning) or moist Micro-Cel E (a treatment termed 'matriconditioning', the soaking seeds in predetermined amounts of water and a solid

Table 3.1 Factors inducing and releasing dormancy compared with factors preventing and inducing germination.

DIF	DRF or DPF	GPF	GIF
Unavailability of GA	GA availability	Low temperature	Moderate temperature
GA synthesis inhibition	GA synthesis	Low ψ	High ψ
Prolonged exposure in darkness to GPF	Moist-chilling	Anaerobiosis	
	Irradiation	Inhibitors (e.g. ABA)	Cytokinin
		High temperature	Ethylene
		Seed-coat restraint	Oxygen

ψ = Water potential; DIF = dormancy-inducing factors; DRF or DPF = dormancy-releasing or dormancy-preventing factors; GPF = germination-preventing factors; GIF = germination-inducing factors.

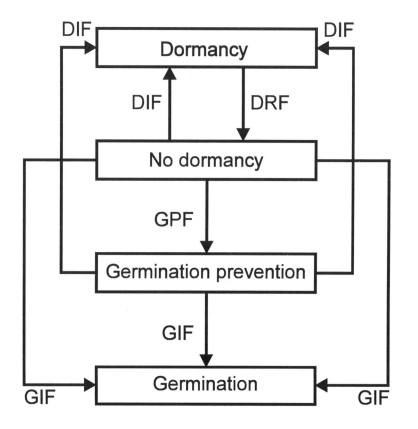

Fig. 3.4. A scheme showing the various physiological states of a seed in a field or a simulated field environment and their ability to switch from one state to another. DIF = dormancy-inducing factors; DRF = dormancy-releasing or dormancy-preventing factors; GPF = germination-preventing factors; GIF = germination-inducing factors.

carrier, such as Micro-Cel E, a synthetic calcium silicate, with matric properties and a high water-holding capacity), prevents dormancy induction (Khan and Samimy, 1982; Khan, 1992; A.A. Khan, unpublished). In addition, these treatments enhance embryo growth potential, reduce subsequent germination time and synchronize germination and seedling emergence. In several cultivars of carrot, onion, tomato and pepper, these effects are manifested in both light and darkness.

In seeds such as tomato and pepper, GA-mediated events participate in removing the endosperm barrier to embryo growth (Watkins *et al.*, 1985; Groot *et al.*, 1988). As treatment with tetcyclacis induces dormancy and is reversible by GA (Khan, 1994a), it is possible that GA produced during soaking may control dormancy via GA-mediated production of hydrolases and consequent digestion of the endosperm layer surrounding the embryo. It is also possible that changes in embryo growth potential during seed hydration contribute to the release and induction of dormancy. Matriconditioning of tomato and pepper seeds enhanced the embryo growth potential as shown by increased ability to germinate at lower water potential (Ilyas, 1993; Khan *et al.*, 1993). Osmotic conditioning of light-treated lettuce seeds in 0.5 molal mannitol is reported to mobilize storage reserves in the embryo, which seemingly provide nutritional and osmotic factors that overcome the resistance of surrounding layers (Nabors *et al.*, 1974).

These studies indicate that a treatment with GA or light during suspension of germination of many seed types provides an effective means to prevent dormancy induction, enhance embryo growth potential and remove barriers to germination, leading to improved performance of seeds. In the absence of these treatments, the embryo growth potential is reduced and seeds germinate at a slower rate (which can be regarded as a measure of dormancy) and may even become dormant.

Combining dormancy prevention and release with stress alleviation

The release of dormancy by itself, particularly in the field, is not always sufficient to bring about germination. Additional measures are usually required to effect germination and emergence (Khan and Andreoli, 1993). Separating dormancy from germination has permitted a concerted study on the factors controlling the two processes and the development of procedures for affecting them jointly. Release or prevention of dormancy can be combined with the alleviation of germination stress to maximize seed performance. Mechanical stress, imposed by embryo covering structures, or physical stresses, caused by high temperature or low water potential imbibing media, may render a non-dormant seed dormant or prevent it from germinating.

Although inclusion of GA or exposure to light in many cases will prevent dormancy induction and thus keep the seed in a non-dormant or germinable state, these factors by themselves are not adequate, under stressful conditions, to bring about germination. The restraining effect of low water potential and covering structures, or the growth potential-lowering effect of high temperature on embryos, can be overcome or reduced by applying cytokinin and/or ethylene to the seed (Takeba and Matsubara, 1979; Khan and Samimy, 1982; Hemmat *et al.*, 1985; Khan and Prusinski, 1989).

In a study with lettuce seeds, a combination of GA, kinetin and ethephon not only prevented dormancy induction but successfully alleviated the stressful effect of 35°C or low water potential (e.g. -0.3 MPa PEG or -0.4 MPa NaCl solution) on dark germination (Braun and Khan, 1976). Using GA alone, Khan and Andreoli (1993) found that it released or prevented tetcyclacis-induced dormancy in lettuce seeds, even though the seeds failed to germinate at 35°C. Cytokinin and/or ethephon was needed, in addition, to effect germination of non-dormant seeds at 35°C. Thus, a clearer understanding of processes controlling dormancy and germination, and their interaction with stressful factors, permits an integrated approach to improving seed performance under non-optimal conditions.

Presoaking and thermotolerance

It has long been known that presoaking of seeds in water and salt solution improves subsequent germination performance. These treatments have formed the basis for developing specialized low water potential seed hydration treatments (e.g. osmotic priming or osmoconditioning, and matric priming or matriconditioning) for improving seed performance under a variety of conditions (see Khan, 1992, for a review). Synthesis of GA during the early soaking period is implicated as GA synthesis inhibitors applied during this period induce dormancy and the induction is prevented upon addition of GA (Ilyas, 1993; Khan, 1994a). Thus, GA may initiate changes that lead to enhanced embryo growth potential and to a non-dormant state. This may account for the observation that presoaking lettuce seeds in water or salt solutions prevents or reduces thermodormancy induction (Guedes and Cantliffe, 1980; Cantliffe *et al.*, 1984). Prolonged thermoinhibition leads to induction of thermodormancy (Khan *et al.*, 1980/81).

Studies conducted with lettuce seeds have shown that enhanced thermotolerance may be due to enhanced production of 1-aminocyclopropane-1-carboxylic acid (ACC) during matriconditioning and an increase in ACC utilization following transfer of conditioned seeds to water at high temperature (Khan and Andreoli, 1993). The ability to germinate at higher temperatures following a low water potential seed hydration treatment has been demonstrated in celery (Khan *et al.*, 1990), pepper (Ilyas, 1993) and other seeds (Carpenter, 1990; Khan, 1992; Madakadze et al., 1993). Thus, controlled hydration of seeds via low water potential media is an effective means to prevent or manipulate thermodormancy, reduce germination time and impart thermotolerance during germination.

Interruption of the annual dormancy cycle

An annual dormancy cycle has been observed in many weed seeds buried in the soil (Taylorson, 1972; Karssen, 1980/81; Baskin and Baskin, 1985). In several cases, seeds that are shed from the mother plant in the autumn have a primary dormancy; the dormancy is released by moist-chilling during winter months, and a secondary dormancy is reestablished in the spring and summer months. It has been possible to prevent the induction of dormancy in *Ambrosia artemisiifolia* and to achieve germination by application of a hormone mixture containing GA, ethephon and cytokinin (Samimy and Khan, 1983b). When the mixture was applied prior to winter chilling, dormancy induction in the spring was prevented due to the presence of GA in the mixture and, when applied after the induction of dormancy in the spring, the induced dormancy was released. Cytokinin and ethephon presumably aided in the germination of these seeds in the soil medium following the release of dormancy by GA. Curly dock seeds also respond to GA, cytokinin, and ethylene, but only after scarification with sulphuric acid (Hemmat et al., 1985). Thus, seed-coats may present a formidable barrier to germination and to hormone penetration in many seeds. Interrupting the dormancy induction cycle by GA is a promising area of research, particularly in weed seeds, and, if properly managed, may lead to a reduction of weed seed populations in agricultural soils.

Inducing dormancy in non-dormant weed seeds

Many weed seeds are responsive to GA treatment. A 24 h soak with 50 μM tetcyclacis of broadleaf plantain (*Plantago major* L.) and smartweed (*Polygonum pensylvanicum* L.) seeds induced complete dormancy as seeds failed to germinate in water at 25°C in darkness or light, compared with 70 to 98% germination in the controls. Partial dormancy was induced in wild oat and common ragweed seeds by a similar treatment. A 24 h soak of wild oat seeds with 50 μM tetcyclacis or 120 μM paclobutrazol, followed by washing, 2 h drying and planting in a peat-vermiculite mix at an alternating temperature of 10°C and 20°C, resulted in 50 and 60% emergence, respectively, compared to 98% for the untreated seeds. In addition to the partial dormancy, the rate of emergence of the treated wild oat seeds was greatly reduced. These data indicate that it may be possible to induce dormancy, reduce germination percentage or slow the rate of emergence of weed seedlings by application of appropriate GA synthesis inhibitors, thereby reducing their competitive advantage over crops for nutrients. Together with preventing and breaking dormancy (see above), this strategy of inducing dormancy may lead to a reduction in herbicide use and to an effective management of weeds.

Autumn planting of dormant seeds

Autumn planting may be desirable for crops that are planted early in the spring. In many regions of the USA and Canada, the cold, wet conditions of the field during the early spring make them unworkable and difficult to manage. Laboratory and field studies were designed to test if the seeds that were rendered dormant by GA synthesis inhibitor application could remain unsprouted for a time after planting, have their dormancy released by moist-chilling temperatures during winter, and germinate and produce seedlings on exposure to moderately high temperatures of early spring. Dormancy was induced in 'Mesa 659' lettuce seeds by a 24 h exposure to tetcyclacis at 25°C; induction occurred more readily in darkness than in light (Khan, 1994a). The induced dormancy was released progressively when seeds were held at 5°C for different periods of time, as shown by progressively greater seedling emergence on transfer to 25°C (Table 3.2). A 30-day moist-chilling treatment released dormancy almost completely. Some success has been reported with autumn-planted dormant (induced by various temperature regimes) oat seeds in obtaining a stand in the spring (Fregeau and Burrows, 1989). Changing the planting season so that dormant seeds can be planted in the autumn, with

Table 3.2 Release of tetcyclacis (TCY) induced dormancy in 'Mesa 659' lettuce seeds by progressively longer moist-chilling treatment at 5°C (from Khan, 1994b).

TCY (μM)	Moist-chilling at 5°C (days)[a]			
	0	8	15	30
	% Emergence at 25°C[b]			
0 (Untreated)	76	96	97	97
10 (L)[c]	36	50	67	82
50 (L)	3	8	23	39
100 (L)	0	2	18	13
5 (D)	10	39	72	80
10 (D)	0	45	62	67
50 (D)	0	4	11	9
$LSD_{0.05}$	4	5	5	6

$LSD_{0.05}$ = least significant difference at 5% level of probability.
[a] Seeds were moist-chilled at 5°C in thoroughly wetted peat–lite mix for 0 to 30 days.
[b] After moist-chilling, containers with the seeds were transferred to 25°C in the light and emergence determined after 10 days.
[c] 'L' and 'D' in parentheses refer to light and darkness during 24 h TCY treatment.

prevention of autumn sprouting and achievement of a good plant stand in early spring, may be a viable seed management and crop production strategy for certain crops in regions with cold, wet springs.

Mixed seeding cropping systems

Depending upon the soak duration temperature and concentration, GA biosynthesis inhibitors induced varying degrees of dormancy in seeds of several grasses (climax timothy, *Phleum pratense* L.; red fescue, *Festuca rubra* L.; canarygrass, *Phalaris arundinaceae* L.; wild oat, *Avena fatua* L.; perennial rye grass, *Lolium perenne* L.) and legumes (red clover, *Trifolium pratense* L.; white clover, *T. repens* L.; crimson clover, *T. incarnatum* L.; subterranean clover, *T. subterraneum* L.; hairy vetch, *Vicia villosa* Roth; alfalfa, *Medicago sativa* L.). Reduced germination and growth rate are regarded as a manifestation of dormancy and can be utilized in cover crop turf and forage quality management strategies. In silage maize (*Zea mays* L.), for example, red clover seeds are sown 6 to 8 weeks after planting maize. This involves high labour costs. A 3- to 6-week delay in emergence was observed in various clover seeds treated with paclobutrazol, uniconazol, ancymidol and tetcyclacis. A mixed seeding of maize with legume seeds, in which only legumes are treated with GA synthesis inhibitors, may allow little or no competition to maize and still permit a good cover crop after the maize is harvested. The ability to reduce seedling growth rate by GA inhibitors without influencing seedling density may have several advantages. For example, it may allow seeding of a fast-growing grass (treated with a GA inhibitor) and a slow-growing legume at the same time, obtaining a good stand in both. These approaches are currently being tested with different crop combinations.

Conclusions

The reversibility of dormancy induction/release processes and the separation of dormancy related processes from germination has opened up several new avenues of agricultural seed management research. It has helped focus attention on measures to most efficiently prevent, overcome or induce dormancy to suit differing needs of growers and the seed industry. Planting dormant seeds in the autumn, preventing immediate sprouting, securing the release of dormancy by moist-chilling in the winter months, and establishing a plant stand in the spring, provides a unique way to alter the planting season, at least in some crops. Thus, control and manipulation of dormancy offers wide-ranging possibilities for developing novel seed management and crop production strategies.

References

Abeles, F.B. (1986) Role of ethylene in *Lactuca sativa* cv 'Grand Rapids' seed germination. *Plant Physiology* 81, 780-787.

Baskin, J.M. and Baskin, C.C. (1985) The annual dormancy cycle in buried weed seeds. *BioScience* 35, 492-498.

Braun, J.W. and Khan, A.A. (1976) Alleviation of salinity and high temperature stress by plant growth regulators permeated into lettuce seeds via acetone. *Journal of the American Society for Horticultural Science* 101, 716-721.

Cairns, A.L.P. and De Villiers, O.T. (1989) Effect of sucrose taken up by developing *Avena fatua* L. panicles on the dormancy and a-amylase synthesis of the seeds produced. *Weed Research* 29, 151-156.

Cantliffe, D.J., Fischer, J.M. and Nell, T.A. (1984) Mechanism of seed priming in circumventing thermodormancy in lettuce. *Plant Physiology* 75, 290-294.

Carpenter, W.J. (1990) Priming dusty miller seeds: role of aeration, temperature, and relative humidity. *HortScience* 25, 299-302.

Conner, A.J. and Conner, L.N. (1988) Germination and dormancy of *Arthropodium cirratum* seeds. *New Zealand Natural Sciences* 15, 3-10.

Davidowicz-Grzegorzewska, A. and Maguire, J.D. (1993) The effects of SMP on the ultrastructure of carrot seeds. In: Come, D. and Corbineau, F. (eds) *Fourth International Workshop on Seeds. Basic and Applied Aspects of Seed Biology*, Vol. 3. ASFIS, Paris, pp. 1039-1044.

Fregeau, J.A. and Burrows, V.D. (1989) Secondary dormancy in dormoats following temperature treatment: field and laboratory responses. *Canadian Journal of Plant Science* 69, 93-100.

Gardner, G. (1983) The effect of growth retardants on phytochrome-induced lettuce seed germination. *Journal of Plant Growth Regulation* 2, 159-163.

Geneve, R.L. (1991) Seed dormancy in eastern redbud *Cercis canadensis*. *Journal of the American Society for Horticultural Science* 116, 85-88.

Groot, S.P.C., Kieliszewska-Rokicka, E., Vermeer, E. and Karssen, C.M. (1988) Gibberellin-induced hydrolysis of endosperm cell walls in gibberellin-deficient tomato seed prior to radicle protrusion. *Planta* 174, 500-504.

Guedes, A.C. and Cantliffe, D.J. (1980) Germination of lettuce seeds at high temperature after seed priming. *Journal of the American Society for Horticultural Sciences* 105, 777-781.

Hemmat, M., Zeng, G.W. and Khan, A.A. (1985) Responses of intact and scarified curly dock (*Rumex crispus*) seeds to physical and chemical stimuli. *Weed Science* 33, 658-664.

Hilhorst, H.W.M. and Karssen, C.M. (1992) Seed dormancy and germination: the role of abscisic acid and gibberellins and the importance of hormone mutants. *Plant Growth Regulation* 11, 225-238.

Hoff, R.J. (1987) Dormancy in *Pinus monticola* seed related to stratification time, seed coat and genetics. *Canadian Journal of Forestry Research* 17, 294-298.

Ilyas, S. (1993) Invigoration of pepper seeds by matriconditioning and its relationship with storability, dormancy, ageing, stress tolerance and ethylene biosynthesis. Unpublished PhD thesis, Cornell University.

Karssen, C.M. (1980/81) Patterns of change in dormancy during burial of seeds in soil. *Israel Journal of Botany* 29, 65-73.

Karssen, C.M., Zagorski, S., Kepczynski, J. and Groot, S.P.C. (1989) Key role for endogenous gibberellins in the control of seed germination. *Annals of Botany* 63, 71-80.

Khan, A.A. (1971) Cytokinins: permissive role in germination. *Science* 171, 853-859.

Khan, A.A. (1980/81) Hormonal regulation of primary and secondary seed dormancy. *Israel Journal of Botany* 29, 207-224.

Khan, A.A. (1982) Gibberellins and seed development. In: Khan, A.A. (ed.) *The Physiology and Biochemistry of Seed Development, Dormancy and Germination*. Elsevier Biomedical Press, Amsterdam, pp. 111-135.

Khan, A.A. (1992) Preplant physiological seed conditioning. *Horticultural Reviews* 13, 131-181.

Khan, A.A. (1994a) Induction of dormancy in nondormant seeds. *Journal of the American Society for Horticultural Science* 119, 408-413.

Khan, A.A. (1994b) Inducing dormancy in nondormant seeds. United States Patent. Patent No. 5,294,593.

Khan, A.A. and Andreoli, C. (1993) Hormonal control of dormancy and germination under stressful and nonstressful conditions. In: Come, D. and Corbineau, F. (eds) *Fourth International Workshop on Seeds*, Vol. 2. ASFIS, Paris, pp. 625-631.

Khan, A.A. and Karssen, C.M. (1980) Induction of secondary dormancy in *Chenopodium bonus-henricus* L. seeds by osmotic and high temperature treatments and its prevention by light and growth regulators. *Plant Physiology* 66, 175-181.

Khan, A.A. and Prusinski, J. (1989) Kinetin enhanced 1-aminocyclopropane-1-carboxylic acid utilization during alleviation of high temperature stress in lettuce seeds. *Plant Physiology* 91, 733-737.

Khan, A.A. and Samimy, C. (1982) Hormones in relation to primary and secondary seed dormancy. In: Khan, A.A. (ed.) *The Physiology and Biochemistry of Seed Development, Dormancy and Germination*. Elsevier Biomedical Press, Amsterdam, pp. 203-241.

Khan, A.A. and Zeng, G.W. (1985) Dual action of respiratory inhibitors: inhibition of germination and prevention of dormancy induction in lettuce seeds. *Plant Physiology* 77, 817-825.

Khan, A.A., Braun, J.W., Tao, K.-L., Millier, W.F. and Bensin, R.F. (1976) New methods for maintaining seed vigor and improving performance. *Journal of Seed Technology* 1, 33-57.

Khan, A.A., Peck, N.H. and Samimy, C. (1980/81) Seed osmoconditioning: physiological and biochemical changes. *Israel Journal of Botany* 29, 133-144.

Khan, A.A., Miura, H., Prusinski, J. and Ilyas, S. (1990) Matriconditioning of seeds to improve seedling emergence. In: *National Symposium on Stand Establishment of Horticultural Crops*, April 4-6, Minneapolis, Minnesota, pp. 19-40.

Khan, A.A., Huang, X., Zeng, G. and Prusinski, J. (1992) Integration of hormonal controls of seed dormancy and germination with environmental demands. In: Fu, J. and Khan, A.A. (eds) *Advances in the Science and Technology of Seeds*. Science Press, Beijing and New York, pp. 313-335.

Khan, A.A., Andreoli, C., Prusinski, J. and Ilyas, S. (1993) Enhanced embryo growth potential as a basis for alleviating high temperature and osmotic stress. In: Kuo,

C.G. (ed.) *Adaptation of Food Crops to Temperature and Water Stress*. Asian Vegetable Research and Development Centre, Taipei, pp. 437-451.

Le Page-Degivry, M.T. and Garello, C. (1992) In situ abscisic acid synthesis. *Plant Physiology* 98, 1386-1390.

Madakadze, R., Chirco, E.M. and Khan, A.A. (1993) Seed germination of three flower species following matriconditioning under various environments. *Journal of the American Society for Horticultural Sciences* 118, 330-334.

Nabors, M.W., Kurgens, P. and Ross, C. (1974) Photodormant lettuce seeds: phytochrome induced protein and lipid degradation. *Planta* 117, 361-365.

Nambara, E., Akazawa, T. and McCourt, P. (1991) Effect of the gibberellin biosynthetic inhibitor uniconazol on mutants of *Arabidopsis*. *Plant Physiology* 97, 736-738.

Negm, F.B. and Smith, O.E. (1978) Effect of ethylene and carbon dioxide on the germination of osmotically inhibited lettuce seeds. *Plant Physiology* 62, 473-476.

Ni, B.R. and Bradford, K.J. (1993) Germination and dormancy of abscisic acid and gibberellin-deficient mutant tomato seeds. *Plant Physiology* 101, 607-617.

Nikolaeva, M.G. (1977) Factors controlling the seed dormancy pattern. In: Khan, A.A. (ed.) *The Physiology and Biochemistry of Seed Dormancy and Germination*. North-Holland Publishing Company, Amsterdam, pp. 51-74.

Powell, A.D., Dulson, J. and Bewley, J.D. (1984) Changes in germination and respiratory potential of embryos of dormant Grand Rapids lettuce seeds during long-term imbibed storage, and related changes in the endosperm. *Planta* 162, 40-45.

Rademacher, W. (1991) Biochemical effects of growth retardants. In: Gausman, H.W. (ed.) *Plant Biochemical Regulators*. Marcel Dekker, New York, pp. 169-200.

Samimy, C. and Khan, A.A. (1983a) Secondary dormancy, growth regulator effects and embryo growth potential in curly dock (*Rumex crispus*) seeds. *Weed Science* 31, 153-158.

Samimy, C. and Khan, A.A. (1983b) Effect of field application of growth regulators on secondary dormancy of common ragweed (*Ambrosia artemisiifolia*) seeds. *Weed Science* 31, 299-301.

Sharples, G.C. (1973) Stimulation of lettuce seed germination at high temperature by ethephon and kinetin. *Journal of the American Society for Horticultural Sciences* 98, 209-212.

Takeba, G. and Matsubara, S. (1979) Measurement of growth potential of the embryo in New York lettuce seed under various combinations of temperature, red light and hormones. *Plant and Cell Physiology* 51, 51-61.

Taylorson, R.B. (1972) Phytochrome controlled changes in dormancy and germination of buried weed seeds. *Weed Science* 20, 417-422.

Taylorson, R.B. and Hendricks, S.B. (1980/81) Anaesthetic release of seed dormancy - an overview. *Israel Journal of Botany* 29, 273-280.

Watkins, J.T., Cantliffe, D.J., Huber, D.J. and Nell, T.A. (1985) Gibberellic acid stimulated degradation of endosperm in pepper. *Journal of the American Society of Horticultural Science* 110, 61-65.

West, S.H. and Marousky, F. (1989) Mechanism of dormancy in Pensacola bahiagrass. *Crop Science* 29, 787-791.

Weston, L.A., Geneve, R.L. and Staub, J.E. (1992) Seed dormancy in *Cucumis sativus* var. harwickii (Royle) Alef. *Scientia Horticulturae* 50, 35-46.

Williams, R.R., Holliday, K.C. and Bennell, M.R. (1989) Cultivation of the pink mulla mulla *Ptilotus exaltatus* Nees. 1. Seed germination and dormancy. *Scientia Horticulturae* 40, 267-274.

4 A Physiological Comparison of Seed and Bud Dormancy

Frank G. Dennis, Jr
Department of Horticulture, Michigan State University, East Lansing, MI 48824-1325, USA

Introduction

Seed and bud dormancy have many features in common. Some characteristics common to both seeds and buds for induction and release of dormancy include: the temperature optima and range for breaking dormancy; similar exposure times (chilling requirements) for buds and seeds of a given genotype; reduced growth potential when chilling is interrupted with high temperature; and the stimulation or inhibition of germination/bud break by growth regulators (Vegis, 1964; Powell, 1987). In some species (e.g. birch), light quality (red/far red) and/or photoperiod also influences both bud and seed dormancy (Wareing and Black, 1958). Thus, 'indiscriminate' comparisons of dormancy mechanisms in seeds and buds tend to emphasize their similarities.

These parallel responses have led scientists to propose that common mechanisms control these two phenomena (e.g. Vegis, 1964; Wareing *et al.*, 1964; Powell, 1987). As a result, information gained from studies with seeds is often assumed to apply to buds as well, and some investigators have suggested using seeds as model systems. Powell (1987), for example, stated, 'Since seeds are far more amenable to manipulation than buds, it would seem that basic studies on the chilling mechanism would be facilitated by using seeds.' Pollock and Olney (1959) noted the advantages of using seeds because they represent a 'closed system'. Indeed, the number of laboratories studying bud dormancy decreased during the 1960s and 1970s (Champagnat, 1983), although few problems in bud dormancy had been solved and general interest remained strong.

Can information based on studies of seeds indeed be extrapolated to phenomena occurring in buds and vice versa? This chapter will review the similarities and differences between dormancy in seeds vs. buds, confining the discussion to endodormancy to avoid some of the complications of correlative effects. Emphasis will be on woody perennials, in most cases, and on environmental effects, since limited understanding of the physiological mechanisms involved makes good comparisons difficult. Two comparative approaches will be used to examine specific similarities and differences between buds and seeds with respect to dormancy phenomena. The first will be to compare dormancy of seeds vs. buds within species, ecotypes and genotypes. The second will compare dormancy of seeds vs. that of the seedlings to which they give rise or of the trees that produced the seeds.

Some investigators note major differences between seeds and buds and believe that their dormant states cannot be compared easily. Crabbé (1994) notes that a seed is a whole plant, whereas a bud consists of a shoot only. Furthermore, the seed is autonomous, whereas a bud is attached to a stem, and other parts of the plant can therefore influence its activity via correlative effects. Crabbé and coworkers (e.g. Crabbé, 1968; Arias and Crabbé, 1975) have used the one-node cutting system of Nigond (1967) to reduce correlative effects; even then, however, such effects can still be demonstrated.

Lang et al. (1987) proposed a consolidation of the terms used to describe the many recognized types of dormancy into three – eco-, para- and endo-dormancy, representing dormancy controlled by conditions outside the plant, outside the affected organ but within the plant and within the organ, respectively. Chouard (1956) proposed a similar classification, using the terms quiescence, correlative inhibition and rest, respectively. All three types of dormancy occur in buds, and seeds definitely exhibit endodormancy, but seed physiologists often reject the term ecodormancy as applying to seeds that are non-dormant (Junttila, 1988). This is a matter of semantics, however, for buds that are ecodormant are also 'non-dormant' in the sense that they are fully capable of growth when external conditions permit. Chouard's (1956) term, quiescence, may be more acceptable, as it does not contain the word 'dormancy'. A more important difference between buds and seeds applies to paradormancy, which occurs in buds, but not in seeds, the seed being a self-contained entity. Lang et al. (1987) suggested that the seed-coat may exert a paradormant effect upon the embryo; this would be valid for embryo dormancy perhaps, but not for seed dormancy. The problem of semantics – what is a seed? – arises here. Bewley and Black (1982) note cases in which seed dormancy is influenced by position on the plant or inflorescence, possibly as a result of competition among fruits. Another classic example is cocklebur (*Xanthium strumarium*), in which the lower seed in the fruit is less dormant than the upper seed (Crocker, 1906). Crocker (1916) recognized several types of seed dormancy, including that caused by hard seed-coats, seed-coats impermeable to water and/or gases, immature embryos and physiologically immature embryos. Additional causes include the presence of chemical inhibitors, and the induction of a secondary dormancy in non-dormant seeds when exposed to certain conditions, such as high temperatures (Mayer and Poljakoff-Mayber, 1982). Of these types of dormancy, only physiological immaturity, secondary dormancy and possibly chemical inhibition apply to buds. Secondary dormancy has been used to describe the partial or total reversal of the chilling effect on buds by high temperature (Vegis, 1964). Strictly speaking, however, this is not secondary dormancy, for the buds were not capable of growing when chilling was applied.

Dormancy in Seeds vs. Buds of Herbaceous Species

Simmonds (1963) reported high correlations between tuber and 'true seed' dormancy in potato lines selected for early vs. late germination or sprouting, and concluded that 'seed and tuber dormancy are under a common biochemical control'. Okagami (1986) compared the temperature requirements for the breaking of dormancy in seeds, bulbils and the buds of subterranean organs of ten species of *Dioscorea* collected from subarctic to tropical climates in Japan. Differences were noted in chilling requirements, and in optimum temperatures for both chilling and germination/bud break. However, Okagami's general conclusion was that seeds of species adapted to the northernmost climates had shorter chilling requirements than did buds on their subterranean organs, whereas the opposite was true for species from the southern climates (Fig. 4.1). Thus, seed and bud dormancy were not correlated.

Dormancy in Seeds vs. Buds of Woody Species

Frisby and Seeley (1993) compared bud, seed and seedling dormancy in 'Johnson Elberta' peach, using seeds from fruits that developed from openpollinated flowers. Based on the similar response of seeds, seedlings and cuttings to chilling, they concluded that 'their dormancy-release and growth-promoting mechanisms due to chilling are similar'.

Both species and ecotypes within species differ with respect to the need for chilling by seeds and buds (Table 4.1). Comparisons are difficult, for sources often differ in evaluating the need for chilling. For example, Sondheimer *et al.* (1968) considered seeds of *Fraxinus ornus* to be non-dormant. Other investigators, however, have reported them to be partially or wholly dormant (e.g. Arrillaga *et al.*, 1992). In some tropical and subtropical species (e.g. orange), neither buds nor seeds require chilling for subsequent growth. In olive, a brief exposure to low temperature (15°C) hastens germination, and the seeds have some requirement for 'chilling' (Istanbouli and Neville, 1977; Lagarda *et al.*, 1983). However, the chilling requirement for buds varies with the type of bud. Rallo and Martin (1991) presented evidence that floral buds require chilling, but not vegetative buds.

In many cases (e.g. *Acer saccharum*), both seeds (shed in the autumn) and buds require chilling (Webb and Dumbroff, 1969; Schopmeyer, 1974). Although buds of most *Acer saccharinum* trees require chilling, those of some individual trees may not (Ashby *et al.*, 1991); all seeds (shed in the early summer), however, are non-dormant (Rudnicki and Suszka, 1969; Ashby *et al.*, 1991). In northern ecotypes of *Acer rubrum*, buds must be chilled; however, only a portion of the seeds are dormant (Wang and Haddon, 1978; Farmer and Cunningham, 1981; Farmer and Goelz, 1984). Some southern ecotypes exhibit neither bud nor seed dormancy (Perry, 1960, 1970, 1971, and pers. comm.). Seeds of *Fagus sylvatica* definitely require chilling (Schopmeyer, 1974), but sources differ as to the buds' need for chilling. Howard (1910) found the species to be among those requiring the most exposure to low temperatures before cuttings would break bud, although he did not appreciate the temperature effect at the time. However, Wareing (1953) reported that bud dormancy was controlled by photoperiod, rather than

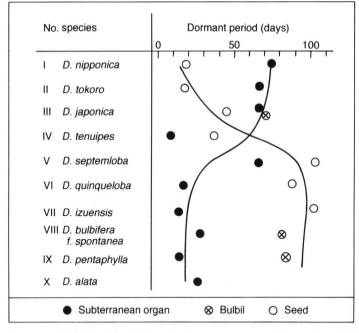

Fig. 4.1. Dormant periods of seeds, bulbils and subterranean organs of ten species of *Dioscorea* in relation to their north–south distribution. The periods required for 50% germination or sprouting (sum of days of prechilling and days of incubation for germination or sprouting) under the most suitable experimental conditions for breaking dormancy are indicated. (From Okagami, 1986, by permission.)

temperature. More recently, Heide (1993) and Lang (1994) have presented evidence that long photoperiods and chilling temperatures interact to break bud dormancy in *Fagus*.

In 'evergreen' peach (Rodriguez-A. *et al.*, 1994), terminal bud dormancy is controlled by a single gene; in homozygous recessive plants, growth of the terminal bud continues whenever temperature permits. However, both lateral buds and seeds require ~100 to 200 chilling units for resumption of growth or germination (Diaz and Martin, 1972; Rodriguez-A. *et al.*, 1994; R. Scorza, pers. comm.; W.B. Sherman, pers. comm.). Thus, seed dormancy parallels the dormancy of the lateral buds, rather than that of the terminals. Bud dormancy in filbert (*Corylus avellana*) is also controlled by a single recessive gene (Thompson *et al.*, 1985). Unfortunately, no data on seed dormancy are available for non-dormant genotypes of this species (M. Thompson, pers. comm.).

On the basis of the data in Table 4.1, the factors controlling dormancy in seeds and buds of the same species may differ considerably. A similar question is whether a relationship exists between the chilling requirements of buds and seeds within species in which both organs require chilling. Westwood and Bjornstad (1968) were among the first to determine the seed chilling requirements of species, within a genus, that have different bud chilling requirements. Temperature optima for seed stratification were higher for *Pyrus* species with short bud chilling requirements than for those with long chilling requirements, in agreement with the climatic conditions in the regions of origin. Individual *P. calleryana* seeds that germinated only after extended periods of chilling produced seedlings with longer than average bud chilling requirements. From these and other observations, they

Table 4.1. Chilling requirements of buds vs. seeds of woody perennials (see text for references).

Chilling required for subsequent growth		
Seeds	Buds	Species
No	No	*Citrus sinensis*
No?	Floral only	*Olea europea*
Yes	Yes	*Acer saccharum*
Yes	Yes*	*Fagus sylvatica*
No	Yes	*Acer saccharinum*
No (hard)	Yes	*Robinia pseudoacacia*
Some	Yes	*Acer rubrum* (northern ecotypes)
No	No	*Acer rubrum* (southern ecotypes)
Yes	No, Yes†	*Prunus persica* genotypes
?	No	*Corylus avellana* genotypes

* Also require long photoperiods for expansion (see text).
† Terminal buds are non-dormant, lateral buds require chilling.

concluded that the chilling requirements of *Pyrus* seeds paralleled those of buds of the same species and of interspecific crosses.

Breeders have made similar comparisons, using various ways of expressing the results (Table 4.2). In many cases, highly significant correlations were obtained between the chilling requirements of individual seeds and those of the seedlings arising from them, or, in the case of hybridizations, of the midparent mean for bud chilling requirement vs. the chilling requirement of the resultant seeds or seedlings. Several exceptions are noteworthy, however. In almond, although correlations were highly significant statistically, the absolute r values were often low (Kester et al., 1977). In low-chilling peaches, r values were both low and, in two of three cases, non-significant (Rodriguez-A. and Sherman, 1985). The latter observation may reflect the fact that chilling requirements of the parent trees were low and varied little

Table 4.2. Correlations between seed and bud dormancy in several woody species.

Species	Time required for		r value	Reference
	Bud break/bloom	Germination		
Peach, almond	Bloom date (midparent mean)	Days to germ. at 10°C	0.803* (1965) 0.863* (1967)	Kester (1969)
Almond	Days from 20 Feb. to 50% shoot emerg.	Same	0.154*	Kester et al. (1977)
	Days from 20 Feb. to 5–10% bloom	Same	0.129*	
	Same (family mean)	Same	0.854*	
	Same (midparent mean)	Same	0.686*	
Peach	Bud break of parent tree (1 = early to 10 = late)	Days to 80% germ. at 6°C	0.73	Perez et al. (1993)
Peach	Bud chilling units (CU) (midparent mean)	Hours to 80% germ. at 5–6°C (family mean)	0.08	Rodriguez-A. and Sherman (1985)
	Same (family mean)	Same	0.01	
	CU of resulting seedling	Hours to 80% germ. at 5–6°C	0.21*	
Apple	Bloom date of parent tree	Weeks to 50% germ. at 4°C	0.85*	Mehlenbacher and Voordeckers (1991)
	Average seedling bud break	Same	0.26–0.66*	

* Significant at $P < 0.01$.

among genotypes. In apple, although bloom date of the parent tree was highly correlated with chilling requirement of the seed, r values for seedling bud break vs. seed chilling requirement were low (Mehlenbacher and Voordeckers, 1991). These data provide some support for a relationship between chilling requirements of seeds and buds, but they also indicate that the relationship often is not a close one. One of the problems involved in these comparisons is determining what to compare. For example, family means generally yield higher r values than do data for individual seedlings or seeds.

Given the accumulated evidence, bud dormancy and seed dormancy are not inseparably linked within species. But does this mean that they are not controlled by the same mechanisms? If genes controlling dormancy exist in the bud, they must exist in the seed as well, and vice versa. However, a hard seed-coat could obviate the need for expression of such genes in seeds. One might test this hypothesis by determining if dormancy can be induced in such seeds following scarification. Seeds of many pine species are not dormant at harvest, but become dormant on drying (Krugman and Jenkinson, 1974). Thus, drying appears to 'turn on' genes responsible for dormancy in such species. A similar situation occurs in filbert (Jarvis, 1975), except that embryo dormancy is involved; embryos excised from freshly harvested seeds can germinate without chilling, but drying the seeds induces embryo dormancy. In other species, expression of genes controlling dormancy may be totally blocked. The ability to synthesize certain pigments, for example, is often confined to the petals of flowers. By analogy, then, certain mechanisms of dormancy could exist in buds of a given species, but not in the seeds, or vice versa.

As in most physiological processes, difficulties arise in drawing conclusions concerning bud vs. seed dormancy. Romberger (1963: p. 177) states the problem well: 'The reader who expects a pithy summary, replete with sweeping truths about the ... physiological basis of dormancy, will be disappointed.' Nevertheless, some mechanisms that prevent development in seeds (e.g. hard seed-coats) do not apply to buds, and vice versa. Within species, bud dormancy and seed dormancy may be similar or may differ considerably. Although the chilling requirements of buds and seeds of a given genotype are often similar, correlations between the two often leave much to be desired. These facts suggest that bud dormancy and seed dormancy are similar, but not identical, phenomena. Hopefully, genetic analysis, using molecular genetics and other tools, will provide definitive evidence as to both the genetic and the physiological basis of dormancy and permit a 'final answer' to this question. Non-dormant mutants have been used to explore the role of hormones in controlling seed dormancy in herbaceous species, e.g. *Arabidopsis* and tomato (Karssen *et al.*, 1987). The use of non-dormant genotypes of *Prunus*, *Corylus* and other species gives promise of providing new answers to the riddle of dormancy in woody plants as well.

References

Arias, O. and Crabbé, J. (1975) Les gradients morphogénétiques du rameau d'un an des végétaux ligneux en repos apparent. Données complémentaires fournies par l'étude de *Prunus avium* L. *Physiologie Végétale* 13, 69-81.

Arrillaga, I., Marzo, T. and Segura, J. (1992) Embryo culture of *Fraxinus ornus* and *Sorbus domestica* removes seed dormancy. *HortScience* 27, 371.

Ashby, W.C., Bresnan, D.F., Huetteman, C.A., Preece, J.E. and Roth, P.L. (1991) Chilling and bud break in silver maple. *Journal of Environmental Horticulture* 9, 1-4.

Bewley, J.D. and Black, M. (1982) *Physiology and Biochemistry of Seeds in Relation to Germination. Vol. 2. Viability, Dormancy and Environmental Control.* Springer-Verlag, New York.

Champagnat, P. (1983) Quelques réflexions sur la dormance des bourgeons des végétaux ligneux. *Physiologie Végétale* 21, 607-618.

Chouard, P. (1956) *Dormance et inhibition des graines et des bourgeons. Préparation au forçage et au thermopériodisme.* Centre de Documentation Universitaire, Paris.

Crabbé, J. (1968) Evolution annuelle de la capacité intrinsèque de la pousse de l'année chez le pommier et le poirier. *Bulletin de la Société Royale Botanique Belgique* 101, 195-204.

Crabbé, J. (1994) Dormancy. In: Arntzen, C.J. and Ritter, E.M. (eds) *Encyclopedia of Agricultural Science*, Vol. 1. Academic Press, New York, pp. 597-611.

Crocker, W. (1906) Role of seed coat in delayed germination. *Botanical Gazette* 42, 265-291.

Crocker W. (1916) Mechanism of dormancy in seeds. *American Journal of Botany* 3, 99-120.

Diaz, D.H. and Martin, G.C. (1972) Peach seed dormancy in relation to endogenous inhibitors and applied growth substances. *Journal of the American Society for Horticultural Science* 97, 651-654.

Farmer, R.E., Jr and Cunningham, M. (1981) Seed dormancy of red maple in east Tennessee. *Forest Science* 27, 446-448.

Farmer, R.E. and Goelz, J.C. (1984) Germination characteristics of red maple in northwestern Ontario. *Forest Science* 30, 670-672.

Frisby, J.W. and Seeley, S.D. (1993) Chilling of endodormant peach propagules: V. Comparisons between seeds, seedlings and cuttings. *Journal of the American Society for Horticultural Science* 118, 269-273.

Heide, O.M. (1993) Dormancy release in beech buds (*Fagus sylvatica*) requires both chilling and long days. *Physiologia Plantarum* 89, 187-191.

Howard, W.L. (1910) An experimental study of the rest period in plants: the winter rest. First report. *Missouri Agricultural Experiment Station Research Bulletin* 1.

Istanbouli, A. and Neville, P. (1977) Etude de la 'dormance' des sémences d'Olivier (*Olea europea* L.). Mise en évidence d'une dormance embryonnaire. *Comptes Rendus Academie des Sciences, Paris, Ser. D* 284, 2503-2506.

Jarvis, B.C. (1975) The role of seed parts in the induction of dormancy of hazel (*Corylus avellana* L.). *New Phytologist* 75, 491-494.

Junttila, O. (1988) To be or not to be dormant: some comments on the new dormancy nomenclature. *HortScience* 23, 805-806.

Karssen, C.M., Groot, S.P.C. and Koornneef, M. (1987) Hormone mutants and seed dormancy in *Arabidopsis* and tomato. *Society for Experimental Botany Seminar Series* 32, 119-133.

Kester, D.E. (1969) Pollen effects on chilling requirements of almond and almond-peach hybrid seeds. *Journal of the American Society for Horticultural Science* 94, 318-321.

Kester, D.E., Raddi, P. and Asay, R. (1977) Correlations of chilling requirements for germination, blooming and leafing within and among seedling populations of almond. *Journal of the American Society for Horticultural Science* 102; 145-148.

Krugman, S.L. and Jenkinson, J.L. (1974) *Pinus* L.: pine. In: Schopmeyer, C.S. (ed.) *Seeds of Woody Plants in the United States*. Agriculture Handbook No. 450, Forest Service, US Department of Agriculture, Washington, DC, pp. 598-638.

Lagarda, A., Martin, G.C. and Kester, D.E. (1983) Influence of environment, seed tissue and seed maturity on 'Manzanillo' olive seed germination. *HortScience* 18, 868-869.

Lang, G.A. (1994) Dormancy: the missing links - integrating molecular information into complex plant/environment systems. *HortScience* 29, 1255-1263.

Lang, G.A., Early, J.D., Martin, G.C. and Darnell, R.L. (1987) Endo-, para- and eco-dormancy: physiological terminology and classification for dormancy research. *HortScience* 22, 371-377.

Mayer, A.M. and Poljakoff-Mayber, A. (1982) *The Germination of Seeds*, 3rd edn. Pergamon Press, New York.

Mehlenbacher, S.A. and Voordeckers, A.M. (1991) Relationship of flowering time, rate of seed germination and time of leaf budbreak and usefulness in selecting for late-flowering apples. *Journal of the American Society for Horticultural Science* 116, 565-568.

Nigond, J. (1967) Recherches sur la dormance des bourgeons de la vigne. I. Caractères généraux de l'évolution des bourgeons. *Annales de Physiologie Végétale* 9, 107-152.

Okagami, N. (1986). Dormancy in *Dioscorea*: different temperature adaptation of seeds, bulbils and subterranean organs in relation to north-south distribution. *Botanical Magazine, Tokyo* 99, 15-27.

Perez, S., Montes, S. and Mejia, C. (1993) Analysis of peach germplasm in Mexico. *Journal of the American Society for Horticultural Science* 118, 519-524.

Perry, T.O. (1960) Genetic variation in the winter chilling requirement for date of dormancy break for *Acer rubrum. Ecology* 41, 790-794.

Perry, T.O. (1970) Immobility of dormancy regulators in interracial grafts of *Acer rubrum* L. *Plant Physiology* 46 (Suppl.), 175.

Perry, T.O. (1971) Dormancy of trees in winter. *Science* 171, 29-36.

Pollock, B.M. and Olney, H.O. (1959) Studies of the rest period. I. Growth, translocation and respiratory changes in the embryonic organs of the after-ripening cherry seed. *Plant Physiology* 34, 131-142.

Powell, L.E. (1987) Hormonal aspects of bud and seed dormancy in temperate-zone woody plants. *HortScience* 22, 845-850.

Rallo, L. and Martin, G.C. (1991) The role of chilling in releasing olive floral buds from dormancy. *Journal of the American Society for Horticultural Science* 116, 1058-1062.

Rodriguez-A., J. and Sherman, W.B. (1985) Relationships between parental, seed and seedling chilling requirement in peach and nectarine. *Journal of the American Society for Horticultural Science* 110, 627-630.

Rodriguez-A., J., Sherman, W.B., Scorza, R., Wisniewski, M. and Okie, W.R. (1994) 'Evergreen' peach, its inheritance and dormant behavior. *Journal of the American Society for Horticultural Science* 119, 789-792.

Romberger, J.A. (1963). *Meristems, Growth and Development in Woody Plants.* Technical Bulletin No. 1293, US Department of Agriculture, Washington, DC.

Rudnicki, R. and Suszka, B. (1969) Abscisic acid in non-dormant seeds of silver maple (*Acer saccharinum* L.). *Bulletin de l'Académie Polonaise des Sciences, Series 5* 17, 325-331.

Schopmeyer, C.S. (1974) *Seeds of Woody Plants in the United States.* Agriculture Handbook No. 450, Forest Service, US Department of Agriculture, Washington, DC.

Simmonds, N.W. (1963) Correlated seed and tuber dormancy in potatoes. *Nature* 197, 720-721.

Sondheimer, E., Tzou, D.S. and Galson, E.C. (1968) Abscisic acid levels and seed dormancy. *Plant Physiology* 43, 1443-1447.

Thompson, M.M., Smith, D.C. and Burgess, J.E. (1985) Non-dormant mutants in a temperate tree species, *Corylus avellana* L. *Theoretical and Applied Genetics* 70, 687-692.

Vegis, A. (1964) Dormancy in higher plants. *Annual Review of Plant Physiology* 15, 185-224.

Wang, S.P. and Haddon, B.D. (1978) Germination of red maple seed. *Seed Science and Technology* 6, 785-790.

Wareing, P.F. (1953) Growth studies in woody species. V. Photoperiodism in dormant buds of *Fagus sylvatica* L. *Physiologia Plantarum* 6, 692-706.

Wareing, P.F. and Black, M. (1958) Photoperiodism in seeds and seedlings of woody species. In: Thimann, K.V. (ed.) *The Physiology of Forest Trees.* Ronald Press, New York, pp. 539-556.

Wareing, P.F., Eagles, C.F. and Robinson, P.M. (1964) Natural inhibitors as dormancy agents. In: *Régulateurs Naturels de la Croissance Végétale*, Fifth International Conference on Plant Growth Substances, Gif-sur-Yvette, France. Centre National de la Recherche Scientifique, Paris, pp. 377-386.

Webb, D.P. and Dumbroff, E.B. (1969) Factors influencing the stratification process in *Acer saccharum. Canadian Journal of Botany* 47, 1555-1563.

Westwood, M.N. and Bjornstad, H.O. (1968) Chilling requirements of dormant seeds of 14 pear species as related to their climatic adaptation. *Proceedings of the American Society for Horticultural Science* 92, 141-149.

II Bud Dormancy Systems and Concepts

5 Dormancy and Symplasmic Networking at the Shoot Apical Meristem

CHRIS VAN DER SCHOOT
Agrotechnological Research Institute (ATO-DLO), Bornsesteeg 59, PO Box 17, 6700 AA Wageningen, The Netherlands

Introduction

The phenomenon of dormancy is defined by the lack of visible growth (Romberger, 1963; cited by Lang *et al.*, 1987). This notion is essentially phenomenological since it describes something – suspended visible growth – without reference to a specific mechanism; dormancy is just 'the state of being dormant' (Webster, 1966). In physiological terms, dormancy has been defined as a state in which the metabolic processes are slowed (Usher, 1965). It has been recognized that the dormant state might be described more precisely in terms of complex physiology (Lang *et al.*, 1987). It is not clear how this could be done, due to the lack of any mechanistic model to which real physiogical states could be compared.

Since the dormant state is a temporary absence – of visible growth – it might be defined alternatively with reference to the active state. Investigations of dormancy have, therefore, focused on when and how a 'presence', i.e. visible growth, is lost and resumed. The phenomenon of dormancy, then, is pinpointed by characterizing transitions rather than 'states'. Like the dormant state, however, the transitions remain elusive in terms of mechanisms (Seeley, 1994). The way the various definitions restrict the use of the word dormancy to particular cases of 'visibly suspended growth' is based on the behaviour of the particular plant structure. This behaviour is thought to be spontaneous or enforced and it is interpreted in terms of hypothetical physiological mechanisms, the existence of which is inferred from physiological changes occurring during this behaviour. There is no consensus about the questions of: (i) whether these hypothetical physiological mechanisms are specific; and (ii) whether or not they originate in the particular plant part itself (see Lang *et al.*, 1987; and comments of Junttila, 1988).

The existence of various dormancy definitions illustrates both the uncertainties concerning the physiological mechanisms and the different views regarding what dormancy is (reviewed by Lang *et al.*, 1987). These uncertainties relate to a wide range of basic questions about dormant systems. Why is there a phenomenon of dormancy at all and how is it mastered? Is dormancy a property of whole plants (e.g. Koller, 1969) or seeds, plant parts (e.g. van Ittersum, 1992), individual growing points (e.g. Rees, 1981; cited by Lang *et al.*, 1987), or all meristem-containing structures (Lang *et al.*, 1987)? How do potentially dormant parts within one plant influence each other and what keeps surveillance over the whole? Are the various cases of dormancy a manifestation of the same kind of physiological mechanism? Is it possible to classify the various cases of dormancy on the basis of a single (set of) physiological mechanism(s), implemented in different ways in the various 'types' of dormancy (see discussions in Lang *et al.*, 1987), or do we have to think of multiple mechanisms (Dennis, 1994)? Are these mechanisms specific (Junttila, 1988) or might they be unspecific too (e.g. as in ecodormancy; Lang *et al.*, 1987)? Are these mechanisms just devices linking the potentially dormant structure to its internal and/or external environment, or are they integral parts of the system? In the case of 'external links', the assumed specificity may not be relevant since it might be a property of the responding system; non-specific stimuli can invoke a specific response due to the restricted response possibilities of the system. In the case of 'internal links', it is still possible that specific and unspecific stimuli invoke an identical endstate via a different trajectory. Is there a physiological basis for a universal descriptive terminology (Lang *et al.*, 1987), or is the phenomenon universal but the underlying mechanism exclusive?

The uncertainty concerning the physiological nature of dormancy is reflected in the way that the phenomenon is studied. Is the aim to give an accurate description of dormancy as a function of the environment and as a function of the physiology of individual systems (e.g. Junttila, 1988), or as a function of the environment and those aspects of the physiology which are more general and universal (Lang *et al.*, 1987)? Are additional experimental approaches to dormancy (e.g. the search for specific genes) essential for our understanding (e.g. Lang, 1994)? What role do growth substances, in particular the 'dormancy hormone' abscisic acid (ABA), play in development towards the dormant state, and can this role be established experimentally without overriding the control systems of the plant (Trewavas and Jones, 1991)? Do we understand dormancy if we are able to experimentally manipulate the system and predict its behaviour?

The answers to these physiological and methodological questions shape our perception and, importantly, affect the formulation of research strategies. They relate to the particular motivation of our interest in dormancy, our desire to either master the phenomenon or explain the nature and behaviour of a system which switches between various active and dormant states. In the latter case, our view on plant organization and on the notion of individuality and autonomy of plant parts is crucial. In the following, attention will be focused on the centres of primary growth and development in shoots, endorsing the intuition of Rees that 'dormancy is a property of the growing point' (Rees, 1981; cited by Lang *et al.*, 1987).

Dormancy will be viewed as a stationary phase in the development of the growing point. Although it is often assumed that development is programmed in the genes, the view is more appropriate that the developing system displays differential gene expression (e.g. Harold, 1990). Since development is a property of an integrated system, it is necessary to understand the nature of the system, and its actual and potential behaviour, in order to understand how and why such a system can display dormancy. It will be argued that, by looking at the global aspects of development, specific transition points in the transformation of the system can be identified. These transition points are open to specific experimental manipulation by molecular technologies.

Physiological Mechanisms, Networking and Morphogenesis

Physiological mechanisms underlie the induction, maintenance and release of dormancy. The conditions internal and external to the plant influence or evoke the emergence of these physiological mechanisms. Therefore, a careful description of dormancy with respect to both environmental control and physiological mechanisms is essential (e.g. Junttila, 1988). It is useful to make an *a priori* distinction between triggers or triggering processes and the transformation of the system (the shoot apical meristem, AM) itself.

Classically, dormancy has been defined and classified with regard to the initial physiological reactions which eventually lead a non-dormant system to a state of dormancy (see for an overview, Lang *et al.*, 1987) as well as with regard to the overall behaviour of the dormant system, e.g. whether or not it can resist favourable conditions (e.g. Junttila, 1988). In the first case it is assumed that the way a non-dormant system develops towards dormancy is determined by the initial triggers, and that the subsequent pattern of behaviour of the resulting dormant state (i.e. when and how it resumes growth) is also determined by the initial triggers (see comments of Lang *et al.*, 1987, and Junttila, 1988). In the second case, the behaviour of the system motivates the classification. When the two classifications conflict, the conclusion seems justified that the behaviour is not determined by the triggers.

With regard to this basic classification problem, Lang (1987) and Lang *et al.* (1985, 1987) aimed to derive a physiologically based descriptive terminology applicable to all the different within-plant and between-plant cases of dormancy. They proposed the use of a single base term 'dormancy', with a further specification into three types of dormancy: eco-, para- and endodormancy. The initial triggers could be different, yet the outcome would be similar: 'the single observable event of dormancy may be a function of multiple processes that are equal, interactive and dynamic' (Lang et al., 1987). Junttila (1988) pointed out that this classification does not accommodate the frequently occurring cases of eco-para-endodormancy; in fact, this is implied already in the notion of 'interactive, multiple processes'. In summary, multiple interactive processes (different physiological pathways), variable in their evolution and propagated through a transforming system, result consistently in an identically organized dormant state (the 'single observable event', Lang et al., 1987). This implies that dormancy is an alternative physiological state, the possibility of which is built into the nondormant state. Therefore, a general identification of the dormant state with a specific triggering mechanism cannot be justified.

The development and alleviation of dormancy can be modelled as events related to catastrophic changes in physiological networking. These catastrophic changes in the network can be brought about by variable combinations of stimuli entering the network at multiple locations. As a result, the network transforms, through a variable trajectory, into a dormant state. Dormancy as a developmental phase can thus be understood, not from physiological changes as such, but from patterns of organization emerging from physiological networking. In fact, the question 'What is dormancy?' is similar to the question 'What is morphogenesis?'. Neither term has a specific physiological content. Morphogenesis denotes 'organization of development' and dormancy denotes its 'temporary absence'. So, just as the concept of morphogenesis does not imply any specific physiological mechanism, neither should the concept of dormancy. Morphogenesis and dormancy similarly denote two different states of a dynamically organized system. During morphogenesis, this dynamic system has a developmental trajectory. During dormancy, the developmental trajectory has a vector of zero or almost zero. Therefore, dormancy can be described as the stationary phase of the physiological networks that pattern the shoot.

Morphogenesis and dormancy can be modelled as alternative states of the AM growth centres (the shoot apical meristems). The state of dormancy is distinct from the mechanisms (triggers) of induction and release, and can be characterized separately. This would be true even if the 'mechanisms' were integral parts of the system. It requires only that we distinguish between the system which is developing towards, or emerging from, the dormant state (i.e. the transforming system) and the dormant state of the system (i.e. the transformed system). Upon dormancy induction, the overall system (the AM) is committed to a developmental trajectory which ends in the metastable state of dormancy. This process is homologous to that of an individual cell in the AM which is committed to a certain developmental path and ends up in the metastable state of a particular cell type. Each cell is in fact a complex system which can be conceived of as gravitating into the 'attractor' of the corresponding cell type during the process of morphogenesis (Kauffman, 1993). During the development of dormancy, the entire tissue of the AM 'can settle to a spatially heterogeneous and more or less ordered attractor pattern' (Kauffman, 1993) that ends in a stable dormant state which has lost its developmental vector: it is trapped into an overall attractor.

This model accommodates classic notions of dormancy at a different conceptual level. Accordingly, the dormant system develops the ability to 'climb out of its attractor' via removal of global or non-specific constraints (ecodormancy), or more specific constraints imposed by other plant parts (paradormancy), or even by spontaneous action of innate cause (endodormancy). It is the AM which displays this behaviour and which is the embodiment of the potentially dormant system. This system is autopoietic (Varela et al., 1974): it is self-referential in its growth and development, a unit of self-organization which produces, in a repetitive way, species-specific patterns during postembryonic development of the plant.

The Study and Classification of Dormancy

The experimental study of dormancy is confronted with two difficulties: the dynamic nature of the system which displays dormancy and the phases through which it transitions. The nature of an integrated system influences its possible responses to an experimental treatment in the sense that its dynamic organization permits certain responses but not others. By just perturbing the system, a commitment to a developmental attractor can be induced, and the specificity of this response is a property of the perturbed system (Kauffman, 1993; compare also Trewavas, 1986, 1987). This implies that the responses obtained following experimental treatment of a potentially dormant system – an AM – reveal which transformations the system can undergo on grounds of its internal organization rather than what mechanisms evoke the same transformation in the unperturbed system. Different transformation processes might result in transformed states identical in organization and resulting morphology (Trewavas, 1986, 1987).

Another difficulty is presented by the fact that potentially dormant structures display network states that change in time. Distinguishing between these different states is important if we are to infer causalities from the behaviour observed in the system which is developing or breaking dormancy. Induction of dormancy depends on the sensitivity of the non-dormant state, whereas breakage of dormancy depends on the sensitivity of the dormant state. Any dormancy-related innate physiological mechanism is integrated into the continuously changing network of the AM, which strongly hampers investigation of such mechanisms. To circumvent this, we might focus on the dynamic organization of the system and thus approach the phenomenon by studying the overall behaviour of the integrated system, not the component parts. The experimental approach described below investigates dormancy and development as functions of alterations in the global morphological development of the network system (i.e. the AM). Explanation, at the global level, of the dynamic properties of the system will shed some light on general properties common to all dormant states.

Self-regulatory Growth at the Apical Meristem

In dormant potato tubers, the bud is the most important structure (Burton, 1966; Rappaport and Wolf, 1969; Cutter, 1978). It is a centre of growth and organization and as such it has a high degree of autonomy (Steeves and Sussex, 1989; Sussex, 1989). Its activity can be modulated by influences from within or without the plant or tuber system, but pattern development is dependent on its own existing organization. The structure of this organization results in a species-specific architecture of the plant, something which is generally a conservative feature. In a certain sense, buds are semi-autonomous structures linked up with the plant body; morphogenesis in buds (i.e. the AMs) is independent of the plant or tuber that carries them. In that the AM has a small repertoire of behavioural responses to unbalancing factors, and that this repertoire is a property of the responding AM, their potential responses are therefore characteristic of the system. Excised buds and even isolated AMs can be grown in a petri dish when they are supplied with sufficient nutrients and hormones (Smith and Murashige, 1970). Their morphogenetic activity must, therefore, be a property of their specific organization. This morphogenetic potential must be located specifically in the AM. When an AM is split, each part regenerates a complete apex in which the regenerating AM models itself upon itself (e.g. Ball, 1955, 1980). The centre of a meristem has the potential to regenerate a complete apex, and even small panels of cells have this potential provided the other tissues are removed (Ball, 1980; Steeves and Sussex, 1989, and references therein). Such observations support an earlier conclusion of Sussex (1952) that the AM (of potato) is a 'self-regulatory field of growth'.

Apical meristems are linked via the signal and supply routes of the stem and the branches. These links shape plant growth form by coordinating the activities of the individual growth centres or meristems, but they do not govern pattern formation at the growth centres. Via these links, morphogenetic activity of the semi-autonomous shoot meristem can be activated or blocked, individually or collectively, from outside or inside the plant body. How is this dynamic network positioned in the AM, and how does it relate to the structures which it generates and by which its activity, in turn, is modified?

In order to produce a shoot, the pattern-generating mechanism of the AM must oscillate between certain basic states. As the network of the meristem is an integration of the individual networks of each cell, it is implied that alterations in cell–cell contacts make this possible. Movement of cells to the periphery of the AM will change the relationships between them, thereby altering the structure of the overall network of the AM. Although plant cells do not move, being fixed in the plant body, their cytoplasm moves. In the organismic view of plant development, the cells are chambers in the symplasm (e.g. Kaplan and Hagemann, 1991). Cell divisions are incomplete and it is the symplasm which expands, differentiates and subsequently consolidates these initial differences by placing perforated cell walls within the membranous compartment of the cell. In this view, it is the symplasm which has primacy: cells or chambers are created within an expanding symplasm. The spatial differentiation of the symplasm is reflected in a continuous stretching and modification of the 'extracellular matrix' (Roberts, 1994).

Apical meristem patterns and dynamics

The AM of a dicotyledonous plant can be described in different and complementary ways (for review, see Sussex, 1989). The tunica–corpus (layer) model describes the patterns of cell production (Fig. 5.1). Cells are produced by anticlinal divisions in centrally located initial cells, resulting in distinctive layers. The number of layers varies among species and may vary periodically (Sussex, 1955). In Fig. 5.1, three layers are represented, which are superimposed on each other to form the tunica. They extend laterally by division in the central mother cells, but even more by division of their derivatives. From the circumference of this layered structure, the primordia are formed, giving rise to laterally attached organs. The second population of meristematic cells is located immediately below the tunica. This so-called corpus gives rise to the secondary meristematic tissues of the stem and to its parenchyma. Its activity pushes the AM 'up'. Since cells propagating horizontally from a mother cell are genetically similar, this 'layer model' often is used in studies of tissue-related gene expression.

The alternative 'zonation' model focuses on the functional patterns of cells. It appears that cytologically distinct zones can be distinguished which differ in metabolism, cell cycling rate and sometimes presence of organelles. The most prominent aspect is the presence of a central zone whose overall activity is less than that of the surrounding peripheral zone (Fig. 5.1b). The central zone includes cells of all tunica layers and the upper corpus layers. Morphologists have commonly preferred this model and have, for instance, stressed the primacy of the symplasm over cell generation (Kaplan and Hagemann, 1991). These two models focus on different aspects of organization, both of which result from, and relate to, the networking activity of the AM. In essence, the AM is not a strictly programmed production centre of cells, but rather a unit of self-organization which functionally compartmentalizes its symplasm in order to divide morphogenetic activity.

Specialization of cells in the AM is relative (Tilney-Bassett, 1986; Steeves and Sussex, 1989; Sussex, 1989). When dead cells in the AM are replaced by divisions in other layers, the new cells attain the specialization of the lost cells. Therefore, cells specialize in a manner which reflects their position in a lineage, not because they have inherited this specialization tendency, but because they have been put spatially in this particular position in the symplasmic network. When they change position in this symplasmic network, they are 'reprogrammed'. For instance, microsurgery has demonstrated that changes in the position of cells changes their individual, but not their collective, determination (Sussex, 1952), indicating that the 'self-referential' organization preserves its dynamic pattern when parts are excised. It is a feature of networks that the parts possess properties characteristic of the entire network. Therefore, it seems that the AM is the actual

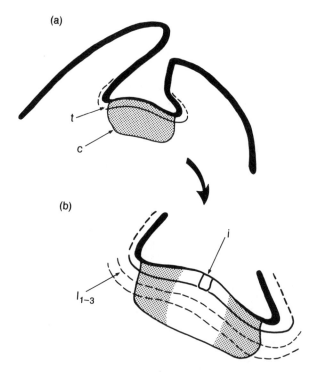

Fig. 5.1. Schematic representation of the apical meristem of potato (not to scale). (a) The apical meristem (AM, shaded) is hidden between overarching young leaves and is composed of a tunica (t) and a corpus (c). (b) Detail of the AM of a resting bud of potato: the initial cell (i, arrow), peripheral zone (shaded) and central zone (not shaded). When the bud assumes morphogenetic activity after the breaking of dormancy, stratification of the tunica increases (stippled layers l_{1-3}).

organizer of the bud and, thus, the shoot. The generator of species-specific organization and architecture is embodied in the AM, which emerges at an early stage of embryogenesis. In conclusion, the behaviour and structure of the cells in the AM are the local expression of the symplasmic networking activity of the integrated AM. Shoot organization is possible through self-organization at the AM.

Cell signalling networks and cellular sensitivity to growth substances

Cells develop cell-cell signalling networks, sense their position in this network, and become committed to one of the possible developmental trajectories. The number of possible trajectories can be thought of as the number of possible cell types, each having their own 'attractor' that pulls the cell into this developmental trajectory (Kauffman, 1993). The commitment to a certain developmental trajectory alters the signal exchange between cells, which in turn results in an overall change of the signal network.

Cell-cell communication can be achieved by various collaborating mechanisms. The two most prominent are membrane-bound mechanisms and the symplasmic continuum. In other words, signals might travel from cell to cell via plasma membranes or via protoplasmic continuities. The first mechanism is open for control via production of growth substances, regulation of receptor densities and modification of the sensitivity of the cells to growth substances. The second mechanism is open to control via modulation of the permeability of plasmodesmata to signals of various kinds.

How do hormones relate to a model of dormancy viewed from the perspective of cell-cell networking and integrated plant systems? Although hormones are required for growth from the AM, Trewavas (1987) has suggested that they are arbitrators of nutrient fluxes rather than controllers of morphogenesis. Their effect is dependent on the presence of other controlling factors and their 'control strength' or 'sensitivity to control'. This 'control strength' is a property not of these growth substances, but of the system which responds to them (Trewavas, 1986, 1987). The internal organization of the system determines how, for instance, second messengers, released after hormone-receptor binding, propagate through the system in numerous parallel and interacting ways. The response will be different each time receptor-binding occurs since the system rearranges itself continuously.

In an integrated system, it is still the internal state of the individual cell that determines its sensitivity. Cell state and corresponding sensitivity are established during the emergence of spatiotemporal patterns in the symplasm, subsequent compartmentalization of the symplasm by (incomplete) cell divisions and differentiation of each individual cell according to the developmental trajectory to which it is committed. Obviously, the extent of signal exchange, e.g. via plasmodesmata, influences which developmental trajectory the cell will follow, partly by modulating its sensitivity to growth substances.

Supracellular organization of the apical meristem

Since dormancy can be regarded as the temporary suspension of morphogenetic activity, attention will be given to those functional changes in the symplasm correlated with the transition from dormancy to activity. The symplasm is 'constricted' at the walls where it has to pass through the plasmodesmata to the next cell. These plasmodesmata are not organelles (Esau, 1977), but continuities of the symplasm. They are equipped, however, with a variable supramolecular complex which permits various kinds of control over the exchange processes they mediate (Lucas *et al.*, 1993). They might also occasionally possess structures, the so-called sphincters (Oleson and Robards, 1990), in the neck region which allow additional control over signal exchange between cells. To better understand the potential role of the symplasm in dormancy, the ultrastructure of plasmodesmata, their basic properties and their supposed role in morphogenesis will be described. Following this, current knowledge on the structure of the symplasm in potato buds will be evaluated.

Plasmodesmata are protoplasmic continuities between individual plant cells (Fig. 5.2). The plasma membrane passes through pores in the cell walls, forming a plasma membrane tube which allows diffusion of small metabolites from cell to cell. The endomembrane system of the connected cells is continuous via the lumen of this protoplasma-filled tube: it contains endoplasmic reticulum (ER), which is appressed into a cylinder (e.g. Lucas *et al.*, 1993) or depicted as a tube (Gunning and Robards, 1976; Botha *et al.*, 1993). Thus, the nuclei of adjacent cells are linked into one supracellular endomembrane system inside the boundaries of the supracellular plasma membrane continuum. Plasmodesmata are engaged in several types of signal transfer from cell to cell, e.g. small metabolites can diffuse through the cytoplasmic sleeve from cell to cell along their chemical gradient. The molecular size exclusion limit (MSEL), which indicates the upper size limit of molecules that can pass through the plasmodesmata, is normally 800–1000 Da, although higher values are reported for specialized cell–cell contacts (e.g. Kempers *et al.*, 1993), and this restricts movement through the sleeve to molecules which are small and water-soluble. Cells can sense the membrane potential of neighbouring cells. Action potentials, triggered in a particular cell or cell groups, propagate through a least-resistance route (i.e. the plasmodesmata if they are present) to other cells. Research on the systemic spread of viruses in plants has revealed the existence of virally encoded movement proteins which facilitate trafficking of viral genetic material from cell to cell. Recent evidence indicates that the movement proteins can also move from cell to cell after targeting to binding sites at the plasmodesmata (e.g. Waigmann *et al.*, 1993). The ER has been shown to provide a lipid signalling pathway for lipids in the endomembrane system (Grabski *et al.*, 1993). This implies that certain lipids might move from the outer nuclear membrane in one cell to that in the next cell. It is clear that plasmodesmata offer a route for intercellular signalling and communication, and various signals which are potentially morphogenetic in their activity use them to traffic to other cells.

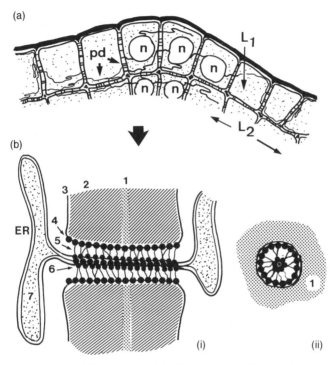

Fig. 5.2. Schematic representation (not to scale) of the position and the ultrastructure of plasmodesmata. (a) Part of the outer two tunica layers, L_1 and L_2, show position of nuclei (n), endoplasmic reticulum (ER) and plasmodesmata (pd, arrows). (b) Generalized scheme of a plasmodesm (redrawn after Ding *et al.*, 1992). (i) Transverse view of the cell walls of two adjacent cells, showing a pore through which the plasma membrane as well as the ER are continuous. 1 = Middle lamella; 2 = cell wall; 3 = plasma membrane; 4 = proteinaceous particles embedded in the plasma membrane tube; 5 = spoke-like extensions, probably partially proteinaceous; 6 = proteinaceous particles embedded in the cylinder of appressed ER; probably involved in macromolecular transfer through plasmodesmata; 7 = lumen of the ER. (ii) Cross-sectional view of the plasmodesm at the level of the middle lamella. The centrally located appressed ER is clothed with particles with are connected by spoke-like extensions to the particles embedded in the plasmamembrane tube. 1 = Middle lamella.

Role of the symplasm in morphogenesis

If plasmodesmata are produced by orchestrating the entrapment of the ER in the developing cell plate (Heplar, 1982), this would yield a population of clonally derived cells in the AM which are symplasmically continuous, but symplasmically isolated from other such groups in the same AM. It is clear from published photographs in ultrastructural studies of apices of various plant species that all cells in any AM are connected by numerous plasmodesmata. This is evidence that between cells of different clonal origin, plasmodesmata are produced *de novo* after the cells have formed. The plasmodesmata that are formed independently of the cell plate are classified as secondary plasmodesmata. The genes involved in the production of these secondary plasmodesmata are continuously active in all cells of the apical meristem.

Returning to the two models of apical organization, the layer model is favoured in molecular analyses of tissue-related gene expression, while the zonation model is favoured in the search for functional physiological zonation related to morphogenesis. The fact that plasmodesmata are present between the various apical layers (Fig. 5.2) is decisive for interpretation of the functional organization of the AM. The obvious structural pattern is the layered structure arising from the clonal propagation of cells within each layer from the centrally located mother cells. In a clonal file of cells, cell types can emerge by means of clonal computation 'in which each cell ignores its neighbours and bases its behaviour on its present state and presumably on its recording of the past history of the clone . . . ' (Kauffman, 1993). When cells can also exchange signals, the situation is more complicated; 'cells carry clonal histories and also consult their neighbours'. In the AM, this signal network is multidimensional. The clonally related cells are symplasmically interconnected in branched patterns, spread out in each tunica layer. Secondary symplasmic connections are made between such clonal groups within and between layers. The production of these secondary plasmodesmata is likely to be very complicated: cell walls must be digested locally, the plasmalemma of both cells must fuse, and the ER of adjacent cells must connect (Kollmann and Glockmann, 1991). The formation of new plasmodesmata in carpel fusion in periwinkle flowers (van der Schoot *et al.*, 1995) can be achieved in a mere 5-6 h, well within the time span of an average cell cycle of 1-3 days in potato AMs (Clowes and McDonald, 1987). The energy requirement and efficiency of this process indicate that this differentiated symplasmic network is vital in morphogenesis at the AM.

An important role for plasmodesmata in morphogenesis has been suggested in many earlier studies (reviewed in Carr, 1976) and is apparent from investigations of a variety of systems (reviewed in Lucas *et al.*, 1993), some of which are simple model systems. Apparently, the symplasm can hold the developmental potential of individual cells in check; removal of symplasmic constraints in lower plants can permit cells to completely redifferentiate and start somatic embryogenesis (Carr, 1976). In gametophyte morphogenesis in *Onoclea sensibilis* (Tilney *et al.*, 1990), the frequency and distribution of plasmodesmata are regulated precisely in relation to future sites of cell division. This suggests that the sites of cell division are controlled via a signal which is distributed through these precisely positioned plasmodesmatal connections. Alternatively, a signal or lack of a signal might result in this pattern of plasmodesmata and cell divisions. In another example, temporary plugging of *Chara* plasmodesmata with electron dense material isolates groups of cells, which synchronizes their divisions independently of other such groups (Kwiatkowska and Maszewski, 1976, 1985, 1986).

In AMs, floral induction is accompanied by mitotic synchronization and restriction of cell–cell communication during the mitotic waves (Goodwin and Lyndon, 1983). *De novo* production of plasmodesmata has been shown to play a role in the production of flower organs (van der Schoot *et al.*, 1995). Very early during carpel fusion in *Catharanthus roseus*, plasmodesmata are formed between the touching carpels, suggesting a role in the symplasmic exchange of signals for the coordinated development of both carpels.

The specific patterns of cell–cell communication through plasmodesmata and the locally variable properties of plasmodesmata are thus a function of the developmental status of the tissue. Alterations in symplasmic integration and selective exchange between cells can completely change the developmental status of the tissue. Complete isolation of individual cells in lower plants might release all developmental constraints on gene expression and lead to a reiteration of developmental potential. In higher plants, more severe constraints are operational. Isolated cell groups perform specialized functions, and single cells are arrested in a dormancy-like state. The opposite formation of a syncytium-like state might synchronize all activities and block differentiation and development.

Supracellular and Cellular Features of AM Organization

Plasmodesmatal connections in the apical meristem

The AM symplasmic network might regulate pattern formation and morphogenesis via selective exchange of signals, as by spatially differentiated characteristics related to the two types of plasmodesmata, i.e. primary and secondary. One distinct type of plasmodesmata, formed by the modification of primary plasmodesmata during the maturation of, for instance, leaves (e.g. Ding *et al.*, 1992), are probably able to transport certain macromolecules from cell to cell: they are sites for binding of viral movement protein, the RNA of which moves between the cells. It is not known if all modified plasmodesmata have the potential to move such large molecules, though there is evidence that this may be the case in the AM (see Lucas *et al.*, 1993). The existence of a mechanism for macromolecular trafficking between cells of different clonal origin could contribute to the control of pattern formation and morphogenesis.

The ability to permit diffusional exchange of small metabolites from cell to cell appears to be similar in primary and secondary plasmodesmata. Changes in MSEL are related to morphogenetic events, but not to plasmodesmata type.

The permeability of the symplasmic continuum for small diffusing metabolites and possible signalling molecules can be investigated with microinjection techniques (described in van der Schoot and Lucas, 1995) based on the iontophoretic injection of fluorescent dyes in individual cells. The dyes, e.g. Lucifer Yellow (Stewart, 1981), are selected because of their specific properties: they have a high quantum yield, are non-toxic, are soluble in water and do not penetrate the plasma membrane. Their net charge permits their injection from a microelectrode by applying a small (-2 to -5 nA) current to it (van der Schoot and Lucas, 1995). The injected dye diffuses through the cytoplasm of the cell and, if the cell is cytoplasmically continuous with neighbouring cells via plasmodesmata, the dye will move to these cells (Fig. 5.3). Such 'dye-coupling groups' (Stewart, 1981) form cytoplasmically united cell groups which are relatively isolated from the surrounding cells.

Metabolites and other dissolved molecules with a molecular mass smaller than the MSEL can diffuse between cells of the coupling group and, therefore, these cells are integrated metabolically to some degree. The degree of coupling can be regulated: the plasmodesmata can, for example, be partially closed, thus lowering the MSEL, and by this means the degree of metabolic integration can be adjusted. Likewise, the coupling group might also be transient because plasmodesmata can re-open to allow cell–cell exchange of metabolites (see Lucas *et al.*, 1993). The significance of this is that the formation and alteration of such symplasma domains is likely to play a main role in morphogenetic processes. The dynamics of the symplasma are expected to correspond to controlled alterations in the physiology of cell coupling groups, which in turn are related to morphogenesis.

Irregular patterns and altered symplasm structure

Shortly after harvest of potato tubers, part of the plasmodesmata between the cells of the AM lack appressed ER in the lumen (Fig. 5.4). This is probably related to the presence of large (up to 5 μm) packages of ER in the cytoplasm of a substantial number of cells. These were observed in various potato cultivars by Shih and Rappaport (1971), who called them 'concentric configurations of endoplasmic reticulum' (CER). They assumed that the presence of CERs was due to hormonal regulation of dormancy. The formation of CER may involve the removal of ER from the cytoplasmic sleeve, resulting in empty plasmodesmata. This would create an enlarged cell – cell continuity which would allow virtually unobstructed and non-selective cell – cell

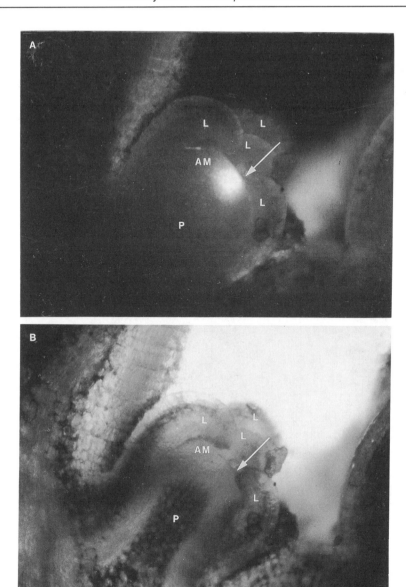

Fig. 5.3. A dye-coupling experiment in which the membrane-impermeable fluorescent dye, Lucifer Yellow CH (LYCH), was microinjected into a single cell of the apical meristem (AM) and moved to other cells, demonstrating the presence of a central group of plasmodesmally linked cells in the AM. (a) Fluorescence of LYCH under blue light. Note presence of LYCH in individual cells. (b) Control photograph with white light indicating the position of the dye-coupling group. AM = apical meristem, L = young leaves, P = pith. Arrow points at injection site.

transport, creating local 'syncytia' within the AM. Injection of Lucifer Yellow CH (LYCH) coupled to 10 kDa dextrans, which cannot pass through normal plasmodesmata (MSEL ca. 900 Da), appeared to traffick from cell to cell in endodormant buds (for schematic representation, see Fig. 5.5a). LYCH (457 Da) trafficked unlimited from cell to cell in similar irregular patterns. This indicates that both molecules might traffick through the same modified plasmodesmata, and that other plasmodesmata are constricted.

During later stages of dormancy (i.e. ecodormancy), the CER dissolved gradually and LYCH-dextran no longer moved from cell to cell when injected in one of the apex cells (e.g. see Fig. 5.5b). Cell – cell movement of LYCH

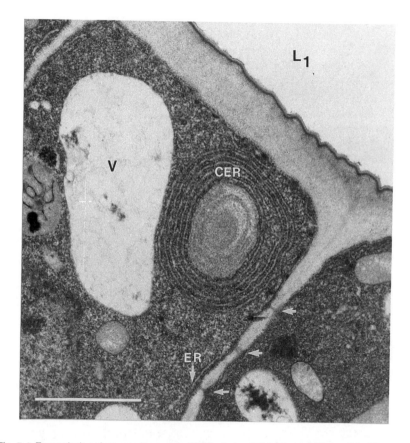

Fig. 5.4. Transmission electron microscopic (TEM) photograph of the outer layer (L_1) of the apical meristem (AM) of the apical eye central bud of an endodormant potato tuber (cv. Desirée), after storage at 4°C for 1 week after harvest. There is only one tunica layer. In the L_1 cell, a concentric configuration of endoplasmic reticulum (CER) is present and endoplasmic reticulum (ER, arrow) sometimes lines the plasma membrane. Arrows point at plasmodesmata. Bar = 4 µm.

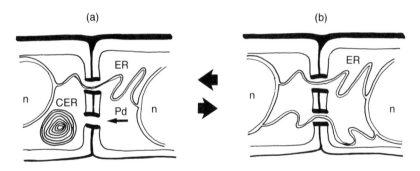

Fig. 5.5. Schematic representation (not to scale) of two adjacent tunica cells in the L_1-layer of a potato tuber bud in different stages of dormancy. (a) Cells of an endodormant bud. Presence of concentric configurations of endoplasmic reticulum (CER) in most cells is correlated with the presence of empty plasmodesmata (Pd) and cell groups dye-coupled by transfer of an enlarged Lucifer Yellow CH molecule (LYCH-dextran). (b) Cells of an ecodormant bud. CER and empty plasmodesmata are not present. No transfer of LYCH-dextran occurs from injected cells to other cells. ER = endoplasmic reticulum; n = nuclei.

became restricted to certain well-defined coupling groups. For instance, when LYCH was injected into a cell of the AM central zone, dye-coupling was restricted to central zone cells (Fig. 5.3). This indicates that these cells are physiologically coupled but isolated from the surrounding cells. The ER connections via plasmodesmata still may constitute a dynamic membrane network across the domain boundary, providing a pathway for signalling (e.g. lipid) molecules.

In the AM of 1 cm potato sprouts, a similar central domain was present. In the late stages of dormancy (at 4°C, imposed or prolonged dormancy), this central domain could still be a static pattern since cell proliferation is minimal, but in vigorously growing sprouts cell production and morphogenetic activity are very high and this central symplasmic domain must be kept in place while the cells are propagating towards the periphery of the AM. As cells pass out of the central zone into the peripheral zone, they must upregulate their constricted plasmodesmata again in order to be integrated physiologically into the symplasm of the peripheral meristem. Since the location of this boundary is approximately constant, there must be a spatially anchored master mechanism which keeps surveillance over plasmodesmata closure. This would serve to keep organ-producing activity confined to the peripheral meristem. Since the only spatially fixed point is the group of initial cells, the mechanism may be set by these cells. Transmission electron microscopy revealed sphincters, which are supposed to be calcium-regulated devices (Oleson and Robards, 1990; Lucas *et al.*, 1993). Therefore, constriction of the plasmodesmata at these domain boundaries might be triggered via a transient release of free calcium and the glucan synthase activator β-furfuryl-β-glucoside at the neck region of the plasmodesmata (Lucas et al., 1993), caused by a signal coming from the initial cell.

Interestingly, the symplasmic domain in the centre of the AM corresponds to the cytological central zone of the classic zonation model. Morphologists have always stressed that zonation is the principal feature of the functional organization in plant apices, and not so much the clonal generation of cells in their respective layers (e.g. Kaplan and Hagemann, 1991). The latter results from compartmentalization of the differentiating symplasmic field as it extends to the AM periphery. Compartmentalization or cell division follows and consolidates symplasma differentiation, rather than causing it, permitting then a further differentiation into specialized cells.

Identification of possible control points in dormancy

These and other data allow the reconstruction of several transitions in AM symplasmic permeability, from the dormant state to the resumption of morphogenetic activity. (i) In early dormancy, part of the plasmodesmata allowed LYCH-dextran$_{10kDa}$ to pass (C. van der Schoot, unpublished results), so their MSEL is at least 10 kDa. This increased MSEL is correlated to the presence of plasmodesmata which lack appressed ER in the lumen. Most plasmodesmata, however, still have their ER and might be downregulated (C. van der Schoot, unpublished results). Both phenomena are correlated with the occurrence of CER in part of the AM cells. It is not yet established that cells which possess CER are identical to those which possess some empty plasmodesmata or to those which exchange the fluorescent 10 kDa dextran. It seems likely, however, that these phenomena are linked. This would suggest that the symplasm is partly in a syncytium-like state, blocking virtually all pattern formation. (ii) In later dormancy, CER has disappeared, LYCH-dextran$_{10kDa}$ is always confined to the injected cell, and no empty plasmodesmata are present (C. van der Schoot, unpublished results). LYCH (mol. wt. 457 Da), however, is transmitted freely between the cells of the central domain. (iii) In the morphogenetically active state, there is no CER, there are no empty plasmodesmata and the central domain is retained (C. van der Schoot, unpublished results). Since cell production is centrifugal, the stability of the central domain demonstrates the dynamic nature of this domain.

This leads to the tentative conclusion that the following happens: the partly syncytium-like state of the AM blocks all processes of specialization. During the breaking of dormancy and the resumption of growth, the exchange possibilities through the plasmodesmata should be restricted by replugging of the ER into the empty plasmodesmata. This is a requirement for the next step to occur, the setting up of pattern in the AM, which precedes morphogenetic activity.

Prevention of sprouting can be achieved by interfering with the breakage of dormancy or by interference with the onset of organogenetic activity needed for sprout growth, and identified points of transition in symplasmic organization might be targeted specifically by molecular techniques. By preventing the replugging of the plasmodesmata by ER, sprouting would be blocked. Alternatively, if the creation of the central symplasmic domain could be prevented, pattern formation would be blocked and, as a result, sprout development.

Conclusions

Dormancy is a phenomenon that is difficult to define in terms of physiological processes. It is a complex, dynamic, multi-faceted phenomenon that probably evolved as a survival strategy. During the emergence of complex plant life, dormancy was allocated to the centres of growth, in particular the terminal growth centres. Since the terminal growth centres have a relative autonomous behaviour with regard to the rest of the plant, they are also relatively autonomous in their development of dormancy. Relative coordination of dormancy and activity patterns between the growth centres provided control over plant architecture and led to an optimization of functional plant form. This optimized plant behaviour in its particular environment.

In the model presented, dormancy is viewed as a stationary phase in plant development. Since primary development is executed by the growing points, dormancy is also a property of the growing points. Dormancy is a phylogenetically significant plant response utilized to optimize architecture as plant form became more complex. The basic feature is that the autopoietic system (Varela *et al.*, 1974) can be arrested in a dormant state. This can happen by non-specific means as a response to unfavourable environments (ecodormancy); the behaviour is adequate and specific, though. This specific response is assimilated phylogenetically 'into the genes', i.e. into the basic networks, such that even tropical trees display periodic dormancy phenomena. The arrest of the autopoietic system can also occur in the setting of internal optimalization processes in the plant body, through which an adequate plant form is created (paradormancy). It can also occur as a response of an individual growing centre to environmental signals, in which only the growing centre is signal-perceptive; this particular receptiveness results in a transition from which the growing centre emerges spontaneously after a certain time, which cannot be shortened by making the conditions more favourable (endodormancy).

In this model, the dormant state can be characterized separately from how it is induced. Endo-, para- and ecodormancy are similar as a state in the sense that they are stationary phases in AM development, resulting in dormancy of part or all of the plant, depending on strategy. Therefore, the mechanisms by which these states are induced should be studied in relation to the context in which they function: survival of the plant, optimization of growth form, etc. The depth of dormancy is different. In ecodormancy, removal of global constraints is sufficient for reactivation of morphogenetic activity. In paradormancy, an alteration of meristem hierarchy is sufficient, and in endodormancy, an elapse of time, with its seasonal progression of environmental conditions, is sufficient. During endodormancy, the network is pushed completely through a phase transition into an alternative, metastable state. Only gradually does it climb out of this 'attractor' to resume growth. Fundamentally, the only difference between these states is the 'control' over their stability. In terms of cell-cell coupling, endodormancy is the most stable state of the AM network because it is trapped into this state by a structurally enforced alteration of the symplasm. In this model, the networking activity of the AM, resulting in different states of dormancy and morphogenetic activity, is sought in catastrophic alterations in the coupling of the subsystems (cells) of the network (AM) and the resulting lability/stability of such altered network states.

Acknowledgements

Thanks are due to Ing. El Bouw for making the TEM figure. I thank the Drs Klaasje Hartmans and Hans van Tol (ATO-DLO, Wageningen, The Netherlands) and Dr Päivi Rinne (University of Oulu, Oulu, Finland) for many stimulating discussions. I am grateful to Dr Jeremy Harbinson (Agrotechnical Research Institute (ATO-DLO), Wageningen, The Netherlands) and Dr Gregory Lang (Washington State University, Prosser, USA) for helpful remarks on the manuscript. This work was partly financed by the Dutch Ministerie van Landbouw, Natuurbeheer en Visserij and by the Dutch potato growers associated in the Nederlandse Aardappel Associatie (NAA).

References

Ball, E. (1955) On certain gradients in the shoot tip of *Lupinus albus*. *American Journal of Botany* 42, 509-521.

Ball, E. (1980) Regenaration of isolated portions of the shoot apex of *Trachymene coerulea* R.C. Grah. *Annals of Botany* 45, 103-112.

Botha, C.E.J., Hartley, B.J. and Cross, R.H.M. (1993) The ultrastructure and computer-enhanced digital image analysis of plasmodesmata at the Kranz mesophyll-bundle sheath interface of *Thermeda trianda* var. *imberbis* (Retz). A.Camus. in conventionally fixed leaf blades. *Annals of Botany* 72, 255-261.

Burton, W.G. (1966) *The Potato*. Veenman, Wageningen.

Carr, D.J. (1976) Plasmodesmata in growth and development. In: Gunning, B.E.S. and Robards, A.W. (eds) *Intercellular Communication in Plants: Studies on Plasmodesmata*. Springer-Verlag, Berlin, Heidelberg, New York, pp. 244-289.

Cutter, E.G. (1978) Structure and development of the potato plant. In: Harris, P.M. (ed.) *The Potato Crop*. Chapman and Hall, London, pp. 70-152.

Clowes, F.A.L. and MacDonald, M.M. (1987) Cell cycling and the fate of potato buds. *Annals of Botany* 59, 141-148.

Dennis, F.G., Jr (1994) Dormancy - what we know (and don't know). *HortScience* 29, 1249-1255.

Ding, B., Turgeon, R. and Parthasarathy, M.V. (1992) Substructure of freeze substituted plasmodesmata. *Protoplasma* 169, 28-41.

Ding, B., Haudenshield, J.S., Hull, R.J., Wolf, S., Beachy, R.N. and Lucas, W.J. (1992) Secondary plasmodesmata are specific sites of localization of the tobacco mosaic virus movement protein in transgenic tobacco plants. *Plant Cell* 4, 915-928.

Esau, K. (1977) *Anatomy of Seed Plants*. 2nd edn. Wiley & Sons, New York.

Goodwin, P.B. and Lyndon, R.F. (1983) Synchronization of cell division during transition to flowering in *Silene* apices is not due to increased symplast permeability. *Protoplasma* 116, 219-222.

Grabski, S., de Feijter, A.W. and Schindler, M. (1993) Endoplasmic reticulum forms a dynamic continuum for lipid diffusion between contiguous soybean root cells. *Plant Cell* 5, 25-38.

Gunning, B.E.S. and Robards, A.W. (1976) *Intercellular Communication in Plants: Studies on Plasmodesmata*. Springer-Verlag, Berlin.

Harold, F.M. (1990) To shape a cell: an inquiry into the causes of morphogenesis of microorganisms. *Microbiological Reviews* 54, 381-431.

Heplar, P.K. (1982) Endoplasmic reticulum in the formation of the cell plate and plasmodesmata. *Protoplasma* 111, 121-133.

Junttila, O. (1988) To be or not to be dormant: some comments on the new dormancy nomenclature. *HortScience* 23, 805-806.

Kaplan, R.D. and Hagemann, H.J. (1991) The relationship of cell and organism in vascular plants. Are cells the building blocks of plant form? *BioScience* 41, 693-703.

Kauffman, S.A. (1993) *The Origins of Order. Self-organisation and Selection in Evolution*. Oxford University Press, New York.

Kempers, R., Prior, D.A.M., van Bel, A.J.E. and Oparka, K.J. (1993) Plasmodesmata between sieve element and companion cell of extrafascicular stem phloem of *Cucurbita maxima* permit passage of 3 kDa fluorescent probes. *Plant Journal* 4, 567-575.

Koller, D. (1969) The physiology of dormancy and survival of plants in desert environments. In: Woolhouse, H.W. (ed.) *Dormancy and Survival*. Symposium of the Society for Experimental Biology (no. XXIII), Cambridge University Press, Cambridge, UK, pp. 449-469.

Kollmann, R. and Glockmann, C. (1991) Studies on graft unions. III. On the mechanism of secondary formation of plasmodesmata at the graft interface. *Protoplasma* 165, 71-85.

Kwiatkowska, M. and Maszewski, J. (1976) Plasmodesmata between synchronously and asynchronously developing cells of the antheridial filaments of *Chara vulgaris* L. *Protoplasma* 87, 317-327.

Kwiatkowska, M. and Maszewski, J. (1985) Changes in ultrastructure of plasmodesmata during spermatogenesis in *Chara vulgaris* L. *Planta* 166, 46-50.

Kwiatkowska, M. and Maszewski, J. (1986) Changes in the occurrence and ultrastructure of plasmodesmata in the antheridia of *Chara vulgaris* L. during different stages of spermatogenesis. *Protoplasma* 132, 179-188.

Lang, G.A. (1987) Dormancy: a universal terminology. *HortScience* 22, 817-920.

Lang, G.A. (1994) Dormancy – The missing links: molecular studies and integration of regulatory plant and environmental interactions. *HortScience* 29, 1255-1263.

Lang, G.A., Early, J.D., Arroyave, N.G., Darnell, R.L., Martin, G.C. and Stutte, G.W. (1985) Dormancy: toward a reduced universal terminology. *HortScience* 20, 809-811.

Lang, G.A., Early, J.D., Martin, G.C. and Darnell, R.L. (1987) Endo-, para-, and ecodormancy: physiological terminology and classification for dormancy research. *HortScience* 22, 371-377.

Lucas, W.J., Ding, B. and van der Schoot, C. (1993) Plasmodesmata and the supracellular nature of plants. *New Phytologist* 125, 435-476.

Olesen, P. and Robards, A.W. (1990) The neck region of plasmodesmata. In: Robards, A.W., Lucas, W.J., Pitts, J.D., Jongsma, H.J. and Spray, D.C. (eds) *Parallels in Cell to Cell Junctions in Plants and Animals*. NATO ASI Series H46, Springer-Verlag, Berlin.

Rappaport, L. and Wolf, N. (1969) The problem of dormancy in potato tubers and related structures. In: Woolhouse, H.W. (ed.) *Dormancy and Survival*. Symposium of the Society for Experimental Biology (no. XXIII), Cambridge University Press, Cambridge, UK, pp. 219-240.

Rees, A.R. (1981) Concepts of dormancy as illustrated by the tulip and other bulbs. *Annals of Applied Biology* 98, 544-548.

Roberts, K. (1994) The plant extracellular matrix: in a new expansive mood. *Current Opinions in Cell Biology* 6, 688-694.

Romberger, J.A. (1963) Meristems, growth, and development in woody plants. *USDA Technical Bulletin* 1293.

Seeley, S.D. (1994) Dormancy – the black box. *HortScience* 29, 1248.

Shih, C.Y. and Rappaport, L.R. (1971) Regulation of bud rest in tubers of potato, *Solanum tuberosum* L. VIII. Early effects of gibberellin A_3 and abscisic acid on ultrastructure. *Plant Physiology* 48, 31-35.

Smith, R.H. and Murashige, T. (1970) *In vitro* development of the isolated shoot apical meristem of angiosperms. *American Journal of Botany* 64, 443-448.

Steeves, T.A. and Sussex, I.A. (1989) *Patterns in Plant Development*, 2nd edn. Cambridge University Press, New York.

Stewart, W.W. (1981) Lucifer dyes – highly fluorescent dyes for biological tracing. *Nature* 292, 17-21.

Sussex, I.M. (1952) Regeneration of the potato shoot apex. *Nature* 170, 755-757.

Sussex, I.M. (1955) Morphogenesis in *Solanum tuberosum* L.: apical structure and developmental pattern of the juvenile shoot. *Phytomorphology* 5, 253-273.

Sussex, I.M. (1989) Developmental programming of the shoot meristem. *Cell* 56, 225-229.

Tilney, L.G., Cooke, T.J., Connely, P.S. and Tilney, M.S. (1990) The distribution of plasmodesmata and its relationship to morphogenesis in fern gametophytes. *Development* 110, 1209-1221.

Tilney-Bassett, R.A.E. (1986) *Plant Chimeras*. Edward Arnold, London.
Trewavas, A. (1986) Understanding the control of plant development and the role of growth substances. *Australian Journal of Plant Physiology* 13, 447-457.
Trewavas, A. (1987) Sensitivity and sensory adaption in growth substance responses. In: Hoad, G.V., Lenton, J.R., Jackson, M.B. and Atkin, R.K. (eds) *Hormone Action in Plant Development*. Butterworth & Co., Bodmin, UK, pp. 19-38.
Trewavas, A. and Jones, H.G. (1991) An assessment of the role of ABA in plant development. In: Davies, W.J. and Jones, H.G. (eds) *Abscisic Acid: Physiology and Biochemistry*. BIOS Scientific Publishers limited, Oxford, UK, pp. 169-188.
Usher, G. (1965) *A Dictionary of Botany*. Constable, London, UK.
van der Schoot, C. and Lucas, W.J. (1995) Microinjection and the study of tissue patterning in plant apices. In: Maliga, P., Klessig, D.F., Cashmore, A.R., Gruissem, W. and Varner, J.E. (eds) *Methods in Plant Molecular Biology*. Cold Spring Harbor Press, New York, pp. 173-189.
van der Schoot, C., Dietrich, M.A., Storms, M., Verbeke, J.A. and Lucas, W.J. (1995) Establishment of a cell-to-cell communication pathway between separate carpels during gynoecium development. *Planta* 195, 450-455.
van Ittersum, M.K. (1992) Dormancy and growth vigour of seed potatoes. Unpublished Doctoral Thesis, Wageningen Agricultural University, Wageningen, The Netherlands.
Varela, F., Maturana, H.R. and Uribe, R. (1974) Autopoiesis: the organization of living systems, its characterization and a model. *Biosystems* 5, 187-196.
Waigmann, B., Lucas, W.J., Citovsky, V. and Zambzyski, P. (1994) Direct functional assay for tobacco mosaic virus cell-to-cell movement protein and identification of a domain involved in increasing plasmodesmal permeability. *Proceedings of the National Academy of Sciences (USA)* 91, 1433-1437.
Webster, N. (1966) *Webster's Third New International Dictionary of the English Language*, unabridged. G. & C. Merrian Company Publishers, Springfield, Massachusetts, USA.

6 A New Conceptual Approach to Bud Dormancy in Woody Plants

Jacques Crabbé[1] and Paul Barnola[2]
[1]*Faculty of Agronomical Sciences, Applied Plant Morphogenesis, B-5030 Gembloux, Belgium;* [2]*University of Nancy I, Biology of Woody Plants, BP 239, F-54506 Vandoeuvre-lès-Nancy, France*

Introduction

The separation of apical dominance, winter dormancy and environmental stress-induced growth arrest, which is common in many reviews and text books, seems to be a shortsighted approach to the study of dormancy. Indeed, the very act of *bud* formation is evidence of the establishment of dormancy, i.e. a result of potentially diverse processes that lead to a common result of suppressed growth. The terms *para-*, *endo-*and *ecodormancy* (Lang *et al.*, 1985, 1987) are supposed to distinguish these different situations while also evoking their likeness. Nevertheless, in all types of dormancy, the primary causes of growth arrest (or absence) are commonly sought outside the bud, even when, as in endodormancy, some endogenous factor within the bud is involved. The inverse approach, that of focusing on the particular morphogenetic events which end in differentiation of a bud, is adopted rarely.

Growth and its absence are not simple situations. Shoot growth activity is complex, consisting of several intra- and extrameristematic subprocesses. Restricting these to the most manifest, i.e. organogenesis, internodal elongation and leaf expansion, it is clear that they interact strongly. Normal growth, i.e. formation of a leafy shoot, results from the perfectly coordinated functioning of these subprocesses. Due to their interplay, any limiting factor for any subprocess can ultimately inhibit normal growth as a whole, without need for a specific factor, internal or external. However, growth limitation or inhibition is not sufficient by itself to create a more or less stable dormant state: the formation of a bud also implies heteroblasty, i.e. changes in the development of foliar primordia, eventually leading to cataphyll formation.

For historical reasons, dormancy was studied first in temperate region species; consequently, winter dormancy and the chilling requirement to break it were important enough to justify intense research efforts. With the remarkable exception of some valuable pioneer works (Alvim, 1964; Alvim and Alvim, 1978; Longman, 1978; Borchert, 1991), much less investigation has been devoted to dormancy in tropical woody species. There has been similar disinterest regarding the short rest phases between growth 'flushes' of many temperate and tropical trees. Thus, enormous gaps in our general understanding of dormancy exist, which may impede further progress. It is clear that studies of the primary cause of any dormancy must consider the numerous factors that control growth activity. Growth regulation is likely to be multifactorial and located at different levels and territories of the meristem.

For over 40 years, the 'linear' hormonal hypothesis, i.e. endodormancy is a physiological state of the bud that is induced and broken by changes in the balance between inhibiting and stimulating endogenous substances, has distracted progress in dormancy research. The multifactorial context and the particular meristematic activity leading to bud differentiation have been overlooked too often in the search for specific hormones. Most recent reviewers (e.g. Saure, 1985; Mauget, 1987; Borchert, 1991; Martin, 1991; Champagnat, 1992) now admit that such a linear view is far too simplistic, and that precise correlations between endogenous hormones and overt progress of dormancy are rare. The possibility of an identical meristematic block for all kinds of dormancy, arising under the influence of different nearby or remote sources, has in fact seldom been taken into account (Guern and Usciati, 1976).

The reconstruction of the chain of events, from remote causes to the possible meristematic 'common denominator(s)', for all types of dormancy in woody species has been the aim of 30 years of physiological and biochemical investigations by the scientific group associated with Professor Paul Champagnat (University of Clermont-Ferrand, France), to whom this review is dedicated. Their testing method includes the measure of dormancy 'depth' and duration in diverse experimental contexts by isolation of buds as one-node cuttings to remove most extrabud influences. The mean time required for bud-burst (MTB) is used for comparisons of endodormancy in experimental or standard forcing conditions. Despite its imperfections (Rageau, 1978; Champagnat, 1983), this method remains very useful and results are confirmed by biochemical tests. Concomitantly, intensive research was initiated to discover the possible metabolic blocks that prevent growth of dormant buds. These investigations were first carried out on tubers of *Helianthus tuberosus* by another team inspired by Champagnat (Gendraud, 1977, 1981) and more recently have been extended to other dormant organs, including buds of woody plants.

Factors Affecting Growth and Dormancy

Internal factors: correlative regulation and interactive meristematic domains

According to Romberger (1963), 'a bud is a non-extended, partly developed shoot having at its summit the apical meristem which produced it'. In fact, an apical meristem is not always at hand (e.g. in some floral buds), but meristematic parts (i.e. those able to grow by meresis and auxesis) remain present. The same author adds that bud break is 'the result of leaf enlargement and subapical meristem activity', but that 'bud formation is not strictly a matter of inhibited internodal and primordial growth': bud scale or cataphyll production 'involves a specific kind of primordial development'. These shrewd remarks have sometimes been forgotten.

Recent works on meristematic activity (Médard, 1989; Médard *et al.*, 1992a, b) show that elongation depends on the initiation and enlargement of an internodal domain, respectively linked to the initiation of a leaf primordium and to the differentiation and expansion of the leaf blade. Internodal growth depends further on water and nutrient supply. Auxin and gibberellin interfere at different steps of this process. On the other hand, bud formation implies not only restricted elongation, but also a complex sequence of correlative effects which proceed from leaves and leaf primordia that accumulate at the apex as a result of shortened internodes (Dostal, 1952; Fulford, 1965, 1966; Neville, 1968; Champagnat *et al.*, 1986b). Their combined effect is heteroblasty, i.e. modified development of leaf primordia in cataphylls and total arrest of elongation. For a while, organogenesis can go on, piling up untransformed leaf primordia within the bud. If the conditions inducing dormancy persist, organogenesis itself will stop and, at the most extreme degree, all the cells of the meristem cease dividing and remain blocked in the G_1 phase of the cell cycle (Cottignies, 1986).

During spring and summer, or even in continuously favourable experimental growth conditions, woody plants quite universally have periods of rapid shoot elongation, which alternate with periods of little or no elongation, a phenomenon known as 'flushing' growth (Romberger, 1963; Lavarenne *et al.*, 1971). The interplay between adjoining meristematic and apical domains creates this behaviour (Champagnat *et al.*, 1986a,b). Causes that are more remote, both correlatively and environmentally, exacerbate these inherent near- and intrameristematic interactions, changing dormancy phases from short to long.

It is often difficult to determine if an environmental cause is perceived directly by the potential dormant bud or by another organ which correlatively inhibits it. The former possibility cannot be excluded (see below), particularly in the winter months after release from endodormancy. However, when inception of dormancy is considered, a prevailing chain of correlative events from summer to winter dormancy is revealed (Champagnat, 1989; see Fig. 6.1).

During normal growth, buds in axillary positions (or terminal on shoots of high order) first appear under *apical dominance*. The current interpretation of this phenomenon is that leading meristems or shoots, thanks to their high auxin production, divert most of the nutrient supply and compete successfully with subordinate ones. Growth of these paradormant buds is restored by suppression of the dominants (e.g. decapitation) or any means that allow some lateral flow of nutrients (Hillman, 1984). Later in the season, with apical growth reduced or even arrested, decapitation is no longer effective and at least partial defoliation is needed to release the buds (Champagnat, 1955; Barlow and Hancock, 1962; Crabbé, 1970). This *foliar inhibition* is probably due to diversion of water and nutrients by still-growing or mature leaves. Still later, neither decapitation nor defoliation is effective, for only isolation of the bud restores growth. The bearing axis itself now exerts control over bud development (Crabbé, 1968; Champagnat *et al.*, 1975; Arias and Crabbé, 1975a). These axis effects maintain the acrotony of the shoot (Meng-Horn *et al.*, 1975; see Fig. 6.2). At the end of the process, even isolated buds placed in proper environmental conditions fail to sprout, because endodormancy has been initiated. The large and abrupt increase in MTB is an indication of this new condition.

Thus, in axillary meristems from spring to autumn, the cause of growth suppression moves from the shoot apex to its leaves and then to the axis tissues, before reaching the *immediate vicinity of the meristem* (i.e. within the lateral bud). Terminal meristems of leading shoots (which exert, instead of undergo, apical dominance) are somewhat different, being the last to stop growth and form buds. Leaf and axis effects probably influence their activity, but the correlative inhibitions seemingly must be aggravated by stressful late-season conditions (see under External factors, below). The result is a different course of dormancy: terminal buds achieve a deeper dormancy, but are released more easily than axillary buds (Champagnat, 1983, 1992; Williams *et al.*, 1979; Mauget and Rageau, 1988).

It is worth noting that all buds in a tree, or even on the same 1-year-old shoot, don't react the same amidst this chain of influences. Each bud has its own correlative context and behaves accordingly (Fig. 6.3). So the course and depth of dormancy have morphogenetic consequences. For example, in autumn, the depth of dormancy on a 1-year-old shoot decreases from the apex to the base, displaying a basitonic gradient of bud growth capacity (Crabbé, 1968; Champagnat *et al.*, 1975). In shrubby species, this *basitony* leads to vigorous basal sprouting in spring (Barnola, 1970, 1972, 1976; Barnola and Crabbé, 1991). In tree species, strong axis effects maintain the potential basitony unexpressed – except under heavy stress – and allow the *acrotony*

Approach to Bud Dormancy in Woody Plants

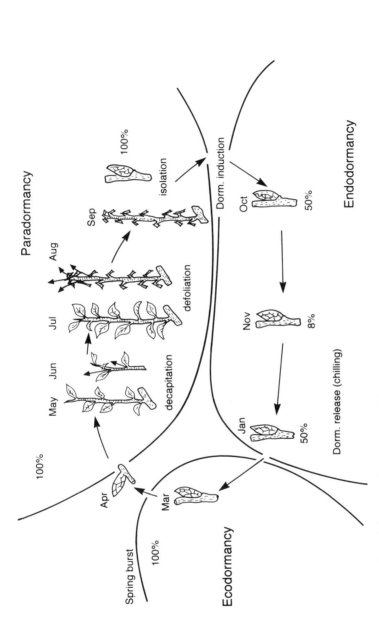

Fig. 6.1. The sequence of correlative events from one type of dormancy to another. Bud break percentages are from one-node cuttings at 25°C. (Redrawn from Champagnat, 1989.)

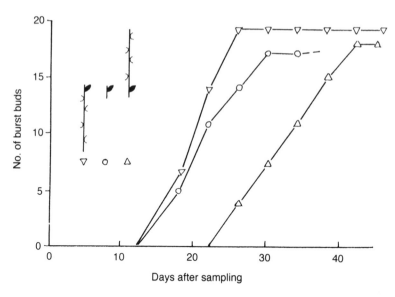

Fig. 6.2. Demonstration of axis effects on the growth capacity of *Rhamnus frangula* buds in March. Compared to a one-node cutting, a disbudded proximal piece of shoot promotes bud break; a distal piece inhibits it. (Redrawn from Meng-Horn *et al.*, 1975.)

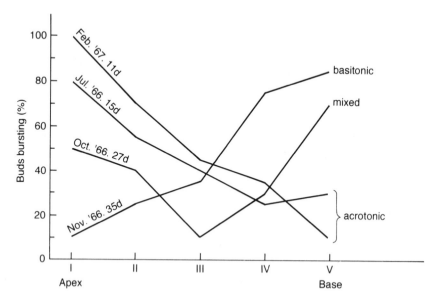

Fig. 6.3. Gradients of bud growth capacity observed along 1-year-old apple shoots at different times of the year. Notice the increase in mean bursting time (in days) from July to November and decrease from November to February. (Redrawn from Crabbé, 1968.)

of spring branching and trunk formation (Champagnat, 1978; Crabbé, 1981, 1984b).

On the other hand, numerous treatments that change the context of bud meristematic domains also modify the course and depth of its subsequent endodormancy (Fig. 6.4). Such effects have been observed after pinching and pruning (Arias and Crabbé, 1975b; Barnola *et al.*, 1976; Mauget, 1987; Bailly and Mauget, 1989), total or partial defoliation (Mauget, 1978; Crabbé, 1984b), partial disbudding (Champagnat *et al.*, 1975) and shoot bending (Crabbé, 1984b). Endodormancy is also influenced by the relative growth circumstances to which buds are exposed during the previous season, e.g. location within the tree (Crabbé, 1968), initial or secondary growth flush (Dreyer and Mauget, 1986a), and latent vs. axillary bud formation (Mauget, 1984; Barnola and Crabbé, 1991).

In conclusion, these facts demonstrate that the ultimate endodormant condition depends on the chain of prior correlative events, i.e. paradormant states, that occur. If we now envision one or more blocks that inhibit meristematic activity - near or within the bud - we should say that the blocks are first labile and under control of remote influences during paradormant phases, finally becoming stable and independent when endodormancy is

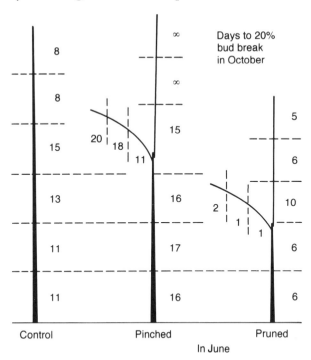

Fig. 6.4. The effect of June pinching or pruning on the subsequent endodormancy of apple buds in various shoot locations. (Redrawn from Arias and Crabbé, 1975b.)

initiated. In this connection, Champagnat (1983) defined endodormancy as 'the last step of a cascade of correlations drawing nearer and nearer to the bud'.

External factors: the multiple effects of temperature

As described above, endodormancy is the result of a long previous history of preformed buds. All along this evolution, buds are exposed to environmental influences that modify or modulate the course of events. For example, conditions favourable to growth can bring about renewed bursting of eco- or paradormant buds, while unfavourable conditions probably reinforce their growth incapacity. On the other hand, experiments have also shown that buds can enter and leave endodormancy without any environmental change, e.g. when maintained in constant conditions in growth rooms or controlled greenhouses (Lavarenne *et al.*, 1975; Mauget, 1983; Mauget and Rageau, 1988). Should this not emphasize the essential role of a physiological evolution of the bud itself and require a profound reconsideration of environmental effects?

One must be aware that temperature is unable to work as a perfect 'signal', since a specific temperature receptor (like phytochrome for light effects) does not exist. Temperature affects the physical state of membranes and the rate of enzymatic reactions, with modified cell compartmentation and upset of metabolic balances as a result. Thus, effects due to temperature changes are not expected to be clear-cut, but rather variable and fluctuating from species to species and according to physiological context. Indeed, precise experimentation demonstrates that chilling temperatures ($\sim$ 0-10°C) have a different effect on buds when applied either during the phase of increasing growth incapacity or afterwards, resulting in increased or decreased endodormancy, respectively (Lavarenne *et al.*, 1975; Mauget, 1981). Moreover, any temperature, with different efficiencies according to its range, can increase or decrease bud growth incapacity, or stimulate bud break after release from dormancy (Lavarenne *et al.*, 1980). In ash (*Fraxinus excelsior* L.), for example, the range from 0 to 10°C is particularly efficient in increasing, and later decreasing, dormancy, whereas it delays subsequent growth. Regimes around 10 to 15°C sustain all phases of the bud cycle. At 18°C and above, release from dormancy is much less efficient, yet bud break is improved greatly. In the experiments cited above, endodormancy in plants held continuously above 15°C became deeper and lasted longer than in natural winter conditions, but it finally decreased to a level that permitted some bud burst (Mauget, 1983; Mauget and Rageau, 1988).

The physiological progression of the bud, from its formation under eco- or paradormant conditions and through these stages toward endodormancy, conditions its response to a given temperature regime. Abrupt variations in this regime moderate or enhance the otherwise spontaneous physiological changes of the buds. Consequences of this situation – both scientific and practical – are numerous and important. First, whatever the temperature treatment and however precise the experiments, it remains difficult to interpret results when the physiological state of the bud is not known exactly and the action of temperature on the metabolism of implicated cells is not elucidated. For example, why is the deep dormancy of a terminal bud broken more easily, by the same chilling treatment, than the relatively shallow dormancy of an axillary bud? Second, there is still no general, sound basis on which to predict the end of endodormancy and date of bud break under any given climate. Models constructed on statistical or theoretical considerations have been developed (Bidabé, 1967; Richardson *et al.*, 1974; Kobayashi *et al.*, 1982; Shaltout and Unrath, 1983), but they work only in restricted climatic areas and in average annual circumstances (Martin, 1991). Third, the course of dormancy in mild winter climates is still understood poorly. The intensity of endodormancy, however, is low in such regions compared with those under more severe winter conditions. Further, under this weak endodormancy, correlative influences may persist throughout the winter and affect the bud break pattern (Herter *et al.*, 1991).

Notwithstanding the shortcomings of our present understanding of temperature influences, this environmental factor is the most important modulator of the course of dormancy, not only for release of buds from rest, but also for acceleration of the shift from eco- and paradormancy to endodormancy.

External factors: day length, water availability and growth regulators

In older literature, short days were often regarded as the factor that induces growth arrest and endodormancy. Some woody species do, in fact, respond to declining photoperiod. Nevertheless, the role of day length in dormancy was later contested because, in natural conditions in temperate regions, shoot growth arrest and bud formation occur mostly in summer under long days. The problem was reexamined recently and intensively in *Gleditsia triacanthos* L. (Aillaud, 1982, 1986; Aillaud *et al.*, 1989; Al-Ibrahem, 1990). Under short days, this species progressively enters into dormancy. Defoliation demonstrated that photoperiodic perception is by the leaves. With less than 30 short days, growth arrest was abolished readily on return to long days; for more than 30–35 short days, terminal meristems stopped growing, senesced and abscised, while axillary buds evolved from para- to endodormancy. This evolution started at the top of the shoot and reached the basal buds after 40–50 short days. Now, if buds formed after 30 short days, i.e. still paradormant, were exposed further to cold, they became endodormant. Prolonged chilling enabled them to grow again (Fig. 6.5). Buds released from

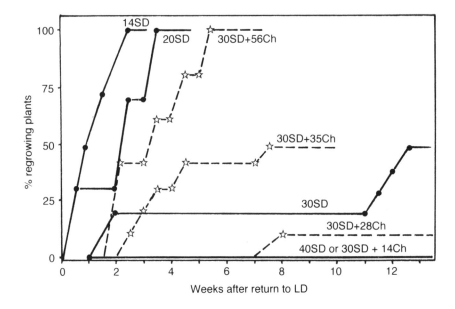

Fig. 6.5. The combined effect of short days (SD) and chilling (Ch) temperature on the course of endodormancy in *Gleditsia triacanthos* L. LD, long days; SD, short days; Ch, chilling days. (Redrawn from G.J. Aillaud, unpublished.)

endodormancy by chilling no longer responded to short days, which accelerated the return to growth. Once again, endodormancy appears as the end-point of a gradual evolution that starts with eco- or paradormancy. Both low temperatures and short days have a dual effect, depending on the physiological state of the receptor bud.

It is beyond doubt that water availability plays a predominant role in bud behaviour. During ecodormant stages, supplying water to the buds is often sufficient to make them grow. In the tropics, seasonal drought and rainfall are the essential factors that induce growth arrest and resumption, with a strong interference of leaf expansion and abscission, which can directly affect the internal water status (Borchert, 1991). Still, it is questionable whether the dry season entails a real endodormancy. On the other hand, Dreyer and Mauget (1986b) indicated that drastic and short (2-3 weeks) dry periods given to walnut trees (*Juglans regia* L.) during spring growth induced an immediate temporary growth arrest. This ecodormant phase further modified the subsequent endodormancy compared with buds of continuously watered trees, i.e. bud dormancy in November was deeper on the shoot parts affected directly by the treatment and more shallow on the parts formed after drought. These effects call to mind those obtained by changes in the correlative context of meristematic domains (e.g. by pruning), a parallelism that is indeed striking and deserves further interest, particularly for tropical and subtropical woody species.

There are other factors known for exerting effects on dormancy onset and release, but they are generally poorly understood. Hormonal relationships and modes of action have yet to be established satisfactorily in repeatable experiments. The efficacy of exogenous growth regulators usually depends on the time of application. For example, sweet cherry buds, isolated and perfused with solutions of regulators, respond according to the physiological state that they have reached on the tree (Arias and Crabbé, 1975a): cytokinins stimulate bud break in the early phase of increasing dormancy, whereas gibberellins do so only when dormancy has begun to decline (Fig. 6.6). Similarly, artificial dormancy breaking agents, like anoxia and ethylene (Regnard, 1985), calcium cyanamide (Shulman et al., 1983), mineral oils and dinitro-o-cresol (Erez and Zur, 1981), only break the residual dormancy left after partial chilling.

Many kinds of dormancy exist and even coexist in a tree. Their primary causes may initiate at a distance from the potentially dormant meristems and relate to unspecific correlative or environmental effects. Whatever their origin, these effects have to be transduced into a morphogenetic sequence of events in the close vicinity of the concerned meristem, and finally into one or more block(s) that prevent the coordinated functioning of different meristematic domains. Meristems and the buds that they eventually form evolve in a

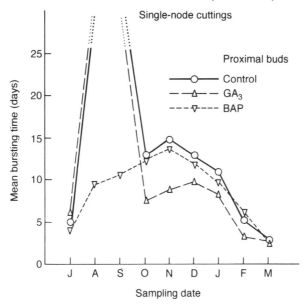

Fig. 6.6. The effect of feeding a cytokinin (6-benzylaminopurine, BAP) or a gibberellin (GA_3) to isolated buds of sweet cherry (*Prunus avium* L., cv. F12/1) during the course of dormancy. (From Arias and Crabbé, 1975a.)

continuously changing context, under both internal and external influences. But correlative and environmental effects – or those of exogenous chemicals – never have a straight and unequivocal action *per se*: response always depends on the state reached by the bud in its individual physiological evolution.

Two fundamental questions are thus to be answered.

1. What is the nature of the involved block(s) to growth?
2. What governs the shift from dependence and lability in eco- and paradormancy, to autonomy and stability in endodormancy?

Biochemical and Metabolic Aspects of Bud Dormancy

Metabolic blocks of growth activity

Knowledge about metabolic aspects of dormancy is inadequate. Many cellular activities begin when a bud passes from rest to growth: reserve carbohydrate mobilization, water content increase, energy support, supply and utilization of metabolites in the growing organs, etc. The reverse change demands that one or more key steps be inhibited. Correlative and environmental factors that are seen as remote causes must eventually trigger one of those inhibitions near to the meristem. Of course, hormones are likely to play a role in the transduction, but they are not by themselves the decisive event.

Gendraud and coworkers used Jerusalem artichoke (*Helianthus tuberosus* L.) tubers to unravel parts of dormancy-related metabolism. Tuber buds are a more suitable material for the biochemist than buds of woody plants, as they are poorly organized – a meristematic dome protected by few scale-like appendages – and are in close contact with a rather homogeneous parenchyma full of stored carbohydrates. Two different traits that coincide with inhibition and onset of tuber sprouting have been discovered.

The first trait concerns nucleotide metabolism (Gendraud, 1975, 1977). Adenylic nucleotides, which store and transfer energy, are produced and turned over continuously to maintain basic cellular activity. Consequently, pieces of parenchyma, taken from dormant as well as from non-dormant material and incubated on a solution of adenosine, increase their adenosine triphosphate (ATP) content. On the other hand, non-adenylic nucleotides (NTP), synthesized by phosphate transfer from ATP to the corresponding nucleosides, are involved in diverse metabolic changes essential for growth: glucose transport, sugar interconversion, ribonucleic acid (RNA) and deoxyribonucleic acid (DNA) synthesis, etc. Parenchyma incubated on adenosine can only increase its NTP content when it is taken from non-dormant tubers, even those that have been chilled and are not yet sprouting. The measure, by bioluminescence, of the ATP and NTP pools in parenchyma cultured on water or fed ^{14}C adenosine has been developed as a biochemical test method: the dormant state is detected by the failure of a material to increase the NTP fraction.

The second trait relates to measurement of the sink strength of dormant or non-dormant tuber tissues, and their subsequent capacity to accumulate metabolites. The prevention of symplastic entry of metabolites into the bud's cells is also indicative of dormancy. In the case of tubers, as long as the parenchyma is a strong sink, 'trapping' all available useful molecules, the buds deprived of these resources remain inhibited: thus, the tuber is dormant (Gendraud, 1981; Gendraud and Lafleuriel, 1983; Tort and Gendraud, 1984; Gendraud and Pétel, 1990). Dormancy is broken only when the sink strength gradient reverses between parenchyma and the bud (Fig. 6.7). Recent work demonstrates that these properties are linked to the nature and characteristics of plasmalemma-bound adenosine triphosphatases (ATPases) (Pétel and Gendraud, 1986, 1993; Pétel et al., 1992). It has been established that the regulation of short distance transport between cells is under control of their pH, which can be evaluated from the absorption of 2(^{14}C) 5,5-dimethyloxazolidine 4-dione (DMO) in the cells (Kurdjan and Guern, 1978). The steeper the upward gradient of pH or DMO concentration, from

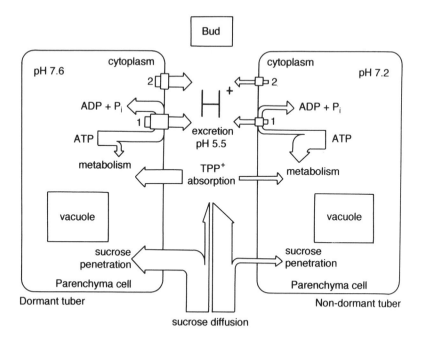

Fig. 6.7. A diagrammatic model of the pH-mediated translocation of sucrose in tubers of Jerusalem artichoke (*Helianthus tuberosus* L.) and its relationship to dormancy. (From Gendraud and Pétel, 1990.)

outside to inside of a tissue (expressed as C_i/C_e), the stronger its sink strength. This is also a convenient test of dormancy. Both of these test methods have been applied successfully to woody plants, with minor modifications.

Biochemical differentiation of bud dormancy types

The bud of a woody plant is morphologically complex and made of organs differing in structure and physiology. Moreover, the bud is connected to the stem through a parenchymatous zone, traversed by a procambial strand: during its early growth, it is fed through the symplasm of this region. Therefore, the transposal of the test methods described above require application to both the living bud tissues and the underlying stem tissues.

The biological (one-node cutting) test and the biochemical tests have been compared in various species, including ash (Lavarenne *et al.*, 1982; Barnola *et al.*, 1986b), oak (Barnola *et al.*, 1986a), apple (Ben Ismail, 1989) and peach (Balandier *et al.*, 1993). This provides a novel illustration (e.g. Table 6.1) of the succession of dormancy states described previously. The growth capacity of buds first declines under the influence of remote correlative or environmental factors that induce the intermittent establishment of a permeability barrier (DMO test) close to the bud, i.e. a pH gradient between stem and bud tissue which opposes the translocation flux required for growth. Throughout most of the winter, this barrier persists. It is withdrawn for a short period right at bud break, to be reinstated sooner or later under renewed para- or ecodormant conditions. Meanwhile, defective NTP synthesis sets in, first in the adjacent stem tissues, when MTB increases sharply, then reaching the bud itself some time later, when endodormancy is at a maximum. The ability to increase the NTP content reappears successively in the stem, and then in the bud at the end of endodormancy when MTB decreases again.

So endodormancy is characterized by the concomitant occurrence of a permeability barrier and an inability to increase the NTP pool in the bud. Chilling is required to rapidly overcome these blocks. Para- and ecodormancy, on the other hand, correspond to a less severely blocked physiological state, which is forced more easily by exposure to favourable conditions, gibberellin or dormancy-breaking agents.

At this state of knowledge, the blocks described above do not fully explain the dormant condition. Other sources of inhibition may be involved and perhaps remain to be discovered. For example, from investigations of paradormant apple bud release by thidiazuron, Faust and coworkers have identified a series of metabolic processes associated with release from dormancy. These include glucidic turnover (Wang *et al.*, 1987), lipid and sterol metabolisms (Wang and Faust, 1988, 1989, 1990), oxidative reactions (Wang *et al.*, 1991a,b,c; Wang and Faust, 1992) and free and bound states of water

Table 6.1. The succession of terminal bud dormancy states of ash (*Fraxinus excelsior* L.) determined by biological (one-node cuttings) and biochemical (nucleotide and DMO) tests. (Adapted from Champagnat, 1989.)

	Sept.	Oct.	Nov.–Dec.	Jan.	Feb.–Mar.	Apr.	May
State of dormancy	Paradormancy (EcoD.)		Endodormancy		Ecodormancy	< Bud break >	(Parad.)
Rate of bud break (one-node cutting test)	Sept. High or beginning of decline	Oct. Rapidly declining	Nov.–Dec. ~0	Jan. ~0	Feb.–Mar. Slowly increasing	Apr. Rapidly increasing	May High, stable or weakly declining
NTP synthesis (nucleotide test)							
In bud tissues	+	+	–	–	+	+	+
In axis tissues	+	–	–	+	+	+	+
Gradient of sink strength (DMO test)							
Bud	⇌	→	→	→	⇌	←	⇌
Axis							

Ecod., ecodormancy; Parad., paradormancy.

(Faust *et al.*, 1991; Liu *et al.*, 1993). At present, none of these events seem as precocious as those described above; only the change from bound to free water could occur as early. However, such studies merit further consideration and could provide missing links to possible hormonal implications.

Several points now seem clear. A 'common denominator' to all kinds of dormancy, composed of two (or more) blocks, impedes basic processes needed to sustain coordinated growth activities at the level of the bud. These blocks are induced first at a distance and then are maintained through correlations emanating from apices, leaves and axis tissues; paradormant bud formation is their consequence. Endodormancy is the last step of this cascade of correlations, which draws progressively nearer to the bud. At endodormancy, the blocks become stable and independent of the remote causes from which they were induced. Their rapid removal requires exposure to chilling temperatures, at least for temperate zone species.

But what is meant here by stability? Does it result from a still unknown additional step? Is it simply the consequence of a multiplication of the blocks? Or is it bound to localization of the NTP synthesis incapacity within the bud? To answer these questions, it seems necessary to redirect attention to meristematic growth activities and to focus on terminal buds, which display alternative phases of growth and rest before finally entering into endodormancy.

Growth Rhythmicity as a Basis of Dormancy Phenomena

Rhythmic growth is almost universal in woody plants (Romberger, 1963; Borchert, 1991). In continuously moist and warm equatorial climates, alternating phases of rapid shoot elongation and little or no growth are commonly observed (Hallé *et al.*, 1978). In seasonal climates, one is struck by the growth arrests and bud breaks that are more or less synchronized with the beginning and end of an unfavourable season. Rhythmic behaviour is expressed in two ways (Lavarenne *et al.*, 1971; Champagnat *et al.*, 1986a). Growth can be *periodic*: active and dormant phases, of rather short and constant duration, alternate quasi-indefinitely, as in many tropical and some temperate (e.g. oak) species maintained in constant environments (e.g. Fig. 6.8). Alternatively, growth can be *episodic*: for most temperate species (e.g. ash, apple) in nature and in constant conditions, a few flushes – sometimes only one – are followed by a long dormant phase. In this latter case, environmental conditions sometimes convert episodic behaviour into periodic, e.g. in apple (Zanette, 1981; Fig. 6.8). Hence, the persistence of growth rhythmicity in controlled constant conditions proves its endogenous origin. A common primary cause of short and long dormant phases may be suspected. Herbaceous plants often lack rhythmic growth, with growth being continuous over a short season:

dormancy is restricted to bud inhibition by apical dominance or is confined to organs which function as propagules (seeds, tubers, bulbs, etc.). The typical rhythmic behaviour is therefore considered 'as an inherent consequence of the developmental constraints of trees as large, long-lived plants pursuing a characteristic adaptative strategy' (Borchert, 1991), an idea also suggested by Romberger (1963).

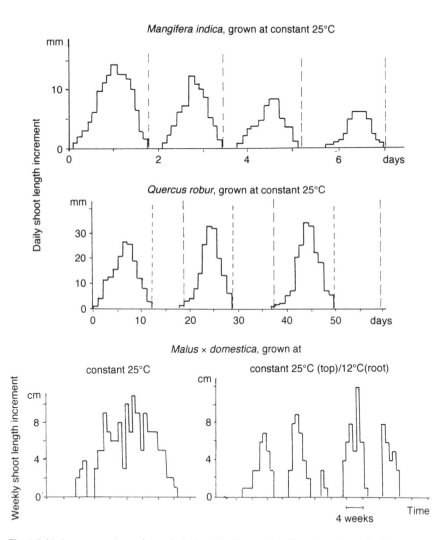

Fig. 6.8. Various expressions of growth rhythmicity. Top and middle: periodic growth of mango (from Parisot, 1988) and oak (from Champagnat et al., 1986b). Bottom left: episodic growth of apple. Bottom right: periodic growth of apple when roots are maintained at a lower temperature than the top. (Redrawn from Zanette, 1981.)

Periodic growth: a perfect coordination of growth processes

The oak, *Quercus robur* L., is a good example of periodic growth when exposed continuously to favourable environmental conditions. At constant 25°C and 16-h-long days, it displays an indefinite succession of flushes, made of 10 to 12 days of elongation separated by 6 to 9 days of rest. The origin of this rhythm is clearly correlative, as demonstrated by defoliation experiments (Lavarenne-Allary, 1965; Champagnat *et al.*, 1986b; Barnola *et al.*, 1990). Complex interactions between apical domains induce periodic variations of the intensity of the main component subprocesses of growth (Fig. 6.9).

During the days before the start of a new flush, the terminal bud is a very strong sink: organogenesis has been enhanced so that most of the leaves of the next flush are already present. These primordia accumulated below the meristem force the plastochron to increase, while, together with the young internodes, they slowly start to grow (leading to bud break). At this moment, the axis below the bud takes over the maximum sink strength and accelerates its elongation. Young, scarcely growing leaves, still close to the apex, induce heteroblasty: the newly forming primordia, at a slower rate due to declining plastochronal activity, fail to develop their blade and thus convert to bud scales. Afterwards, the leaves become the major sink; their rapid expansion maintains the organogenetic activity at a low level: the first leaf primordia of the next flush are formed in the new bud. Meanwhile, internode elongation, first stimulated basipetally by the incipient leaf expansion, now declines progressively and ceases. The rest phase is thus established, with organogenesis continuing slowly until the enhancement that is prelude to a new flush.

The sites of the different component growth processes operate successively as the strongest sinks, demonstrated by the DMO method (Alatou *et al.*, 1989; Fig. 6.9) and the pattern of $^{14}CO_2$ assimilation (Barnola *et al.*, 1990). The low sink activity of the bud that precedes growth arrest leads to a shortage of sucrose, trapped at this time in the axis below the bud (Parmentier, 1993; Alaoui-Sossé *et al.*, 1994). On the other hand, when growth arrest sets in, the ATP and NTP pools of the apical bud are low and the NTP pool cannot be increased by feeding the bud adenosine; near the end of rest, without any change in the environment, both pools can be increased (Barnola *et al.*, 1986a; Fig. 6.9). During a short lapse of time, the terminal bud is thus in a state absolutely reminiscent of endodormancy.

Several other experimental contributions complete this view. Among other things, it has been shown that exogenous regulators, stimulating leaf (cytokinin) or internode (gibberellin) development, can increase the duration of the active vs. rest phase (Champagnat *et al.*, 1986b; Parmentier *et al.*, 1991) and that mineral nutrient deficiency can improve the endodormant characteristics of the rest phase (Parmentier *et al.*, 1991). Thus, periodic behaviour corresponds to a perfectly coordinated functioning of the apex.

Approach to Bud Dormancy in Woody Plants 101

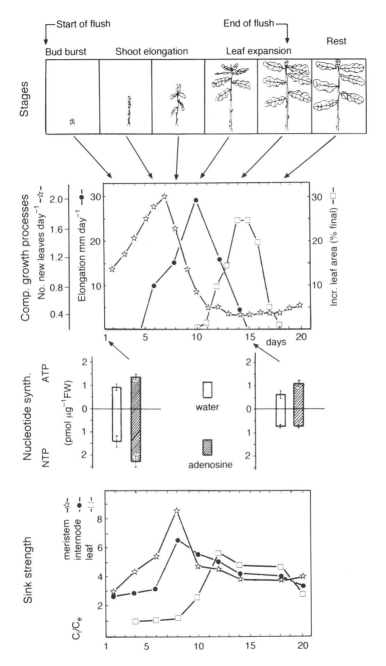

Fig. 6.9. Periodic growth in oak (*Quercus robur* L.) illustrated by the sequence of stages, component growth processes, and nucleotide and DMO test results. (From Alatou *et al.*, 1989, and Barnola *et al.*, 1986a, 1990.)

The relative asynchrony of the component growth processes is perhaps the key by which an 'economy' of their operation is achieved, with internal competition being kept at a minimum. Tropical species displaying periodic growth, like *Hevea* (Hallé and Martin, 1968), mango (Parisot, 1988) and *Terminalia* (Maillard *et al.*, 1987), have the same kind of apical organization.

Episodic growth and endodormancy

Most temperate species that display episodic growth seem less well organized regarding their apical activity. At the onset of a growth phase, organogenesis, internode elongation and leaf expansion start approximately at the same time, possibly leading to increased competition for water and nutrients and resulting in growth limitation. Intermittent unfavourable conditions, to which the plant is exposed in alternate climates, are likely to upset further the widespread growth rhythm. This is illustrated by several studies with ash (Barnola *et al.*, 1986b, c; Lavarenne *et al.*, 1986). Placed in the same constant conditions (25°C, 16-h-long days) that make oak flush indefinitely, young ash plants exhibit one flush and then plunge into an 11- to 13-month-long dormancy, followed by a new flush, and so on. Terminal buds on apical cuttings taken at the start of the dormant phase and placed at 25°C will sprout immediately (Fig. 6.10): isolation is thus sufficient to force these paradormant buds. Six to twelve months later, they still burst at 100%, but with a lag phase of 20 to 30 days. On the contrary, when the young trees are raised at constant 12°C and 16-h-long days, the result is initially the same: one flush followed by an 11- to 13-month-dormancy, etc. Their terminal buds are still able to burst rapidly, if isolated at the beginning of the rest. But, 5 to 10 months later, they cannot be forced to grow (Fig. 6.10). The difference is that, at 12°C, the buds cannot maintain themselves in paradormancy, but enter into endodormancy. This is confirmed by the nucleotide test: buds from plants raised at 25°C remain able to synthesize NTPs, but the axis tissues below them are unable to do so (Fig. 6.10). Buds and axis from plants raised at 12°C have lost this capacity.

Consequently, woody species that historically develop by periodic or episodic flushes may, in a more severe environment, evolve toward an even more disturbed pattern of growth, implying the onset of endodormant characteristics. The comparison of results with various oak species and mango grown in different constant-temperature regimes support this idea (Fig. 6.11). Starting from a regime close to the species optimum, the duration of the active, and especially of the rest, phases increases as the imposed regime moves away from this optimum. A tropical species, like mango, in a range compatible with its original climate, appears less sensitive to changes than do temperate or subtropical (Mediterranean) species.

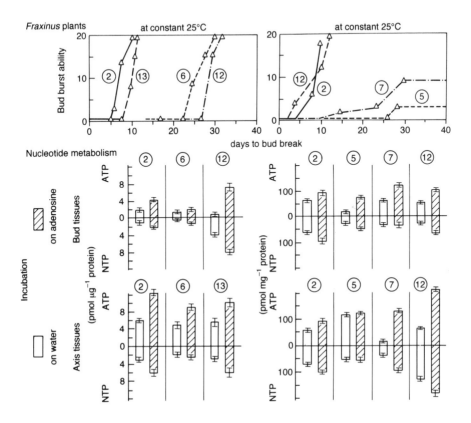

Fig. 6.10. A comparison of the rhythmic behaviours of ash (*Fraxinus excelsior* L.) depending on the temperature regime to which the plants are exposed. At 25°C, the terminal buds remain in a paradormant condition, whereas they become endodormant at 12°C. Circled numbers in or above graphs indicate the number of months after the start of the rest phase. (From Barnola *et al.*, 1986c, and Lavarenne *et al.*, 1986.)

In accordance with Borchert's views, we believe that growth rhythmicity is the unescapable tree response to the constraints generated by their perennial nature and large dimensions. Moreover, it is obvious that a growth rhythm exerts control over the main manifestations of tree development: trunk formation, branching habit, flowering and fruiting. This is the basis of

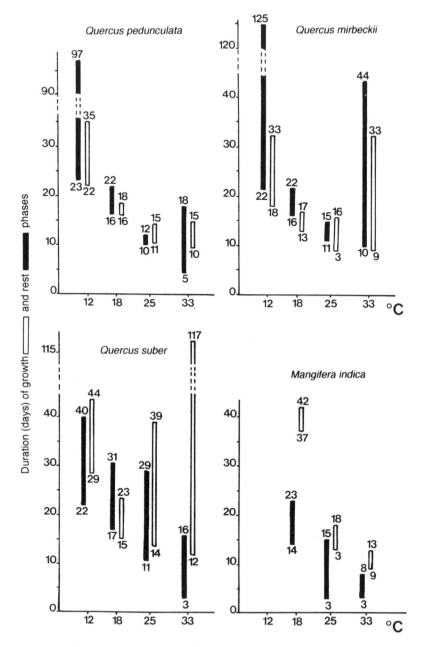

Fig. 6.11. A comparison of the respective durations of growth and rest phases during rhythmic growth of diverse oak species and mango under various temperature regimes. (P. Barnola, 1994, unpublished.)

the temporal organization of the tree. Growth rhythmicity is also the basic mechanism that allows dormancy of terminal meristems. These are the last to grow in late season, while axillary buds are formed long before through correlative events. We have seen how ill-suited environmental factors are – due to their equivocal effects – to ensuring a well-timed arrest. The endogenous rhythm affords periodic opportunities to stop growth, which can lead to long-lasting dormancy with the accessory help of adverse external conditions.

The identical nature of the metabolic blocks, during short stops between flushes as well as long seasonal arrests, reinforces this idea. Endodormancy appears as the extreme point of the rhythm's shift from periodic to episodic, or, in other words, as the result of the destabilization of its endogenous control by climatic constraints (Fig. 6.12).

General Conclusions

With branching as the most efficient response to stationary life, and unlimited branching being insupportable for both energetic and material reasons, plants in the course of evolution had to 'invent' the bud and, at the same time, dormancy. Let us notice that the bud consequently appeared c. 70,000 years earlier than the seed: this does not exclude the possibility that some mechanisms utilized in seed and bud dormancy could be the same, nor does it imply a total similarity of mechanisms (Crabbé, 1994). Simply, at the outset was the bud, in which dormancy evolved at various times to achieve various purposes. Dormancy probably first appeared in subordinated locations of the woody plant to limit expansion (i.e. branching) and economize resources for the main active growth centres. Diverse correlative mechanisms were therefore devised, which are still at work today.

Later, dormancy provided the interruption of ontogeny to allow severance of propagules (e.g. seeds) from the mother plant. When woody species migrated out of the original warm and humid climates or when climate changed, it became necessary to stop growth completely under unfavourable circumstances. To achieve the inhibition of the dominant terminal meristems and maintain their inactivity, another 'invention' was reutilized: growth rhythmicity, whose primary function was to ensure the temporal organization of the plant. That is the reason why endodormancy has retained, beside its adaptive role to an unfavourable seasonal climate, obvious morphogenetic functions: proleptic branching, acrotony and trunk formation, basitony and shrubby habit (Champagnat, 1978; Crabbé, 1981, 1984b; Barnola and Crabbé, 1991) and location of floral sites (Crabbé, 1981). These facts alone might account for enormous physiological differences, which are often overlooked at present.

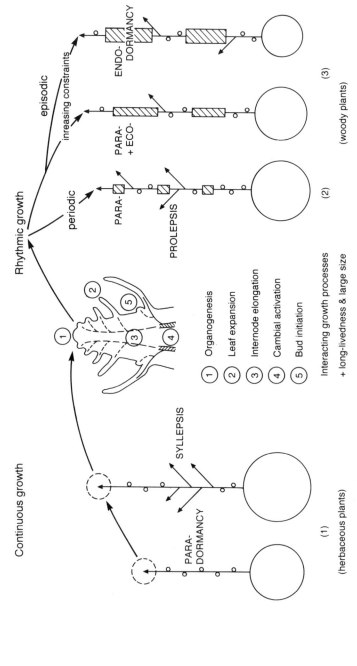

Fig. 6.12. A tentative synthesis of rhythmic growth and dormancy. Rhythmic growth results from internal competition between component growth processes in large, long-lived plants. Dormancy proceeds from still increasing internal and environmental constraints, leading to various physiological expressions. Arrows indicate active meristems, small circles = dormant buds, hatched rectangles = rest phases during rhythmicity, large circles at the bottom represent the amount of available resources. (J. Crabbé and P. Barnola, 1994, unpublished.)

Besides and within the implied correlative endogenous controls, diverse transduction mechanisms conveying information from outside the bud also developed. Among these, hormonal influences have long been suspected, but they are still poorly understood. However, it is clear that some transduction mechanism(s) must necessarily converge at identified, and as-yet unidentified metabolic blocks, working close to the meristem and constituting the true 'common denominators' of all kinds of dormancy. Advancing our understanding of dormancy at the physiological and biochemical level will be hastened if renewed experimental emphasis is placed on the diversity of correlations, the lability and stability of growth incapacity and the barriers to permeability and metabolism in the meristem.

References

Aillaud, G.J. (1982) Etude de la dormance des bourgeons de *Gleditsia*. I. Le cycle annuel de la jeune plante. *Revue Générale de Botanique* 89, 97-109.

Aillaud, G.J. (1986) Etude de la dormance des bourgeons chez *Gleditsia*. II. Comparaison des capacités intrinsèques de débourrement des bourgeons. *Revue de Cytologie et Biologie végétales* 9, 177-191.

Aillaud, G.J., Al-Ibrahem, A. and Neville, P. (1989) Influence de la durée de l'exposition aux jours courts sur la dormance des bourgeons de *Gleditsia triacanthos* L. *Annales des Sciences Forestières* 46 (Suppl.), 236-241.

Alaoui-Sossé, B., Parmentier, C., Dizengremel, P. and Barnola, P. (1994) Rhythmic growth and carbon allocation in *Quercus robur*. I. Starch and sucrose. *Plant Physiology and Biochemistry* 32, 331-339.

Alatou, D., Barnola, P., Lavarenne, S. and Gendraud, M. (1989) Caractérisation de la croissance rythmique du chêne pédonculé. *Plant Physiology and Biochemistry* 27, 275-280.

Al-Ibrahem, A. (1990) Etude de la dormance des bourgeons de jeunes plantes de *Gleditsia triacanthos* L. Unpublished PhD thesis, University of Aix-Marseille III.

Alvim, P.T. (1964) Tree growth periodicity in tropical climates. In: Zimmermann, M.H. (ed.) *Formation of Wood in Forest Trees*. Academic Press, New York, pp. 479-495.

Alvim, P.T. and Alvim, R. (1978) Relation of climate to growth periodicity in tropical trees. In: Tomlinson, P.B. and Zimmermann, M.H. (eds) *Tropical Trees as Living Systems*. Cambridge University Press, Cambridge, pp. 445-464.

Arias, O. and Crabbé, J. (1975a) Les gradients morphogénétiques du rameau d'un an des végétaux ligneux, en repos apparent. Données complémentaires fournies par l'étude de *Prunus avium* L. *Physiologie Végétale* 13, 69-81.

Arias, O. and Crabbé, J. (1975b) Altérations de l'état de dormance ultérieur des bourgeons obtenues par diverses modalités de décapitation estivale, réalisées sur de jeunes plants de *Prunus avium* L. *Comptes-Rendus de l'Académie des Sciences, Paris D* 280, 2449-2452.

Bailly, O. and Mauget, J.C. (1989) Physiological correlations and bud dormancy in the apple tree (*Malus* × *domestica* Borkh.). *Annales des Sciences Forestières* 46 (Suppl.), 220-222.

Balandier, P., Gendraud, M., Rageau, R., Bonhomme, M., Richard, J.P. and Parisot, E. (1993) Bud break delay on single-node cuttings and bud capacity for nucleotide accumulation as parameters for endo- and paradormancy in peach trees in a tropical climate. *Scientia Horticulturae* 55, 249-261.

Barlow, H.W.B. and Hancock, C.R. (1962) The influence of the leaf upon the development of its axillary meristem. *Report of East Malling Research Station for 1961* 71-76.

Barnola, P. (1970) Recherches sur le déterminisme de la basitonie chez le framboisier (*Rubus idaeus* L.). *Annales des Sciences Naturelles, Botanique* 11, 129-152.

Barnola, P. (1972) Etude expérimentale de la ramification basitone du sureau noir (*Sambucus nigra* L.). *Annales des Sciences Naturelles, Botanique* 13, 369-400.

Barnola, P. (1976) Recherches sur la croissance et la ramification du noisetier (*Corylus avellana* L.). *Annales des Sciences Naturelles, Botanique* 17, 223-258.

Barnola P. and Crabbé, J. (1991) La basitonie chez les végétaux ligneux: déterminisme et variabilité d'expression. In: Edelin, C. (ed.) *L'Arbre: Biologie et Développement, Naturalia Monspeliensia*, pp. 381-396.

Barnola, P., Champagnat, P. and Lavarenne, S. (1976) Taille en vert et dormance des bourgeons chez le noisetier. *Comptes- Rendus de l'Académie d'Agriculture de France* 16, 1163-1171.

Barnola, P., Crochet, A., Payan, E., Gendraud, M. and Lavarenne, S. (1986a) Modifications du métabolisme énergétique et de la perméabilité dans le bourgeon apical et l'axe sous-jacent au cours de l'arrêt de croissance momentané de jeunes plants de chêne. *Physiologie Végétale* 24, 307-314.

Barnola, P., Lavarenne, S. and Gendraud, M. (1986b) Dormance des bourgeons apicaux du frêne (*Fraxinus excelsior* L.): évaluation du pool des nucléosides triphosphates et éventail des températures actives sur le débourrement des bourgeons en période de dormance. *Annales des Sciences Forestières* 43, 339-350.

Barnola, P., Lavarenne, S., Gendraud, M. and Jallut, N. (1986c) Etude biochimique d'une dormance rythmique chez le frêne (*Fraxinus excelsior* L.) cultivé en conditions contrôlées (12°C, jours longs). *Comptes-Rendus de l'Académie des Sciences, Paris* 303, 239-244.

Barnola, P., Alatou, D., Lacointe, A. and Lavarenne, S. (1990) Etude biologique et biochimique du déterminisme de la croissance rythmique du chêne pédonculé (*Quercus robur* L.): effets de l'ablation des feuilles. *Annales des Sciences Forestières* 21, 619-631.

Ben Ismail, M.C. (1989) Dormance et reprise de croissance des bourgeons chez les arbres fruitiers. Variabilité et caractère multifactoriel des phénomènes: relation avec la constitution et la mobilization des réserves. Unpublished PhD thesis, Faculty of Agronomy of Gembloux.

Bidabé, B. (1967) Action de la température sur l'évolution des bourgeons de pommier et comparaison des méthodes de contrôle de l'époque de floraison. *Annales de Physiologie Végétale* 9, 65-86.

Borchert, R. (1991) Growth periodicity and dormancy. In: Raghavendra, A.S. (ed.) *Physiology of Trees*. John Wiley, New York, pp. 221-245.

Champagnat, P. (1955) Les corrélations entre feuilles et bourgeons de la pousse herbacée du lilas. *Revue Générale de Botanique* 62, 325-371.

Champagnat, P. (1978) Formation of the trunk in woody plants. In: Tomlinson, P.B. and Zimmermann, M.H. (eds) *Tropical Trees as Living Systems*. Cambridge University Press, Cambridge, pp. 401-422.

Champagnat, P. (1983) Quelques réflexions sur la dormance des bourgeons des végétaux ligneux. *Physiologie Végétale* 21, 607-618.

Champagnat, P. (1989) Rest and activity in vegetative buds of trees. *Annales des Sciences Forestières* 46 (Suppl.), 9-26.

Champagnat, P. (1992) Dormance des bourgeons chez les végétaux ligneux. In: Côme, D. (ed.) *Les Végétaux et le Froid*. Hermann, Paris, pp. 203-262.

Champagnat, P., Lavarenne, S. and Barnola, P. (1975) Corrélations entre bourgeons et intensité de la dormance sur le rameau de l'année pour quelques végétaux ligneux en repos apparent. *Comptes-Rendus de l'Académie des Sciences, Paris D* 280, 2219-2222.

Champagnat, P., Barnola, P. and Lavarenne, S. (1986a) Quelques modalités de la croissance rythmique endogène des tiges chez les végétaux ligneux. *Naturalia Monspeliensia*, special issue, *Colloque international sur l'Arbre*, pp. 279-302.

Champagnat, P., Payan, E., Champagnat, M., Barnola, P., Lavarenne, S. and Bertholon, C. (1986b) La croissance rythmique de jeunes chênes pédonculés cultivés en conditions contrôlées uniformes. *Naturalia Monspeliensia*, special issue, *Colloque international sur l'arbre*, pp. 303-338.

Cottignies, A. (1986) Dormance. *Annales des Sciences Naturelles, Botanique* 8, 93-142.

Crabbé, J. (1968) Evolution annuelle de la capacité intrinsèque de débourrement des bourgeons successifs de la pousse de l'année chez le pommier et le poirier. *Bulletin de la Société Royale de Botanique de Belgique* 101, 195-204.

Crabbé, J. (1970) Influences foliaires sur la croissance de la pousse annuelle du pommier. III. Effets de la suppression de jeunes feuilles sur la levée d'inhibition et le développement des bourgeons axillaires. *Bulletin des Recherches Agronomiques de Gembloux* 5, 136-151.

Crabbé, J.J. (1981) The interference of bud dormancy in the morphogenesis of trees and shrubs. *Acta Horticulturae* 120, 167-172.

Crabbé, J.J. (1984a) Vegetative vigor control over location and fate of flower buds in fruit trees. *Acta Horticulturae* 149, 55-63.

Crabbé, J.J. (1984b) Correlative effects modifying the course of bud dormancy in woody plants. *Zeitschrift für Pflanzenphysiologie* 113, 465-469.

Crabbé, J.J. (1994) Dormancy. In: Arntzen, C.J. and Ritter, E.M. (eds) *Encyclopedia of Agricultural Science*, Vol. 1. Academic Press, New York, pp. 597-611.

Dostal, R. (1952) Experimental morphogenesis of buds in the horsechestnut (*Aesculus hippocastanum* L.). *Moravskoslezske Akademia ved Prirodnich, Prace* 24, 109-146 (in Czech, English summary).

Dreyer, E. and Mauget, J.C. (1986a) Variabilité du niveau de dormance des bourgeons végétatifs suivant les types de rameau d'une couronne de noyer (*Juglans regia* L.): comparaison des cultivars 'Franquette' et 'Pedro'. *Agronomie* 6, 427-435.

Dreyer, E. and Mauget, J.C. (1986b) Conséquences immédiates et différées de périodes de sécheresse estivale sur le développement de jeunes noyers (*Juglans regia* L., cv. 'Pedro'): dynamique de croissance et dormance automno-hivernale des bourgeons. *Agronomie* 6, 639-650.

Erez, A. and Zur, A. (1981) Breaking the rest of apple buds by narrow-distillation-range oil and dinitrocresol. *Scientia Horticulturae* 14, 47-54.

Faust, M., Liu, D., Millard, M.M. and Stutte, G.W. (1991) Bound versus free water in dormant apple buds - a theory for endodormancy. *Hortscience* 26, 887-890.

Fulford, R.M. (1965) The morphogenesis of apple buds. I. The activity of the apical meristem. *Annals of Botany* 29, 167-180.

Fulford, R.M. (1966) The morphogenesis of apple buds. II. The development of the bud. *Annals of Botany* 30, 25-38.

Gendraud, M. (1975) Contribution à l'étude du métabolisme des nucléotides di- et triphosphates de tubercules de topinambour cultivés *in vitro*, en rapport avec leurs potentialités morphogénétiques. *Plant Science Letters* 4, 53-59.

Gendraud, M. (1977) Etude de quelques aspects du métabolisme des nucléotides des pousses de topinambour en relation avec leurs potentialités morphogénétiques. *Physiologie Végétale* 15, 121-132.

Gendraud, M. (1981) Etude de quelques propriétés des parenchymes de pousses de topinambour cultivés *in vitro*, en relation avec leurs potentialités morphogénétiques. *Physiologie Végétale* 19, 473-481.

Gendraud, M. and Lafleuriel, J. (1983) Caractéristiques de l'absorption du saccharose et du tétraphénylphosphonium par les parenchymes de tubercules de topinambour, dormants et non dormants, cultivés *in vitro*. *Physiologie Végétale* 21, 1125-1133.

Gendraud, M. and Pétel, G. (1990) Modifications in intracellular communications, cellular characteristics and change in morphogenetic potentialities of Jerusalem artichoke tubers (*Helianthus tuberosus* L.). In: Millet, B. and Greppin, H. (eds) *Intra- and Intercellular Communications in Plants*. INRA, Paris, pp. 171-175.

Guern, J. and Usciati, M. (1976) Essai de réponse à huit questions concernant la régulation de la croissance des bourgeons axillaires chez *Cicer arietinum* L.. In: Jacques, R. (ed.) *Etudes de biologie végétale dédiées au Professeur P. Chouard*. Masson, Paris, pp. 191-207.

Hallé, F. and Martin, R. (1968) Etude de la croissance rythmique chez l'hévéa (*Hevea brasiliensis* Müll. Arg.). *Adansonia, série 2* 8, 475-503.

Hallé, F., Oldeman, R.A.A. and Tomlinson, P.B. (1978) *Tropical Trees and Forests: an Architectural Analysis*. Springer Verlag, Berlin.

Herter, F.G., Balandier, P., Mauget, J.C., Rageau, R. and Bonhomme, M. (1991) Conséquences des conditions climatiques durant la croissance estivale et la période de repos sur la capacité de croissance des bourgeons chez deux espèces fruitières tempérées: le pommier et le pêcher. In: Edelin, C. (ed.) *L'Arbre: biologie et développement*. *Naturalia Monspeliensia*, Montpelier, pp. 417-431.

Hillman, J.R. (1984) Apical dominance. In: Wilkins, M.B. (ed.) *Advanced Plant Physiology*. Pitman, London, pp. 127-148.

Kobayashi, K.D., Fuchigami, L.H. and English, M.J. (1982) Modeling temperature requirements for rest development in *Cornus sericea*. *Journal of the American Society for Horticultural Science* 107, 914-918.

Kurdjan, A. and Guern, J. (1978) Intracellular pH in higher plant cells. I. Improvement in the use of 5,5 dimethyl-oxazolidine-2(^{14}C),4-dione distribution technique. *Plant Science Letters* 11, 337-344.

Lang, G.A., Early, J.D., Arroyave, N.J., Darnell, R.L., Martin, G.C. and Stutte, G.W. (1985) Dormancy: toward a reduced, universal terminology. *Hortscience* 20, 809-812.

Lang, G.A., Early, J.D., Martin, G.C. and Darnell, R.L. (1987) Endo-, para- and ecodormancy: physiological terminology and classification for dormancy research. *Hortscience* 22, 371-378.

Lavarenne, S., Champagnat, P. and Barnola, P. (1971) Croissance rythmique de quelques végétaux ligneux des régions tempérées cultivés en chambres climatisées à température élevée et constante et sous diverses photopériodes. *Bulletin de la Société Botanique de France* 118, 131-162.

Lavarenne, S., Champagnat, P. and Barnola, P. (1975) Influence d'une même gamme de températures sur l'entrée et la sortie de dormance des bourgeons du frêne (*Fraxinus excelsior* L.). *Physiologie Végétale* 13, 215-224.

Lavarenne, S., Barnola, P. and Champagnat, P. (1980) Climats artificiels et dormance des bourgeons. I. Températures et dormance automnale chez le frêne (*Fraxinus excelsior* L.). *Comptes-Rendus de l'Académie d'Agriculture de France* 20, 92-106.

Lavarenne, S., Champciaux, M., Barnola, P. and Gendraud, M. (1982) Métabolisme des nucléotides et dormance des bourgeons chez le frêne. *Physiologie Végétale* 20, 371-376.

Lavarenne, S., Barnola, P., Gendraud, M. and Jallut, N. (1986) Caractérisation biochimique de la période de repos au cours de la croissance rythmique du frêne cultivé à température élevée et constante. *Comptes-Rendus de l'Académie des Sciences, Paris III* 303, 139-144.

Lavarenne-Allary, S. (1965) Recherches sur la croissance des bourgeons du chêne et de quelques autres espèces ligneuses. *Annales des Sciences Forestières* 22, 1-203.

Liu, D., Faust, M., Millard, M.M., Line, M.J. and Stutte, G.W. (1993) States of water in summer-dormant apple buds determined by proton magnetic resonance imaging. *Journal of the American Society for Horticultural Science* 118, 632-637.

Longman, K.A. (1978) Control of shoot extension and dormancy: external and internal factors. In: Tomlinson, P.B. and Zimmermann, M.H. (eds) *Tropical Trees as Living Systems*. Cambridge University Press, Cambridge, pp. 465-496.

Maillard, P., Jacques, M. and Miginiac, E. (1987) Correlative growth in young *Terminalia superba* in a controlled environment: effect of leaves on internode elongation. *Annals of Botany* 60, 447-454.

Martin, G.C. (1991) Bud dormancy in deciduous fruit trees. In: Steward, F.C. (ed.) *Plant Physiology, A Treatise*, Vol. 10. Academic Press, New York, pp. 183-225.

Mauget, J.C. (1978) Influence d'une ablation totale du feuillage sur l'entrée en dormance des bourgeons du noyer (*Juglans regia* L.). *Comptes-Rendus de l'Académie des Sciences, Paris D* 286, 745-748.

Mauget, J.C. (1981) Modification des capacités de croissance des bourgeons du noyer (*Juglans regia* L.) par application d'une température de 4°C à différents moments de leur période de repos apparent. *Comptes-Rendus de l'Académie des Sciences, Paris III* 292, 1081-1083.

Mauget, J.C. (1983) Etude de la levée de dormance et du débourrement des bourgeons du noyer (*Juglans regia* L., cv. Franquette) soumis à des températures supérieures à 15°C au cours de leur période de repos apparent. *Agronomie* 3, 745-750.

Mauget, J.C. (1984) Comportement comparé des bourgeons de l'année et des bourgeons latents chez le noyer (*Juglans regia* L., cv. Franquette): conséquences sur la morphogenèse de l'arbre. *Agronomie* 4, 507-515.

Mauget, J.C. (1987) Dormance des bourgeons chez les arbres fruitiers de climat tempéré. In: Le Guyader, H. (ed.) *Le Développement des végétaux: aspects théoriques et synthétiques*. Masson, Paris, pp. 133-150.

Mauget, J.C. and Rageau, R. (1988) Bud dormancy and adaptation of apple tree to mild winter climates. *Acta Horticulturae* 232, 101-108.

Médard, R. (1989) Les mécanismes de formation du limbe chez le *Manihot esculenta*: étude microchirurgicale. *Canadian Journal of Botany* 67, 997-1008.

Médard, R., Sell, Y. and Barnola, P. (1992a) Le développement du bourgeon axillaire de *Manihot esculenta*. *Canadian Journal of Botany* 70, 2041-2052.

Médard, R., Walter, J.M.N. and Barnola, P. (1992b) Influence, chez le *Manihot esculenta*, du développement de la lame du limbe sur celui du pétiole et de l'entre-noeud sous-jacent. *Canadian Journal of Botany* 70, 2053-2065.

Meng-Horn, C., Champagnat, P., Barnola, P. and Lavarenne, S. (1975) L'axe caulinaire, facteur de préséances entre bourgeons sur le rameau de l'année de *Rhamnus frangula* L. *Physiologie Végétale* 13, 335-348.

Neville, P. (1968) Morphogenèse chez *Gleditsia triacanthos* L. I. Mise en évidence de corrélations jouant un rôle dans la morphogenèse et la croissance des bourgeons et des tiges. *Annales des Sciences Naturelles, Botanique* 9, 433-510.

Parisot, E. (1988) Etude de la croissance rythmique chez de jeunes manguiers (*Mangifera indica* L.). *Fruits* 43, 175-190, 235-247, 293-312.

Parmentier, C. (1993) Etude physiologique et biochimique de la croissance rythmique endogène du chêne pédonculé: recherche de son déterminisme. Unpublished PhD thesis, University of Nancy I.

Parmentier, C., Barnola, P., Maillard, P. and Lavarenne, S. (1991) Etude de la croissance rythmique du chêne pédonculé: influence du système racinaire. In: Edelin, C. (ed.) *L'Arbre: biologie et développement*. *Naturalia Monspeliensia*, special issue, pp. 327-343.

Pétel, G. and Gendraud, M. (1986) Contribution to the study of ATPase activity in plasmalemma-enriched fractions from Jerusalem artichoke tubers (*Helianthus tuberosus* L.) in relation to their morphogenetic properties. *Journal of Plant Physiology* 123, 373-380.

Pétel, G. and Gendraud, M. (1993) ATP- and NADH-dependent membrane potential generation in plasmalemma-enriched vesicles from parenchyma of dormant and non-dormant Jerusalem artichoke tubers. *Biologia Plantarum* 35, 161-167.

Pétel, G., Lafleuriel, J., Dauphin, G. and Gendraud, M. (1992) Cytoplasmic pH and plasmalemma ATPase activity of parenchyma cells during the release of dormancy of Jerusalem artichoke tubers. *Plant Physiology and Biochemistry* 30, 379-382.

Rageau, R. (1978) Croissance et débourrement des bourgeons végétatifs de pêcher (*Prunus persica* L. Batsch) au cours d'un test classique de dormance. *Comptes-Rendus de l'Académie des Sciences, Paris D* 287, 1119-1122.

Regnard, J.L. (1985) Facteurs artificiels de levée de dormance des bourgeons de végétaux ligneux: analyse des effets de l'anoxie appliquée au peuplier. *5ème Colloque sur les Recherches Fruitières*, Bordeaux, pp. 187-200.

Richardson, E.A., Seeley, S.D. and Walker, D.R. (1974) A model for estimating the completion of rest for 'Redhaven' and 'Elberta' peach trees. *Hortscience* 9, 331-332.

Romberger, J.A. (1963) *Meristems, Growth and Development in Woody Plants*. Technical Bulletin No. 1293, United State Department of Agriculture, Forest Service.

Saure, M.C. (1985) Dormancy release in deciduous fruit trees. *Horticultural Reviews* 7, 239-300.

Shaltout, A.D. and Unrath, C.R. (1983) Rest completion model for 'Starkrimson Delicious' apple. *Journal of the American Society for Horticultural Science* 108, 957-961.

Shulman, Y., Nir, G., Fanberstein, L. and Lavee, S. (1983) The effect of cyanamide on the release from dormancy of grapevine buds. *Scientia Horticulturae* 19, 97-104.

Tort, M. and Gendraud, M. (1984) Contribution à l'étude des pH cytoplasmique et vacuolaire en rapport avec la croissance et l'accumulation des réserves chez le Crosne du Japon. *Comptes-Rendus de l'Académie des Sciences, Paris, III* 299, 431-434.

Wang, S.Y. and Faust, M. (1988) Changes in fatty acids and sterols in apple buds during bud break induced by a plant bioregulator, thidiazuron. *Physiologia Plantarum* 72, 115-120.

Wang, S.Y. and Faust, M. (1989) Changes in membrane polar lipids associated with bud break in apple induced by nitroguanidines. *Journal of Plant Growth Regulation* 8, 153-161.

Wang, S.Y. and Faust, M. (1990) Changes of membrane lipids in apple buds during dormancy and bud break. *Journal of the American Society of Horticultural Science* 115, 803-808.

Wang, S.Y. and Faust, M. (1992) Ascorbic acid oxidase activity in apple buds: relation to thidiazuron-induced bud break. *Hortscience* 27, 1102-1105.

Wang, S.Y., Ji, Z.L. and Faust, M. (1987) Metabolic changes associated with bud break induced by thidiazuron. *Journal of Plant Growth Regulation* 6, 85-95.

Wang, S.Y., Jiao, H.J. and Faust, M. (1991a) Changes in the activities of catalase, peroxidase and polyphenoloxidase in apple buds during bud break induced by thidiazuron. *Journal of Plant Growth Regulation* 10, 33-39.

Wang, S.Y., Jiao, H.J. and Faust, M. (1991b) Changes in superoxide dismutase activity during thidiazuron-induced lateral bud break of apple. *Hortscience* 26, 1202-1204.

Wang, S.Y., Jiao, H.J. and Faust, M. (1991c) Changes in ascorbate, glutathione, and related enzyme activities during thidiazuron-induced bud break of apple. *Physiologia Plantarum* 82, 231-236.

Williams, R.R., Edwards, G.R. and Coombe, B.G. (1979) Determination of the pattern of winter dormancy in lateral buds of apples. *Annals of Botany* 44, 575-581.

Zanette, F. (1981) Recherches descriptives et expérimentales sur la morphogenèse des systèmes aériens et racinaires de quelques porte-greffe de pommier. Unpublished PhD thesis, University of Clermont-Ferrand.

7 Development of Dormancy in Tissue-Cultured Lily Bulblets and Apple Shoots

GEERT-JAN M. DE KLERK AND MEREL M. GERRITS
Centre for Plant Tissue Culture Research, PO Box 85, 2160 AB Lisse, The Netherlands

Introduction

Experimental manipulations in dormancy development studies are usually applied to whole plants. This can hinder progress in understanding dormancy mechanisms for at least three reasons. First, complex interactions may occur between the dormant organ and the other organs of the plant. Second, it is difficult to administer compounds to the organ when it is attached to the plant and application in a quantitative manner is almost impossible. Third, since whole plants often require a lot of space, the effects of environmental factors like temperature or day length can be studied only in an expensive phytotron.

Occasionally, dormancy has been studied in simple systems. For example, the development of dormancy in seeds has been examined in detached pod culture under non-sterile conditions (Cairns and De Villiers, 1989) or in excised zygotic embryos cultured *in vitro* (Le Page-Degivry and Garello, 1990, 1992). However, more sophisticated *in vitro* systems are also available. Embryo dormancy could be studied in somatic embryo cultures and bud, bulb, tuber or corm dormancy in shoot or plantlet cultures. In spite of much progress in tissue culture during the past three decades, these possibilities have hardly been explored. For several years, our group has been investigating dormancy development in lily bulblets regenerated *in vitro*. This chapter reviews progress and results thus far, and explores some preliminary work on bud dormancy in apple cultures.

Dormancy in Lily Bulblets Regenerated *In Vitro*

The experimental system

In vitro lily bulblet experiments have been carried out with *Lilium speciosum* 'Rubrum No. 10', *L. longiflorum* 'Snow Queen', the oriental hybrid 'Star Gazer' and the asiatic hybrid 'Connecticut King'. The bulbs were obtained from a local grower and stored at $-1°C$ until use. The protocol for adventitious regeneration of lily bulblets from scale explants has been described previously in detail (Aguettaz *et al.*, 1990). In short, 7×7 mm explants are cut from surface-sterilized scales of field-grown bulbs and cultured *in vitro* at 20°C and 30 µE m^{-2} s^{-1} for 16 h day^{-1}. The explant is placed with its abaxial side down on a medium of Murashige-Skoog (MS) nutrients (Murashige and Skoog, 1962), 3% (w/v) sucrose, 100 mg l^{-1} myoinositol, 0.4 mg l^{-1} thiamine-HCl, 0.25 µM 1-naphthaleneacetic acid (NAA) and 0.6% (w/v) agar. After 2 to 3 weeks, adventitious buds become visible at the basal edge of the explant. Regeneration occurs at this position because of the basipetal transport of auxin (Van Aartrijk and Blom-Barnhoorn, 1984; Smulders *et al.*, 1988). In our standard protocol, bulblets were harvested 11 weeks after the start of culture (Fig. 7.1) at a fresh weight of *c.* 50 mg.

Fig. 7.1. Plantlets of *L. speciosum* regenerated *in vitro* from scale explants, 11 weeks after the start of tissue culture.

After harvest, the bulblets were transplanted to wooden boxes with steam-sterilized potting soil, covered with *c.* 2 cm soil, and kept at 17°C with 30 µE m^{-2}s^{-1} for 16 h day^{-1}. In some experiments, bulblets were also planted *in vitro* on medium containing only MS salts and agar. Leaf emergence was scored weekly. Depending on the genotype, 20 to 40% of the bulblets that had been regenerated under standard conditions sprouted in soil. After a 6-week cold treatment at 2°C, 90 to 100% sprouted (Fig. 7.2). This demonstrates that dormancy had developed during *in vitro* culture. It should be noted that, at first, only leaves emerged. In *L. longiflorum*, a stem developed during the first growing season, a few months after foliar emergence. In the other lilies, stems developed only after one or two growing seasons.

Factors that may influence dormancy were examined by varying standard culture conditions during bulblet regeneration, e.g. addition of the appropriate compound or changing an environmental condition of the culture. After each treatment, the dormancy status of regenerated bulblets was measured by planting samples of 30 to 50 bulblets in soil and determining the percentage of sprouted bulblets after 10 weeks.

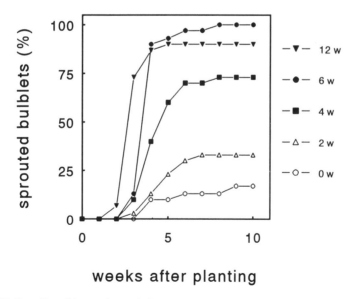

Fig. 7.2. Sprouting of *L. speciosum* bulblets regenerated from scale explants under standard conditions. The bulblets were either planted directly (0 weeks) or after a cold treatment of 2, 4, 6 or 12 weeks at 2°C. w, Weeks.

Effect of physical and nutritional factors

Temperature during regeneration had a major effect on the development of dormancy. In *L. speciosum*, 'Star Gazer' and 'Connecticut King', little dormancy was apparent at 15°C, but dormancy was strong at 20, 25 or 30°C (Fig. 7.3). The behaviour of *L. longiflorum* was different, exhibiting very weak or intermediate dormancy at 30 or 15°C, respectively. Weak dormancy in bulblets of *L. longiflorum* at 30°C also has been reported by Stimart and Ascher (1982). It is interesting to note that high temperature (1 h 45°C water treatment) breaks dormancy in *L. longiflorum* (Stimart *et al.*, 1982) but not in 'Star Gazer' (S. Nashimoto, pers. comm.).

Of the physical factors that have been examined previously for *L. speciosum*, i.e. dark, constant light or photoperiod, wavelength (red, far-red or blue light), osmotic value of the medium or duration of the tissue-culture period, only the latter had a significant effect on the dormancy level (Aguettaz *et al.*, 1990; Delvallée *et al.*, 1990). In all four lilies, dormancy did not develop

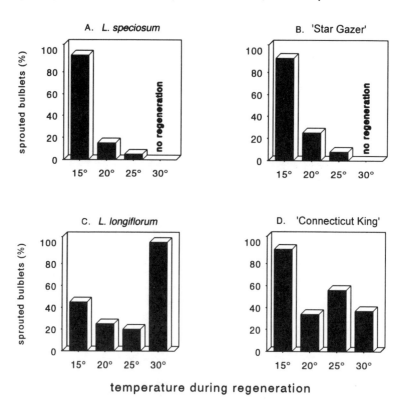

Fig. 7.3. Sprouting of lily bulblets 10 weeks after transplanting into soil. Bulblets were regenerated *in vitro* under standard conditions but at various temperatures and excised after 11 weeks of culture. *L. speciosum* and 'Stargazer' did not regenerate at 30°C.

immediately with the appearance of bulblets on the explants, but only some weeks later, as occurs in seeds several weeks after anthesis (e.g. Borriss and Arndt, 1956). The rate of dormancy development varied with lily genotype (data not shown).

Of the nutritional factors that have been examined, only sucrose concentration was found to influence the establishment of dormancy. For example, in 'Connecticut King' a low sucrose concentration resulted in low dormancy (Fig. 7.4). The same has been reported for *L. auratum* (Takayama and Misawa, 1980) and *L. speciosum* (Aguettaz *et al.*, 1990). Other changes in medium composition, i.e. the type of carbohydrate (fructose or glucose) and the concentration of salts, had no effect.

Effect of plant growth regulators

Addition of cytokinin (benzylaminopurine (BAP), thidiazuron), auxin (NAA, indoleacetic acid (IAA)), the ethylene precursor 1-amino-cyclopropane-1-carboxylic acid (ACC) or the ethylene-producing compound 2-chloroethylphosphonic acid had no significant effect on dormancy development (Aguettaz *et al.*, 1990). Fluridone, an inhibitor of abscisic acid (ABA) synthesis (Zeevaart and Creelman, 1988), prevented dormancy development in all four

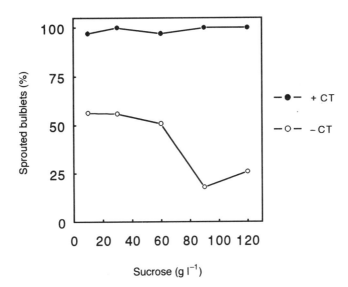

Fig. 7.4. Sprouting of 'Connecticut King' bulblets 10 weeks after transplanting into soil. Bulblets were regenerated *in vitro* at 20°C on various concentrations of sucrose and were harvested after 11 weeks of culture. The bulblets were either planted directly (−CT) or after a cold treatment of 6 weeks at 2°C (+CT).

lilies (e.g. see Fig. 7.5a for *L. longiflorum*). In *L. speciosum*, when ABA is added together with fluridone, dormancy development is restored (Kim *et al.*, 1994). This demonstrates that fluridone acts by inhibiting ABA synthesis. The role of ABA in the development of dormancy in seeds has been well established (Karssen *et al.*, 1983; Koornneef *et al.*, 1989; Le Page-Degivry and

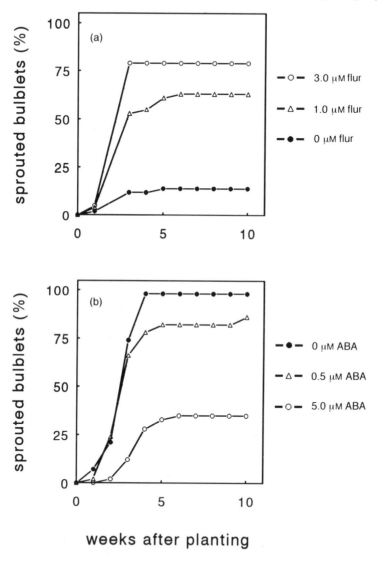

Fig. 7.5. Sprouting of bulblets of *L. longiflorum* after transplanting into soil. (a) Bulblets were regenerated for 11 weeks *in vitro* at 20°C with addition of various concentrations of fluridone and planted directly. (b) Bulblets were regenerated for 11 weeks at 20°C with addition of various concentrations of ABA and planted after a cold treatment of 4 weeks at 2°C.

Garello, 1990; De Klerk, 1992). It should be noted, though, that added ABA did not restore the ability to develop dormancy in *Arabidopsis* seeds that have a low ABA synthesis due to mutation (Karssen *et al.*, 1983).

The results obtained with fluridone suggest that weak dormancy development, for example during regeneration at 15°C, is caused by a low level of endogenous ABA. Indeed, when ABA was added at 15 or 20°C, it sometimes induced dormancy (Fig. 7.5b). However, the effect of added ABA was usually small. Addition of ABA to *L. speciosum* had no effect (Kim *et al.*, 1994). This indicates that the absence of dormancy development is not due solely to a low ABA level, but rather may also involve an ABA insensitivity. In seeds, differences between dormancy levels of various cultivars are related to differences in ABA sensitivity (Walker-Simmons, 1987; Morris *et al.*, 1989, 1991). Therefore, ABA sensitivity in *L. speciosum* bulblets regenerated at 15, 20 or 25°C was determined by bulblet excision from the explant and subculture on to media with only MS salts, agar and different concentrations of ABA. The delay in emergence caused by ABA and the number of bulblets in which sprouting was inhibited by ABA were taken as a measure of ABA sensitivity. During the period in which dormancy was developing, ABA sensitivity was high, but the expected relationship between ABA sensitivity and regeneration temperature was not observed. In fact, the bulblets that regenerated at 15°C had the highest sensitivity to ABA, rather than the lowest as expected (Djilianov *et al.*, 1994).

Gibberellins (GAs) break dormancy in lily bulbs (Ohkawa, 1979) and *in vitro*-produced bulblets (Niimi *et al.*, 1988). Immersion of bulblets for 24 h in a GA_{4+7} solution completely broke dormancy, whereas a similar treatment with fluridone resulted in only 50% sprouted bulblets (Gerrits *et al.*, 1992). When GA_{4+7} or GA_3 was added during culture, higher sprouting percentages were observed occasionally (Aguettaz *et al.*, 1990), yet addition of a GA synthesis inhibitor, paclobutrazol, only occasionally reduced sprouting (data not shown). Since paclobutrazol may be blocking synthesis of ABA as well as GA (Al-Nimri and Coolbaugh, 1990), its addition may act to both increase and decrease the percentage of sprouting bulblets.

The nature of lily bulblet dormancy

During the regeneration of *L. speciosum* bulblets at 25 or 20°C, dormancy developed after 4 or 6 weeks of culture, respectively. At 15°C, dormancy hardly developed, even with prolonged culture (Delvallée *et al.*, 1990). Growth (measured as dry weight increase) at 20 and 25°C continued at the same rate before and after the induction of dormancy (Fig. 7.6a). The increases in weight at 15, 20 and 25°C were very similar (Fig. 7.6a) in spite of the different dormancy levels that were induced at the different temperatures (Fig. 7.3). The increases in the number of scales in bulblets cultured at the three temperatures (Fig. 7.6b) show that the gain in dry weight was due to the formation of new organs, as well as the accumulation of storage products. An increase in the number of scales has been observed previously in pear buds that fail to break dormancy after transfer to a warm glasshouse during winter (Young *et al.*, 1974). The increase of scale number in lily bulblets and pear buds and the high 'normal' levels of metabolism in dormant seeds (De Klerk,

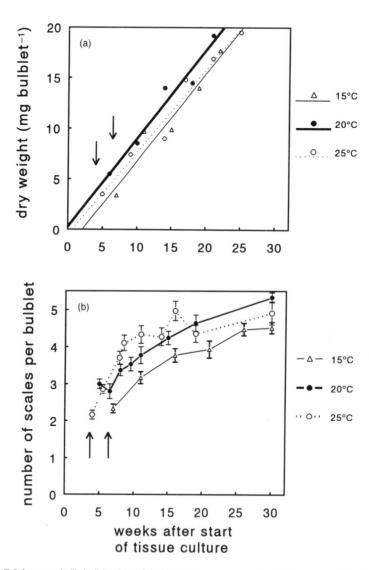

Fig. 7.6. Increase in lily bulblet dry weight (a) and scale number (b) with the course of time in culture at 15, 20 or 25°C. The arrows indicate the start of dormancy development at 25°C (left arrow) or 20°C (right arrow).

1981) and buds (McDonald and Osborne, 1988) demonstrate that the traditional view of dormancy as a cessation of metabolic activity (Bonner, 1965; Kamerbeek *et al.*, 1970, for bulbs) does not hold true.

What is the nature of dormancy in lily bulblets if it does not correspond to an arrest of growth? Delvallée *et al.* (1990) concluded that the induction of dormancy in lily bulblets corresponded to a switch in the developmental pattern of new primordia so that they become incapable of forming leaves and can only form scales. Abundant *in vitro* leaf formation usually correlates with weak dormancy. In particular, the leafy status of the inner scales is related to the dormancy level (Fig. 7.7). There are, however, some obvious exceptions to this rule: both darkness and ABA almost completely prevent leaf formation *in vitro* but have no or very little effect on the dormancy status of *L. speciosum* (Aguettaz *et al.*, 1990; Kim *et al.*, 1994).

In seeds and buds, dormancy is usually relative: growth occurs under a narrow range of conditions, whereas dormancy breaking treatments result in a widening of this range (Borriss, 1940; Vegis, 1964). Dormant bulblets (regenerated at 20°C) did not sprout in soil but did so *in vitro*. Non-dormant

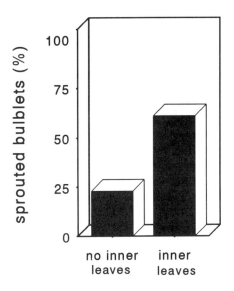

Fig. 7.7. Sprouting of *L. speciosum* bulblets 10 weeks after transplanting to soil. The bulblets were regenerated for 11 weeks at 20°C. At harvest they were divided into two groups: bulblets with a leaf-bearing inner scale and bulblets without an inner leaf. All bulblets were given an incomplete cold treatment of 3 weeks at 2°C and then planted in soil.

bulblets (regenerated at 15°C) sprouted well in soil and also at relatively high temperature (Fig. 7.8). From this, it may be concluded that dormant bulblets will sprout within a narrow range of conditions that includes light and low temperature. The wider range of conditions under which non-dormant bulblets will sprout includes darkness and higher temperatures.

Dormancy in Apple Shoots Proliferated *In Vitro*

The experimental system

In vitro apple shoot experiments were carried out with cultures of *Malus* 'Jork 9' that had been initiated in 1987 and cultured at 20°C as described previously (De Klerk *et al.*, 1990). The carbohydrate source was sorbitol (30 gl^{-1}). About 1-cm-long shoot tips were regularly subcultured onto 15 ml proliferation medium. Due to the presence of 4 μM BAP in the proliferation medium, apical dominance was broken and, after a cycle of 6 weeks, each initial shoot yielded a cluster of 5 to 15 shoots (Fig. 7.9). The standard

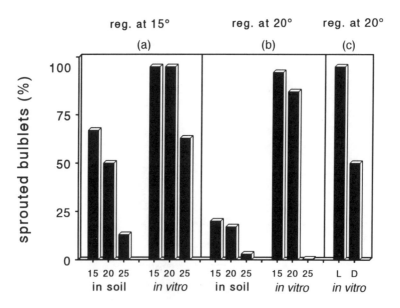

Fig. 7.8. Sprouting of bulblets of *L. speciosum* 10 weeks after transplanting to soil or to solid medium *in vitro*. The bulblets were regenerated for 11 weeks at 15°C (a) or 20°C (b,c) and planted at 15, 20 or 25°C. Bulblets regenerated at 20°C and planted *in vitro* at 15°C were also planted either in the light [L] or in the dark [D] (c).

Fig. 7.9. Shoot clusters of apple 'Jork 9' 2, 4 or 6 weeks (from left to right) after the transfer of a shoot tip of c. 1 cm to standard proliferation medium.

protocol was modified according to the experimental factor being examined for its effect on dormancy development. At the end of the cycle, during which the standard protocol had been modified, 1-cm-long shoot tips or axillary buds (together with 0.5 cm of the stem) were excised. The dormancy levels of shoot tips or axillary buds were measured by their ability to grow on media without hormones or on media with increasing concentrations of indolebutyric acid (IBA) (added to maintain apical dominance), respectively.

The occurrence of dormancy in apple shoot cultures

Once established, apple shoot cultures proliferate indefinitely when subcultured regularly. It therefore seems unlikely that dormancy develops in these cultures. Moreover, it has been reported that dormancy in woody plants is triggered by changes in environmental conditions and that plants grown in glasshouses do not develop dormancy (Vegis, 1964). However, it has also been observed that a cold treatment enhances the growth of some tissue-cultured crops (Cornu and Chaix, 1981; Monette, 1986). An apple rootstock, 'P2', even requires a cold treatment to achieve proliferation (Orlikowska, 1992). When 10-week-old shoot-tip cultures of 'Jork 9' were subjected to a cold treatment prior to excision and subculturing, growth was enhanced (De

Klerk, 1992). Axillary buds that were excised from shoots of 10-week-old cultures doubled their growth when subjected to a cold treatment prior to excision (Fig. 7.10). This demonstrates that dormancy occurs in tissue-cultured apple shoots and that it is superimposed over apical dominance.

Dormancy-influencing factors in apple cultures

It is believed that day length induces bud dormancy (Vegis, 1964). However, shoot tips excised from clusters that had proliferated under short- or long-day conditions had a similar ability to grow on hormone-free medium (Fig. 7.11), indicating a similar dormancy status. A significant increase in ability to grow was observed in shoot tips excised from cultures that had proliferated at low carbohydrate concentration (Fig. 7.11). This suggests that culture-medium carbohydrate level influences dormancy development in apple shoots similarly to that in lily bulblets. Another similarity is that the ability to grow increases sharply when fluridone is added to the medium (De Klerk, 1992). This may be due to the involvement of ABA in dormancy or in apical dominance. Since recent work has shown that the involvement of ABA in

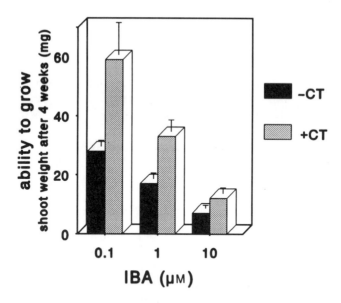

Fig. 7.10. Ability to grow of axillary buds at different concentrations of IBA. The axillary buds had been excised together with 0.5 cm stem from shoot clusters that had (+CT) or had not (−CT) received a cold treatment of 6 weeks at 5°C. The shoot clusters had been cultured from shoot tips on standard proliferation medium for 10 weeks. The ability to grow was determined as the fresh weight after 4 weeks at 25°C.

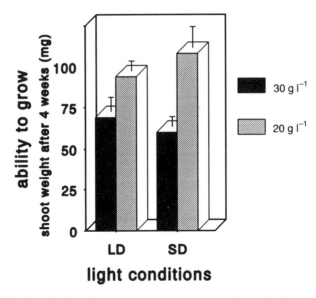

Fig. 7.11. Ability of apple shoot tips to grow on hormone-free medium. The shoot tips had been excised from apple shoot clusters that been cultured for 4 weeks at 25°C at different light (70 µE m^{-2} s^{-1} for 8 (SD) or 16 (LD) h per day) regimes and at 20 or 30 g l^{-1} sorbitol. The ability of the excised apical buds to grow was determined as the fresh weight after 4 weeks at 25°C.

apical dominance of *Rosa*, a related species, is doubtful (H.J. van Telgen, pers. comm.), these results suggest that ABA is involved in the development of dormancy in apple shoots.

Conclusions

Organs that develop dormancy usually form part of a complete plant. This can hinder study of the mechanisms involved in the development of dormancy in these organs. Tissue culture enables the culture of isolated organs and is therefore an excellent tool for the study of dormancy. The results reported above indicate that dormancy develops during tissue culture of both lily bulblets and apple shoots. The effect of various factors on dormancy development in lily bulblet cultures are summarized in Table 7.1. As in seeds (e.g. *Arabidopsis*), ABA and GA appear to play key roles in induction and alleviation of dormancy, respectively (Table 7.1; Karssen and Lacka, 1986; Gerrits *et al.*, 1992). Among the other factors, the sucrose concentration seems to exert a regulatory influence on dormancy status as well. *In vitro* research systems provide a unique ability to manipulate this factor. Comparably to lily bulblets, dormancy development in apple shoots was also affected by carbohydrate concentration and the presence of ABA.

The developmental state of the culture at which the dormancy-influencing factor is received can be important. For example, during *in vitro* culture of tulip, a cold treatment is required to induce bulb formation (bulb formation and dormancy development are concurrent processes) and, at a later stage, when the bulbs are mature, a second cold treatment is required to break dormancy and induce sprouting (Le Nard *et al.*, 1987). In apple shoot cultures there may be great fluctuations from one subculture cycle to the next in shoot proliferation and rooting capability (Denissen *et al.*, 1992). This indicates the occurrence of different developmental states during successive subculture cycles and may be the reason that in apples effects on shoot dormancy could not always be repeated in consecutive experiments. Since in adventitiously formed organs such fluctuations between experiments are unlikely to occur, adventitiously formed bulblets and shoots or somatic embryos may be preferable for studies of dormancy development.

Table 7.1. The effect of various physical, nutritional and hormonal factors on the development of dormancy in lily bulblets regenerated *in vitro* from scale explants.

Factors	Effect on dormancy development	Remarks
Physical		
Temperature	+++	Low dormancy at low temperature
Culture duration	+++	Low dormancy during the first period of culture
Light/dark	0	
Wavelength	0	
Short/long day	0	
Medium osmotic value	0	
Nutritional		
Sucrose concentration	++	Low dormancy at low sucrose concentration
Carbohydrate source	0	
MS concentration	0	
Hormonal		
Cytokinin	0	
Auxin	0	
Ethylene	0	
Abscisic acid	+++	Main effect after inhibition synthesis
Gibberellin	–	Not always reproducible

Acknowledgements

We want to thank Hannah Lilien-Kipnis for the critical reading of the manuscript. The results that are presented here were obtained in collaboration with Pierre Aguettaz, Isabelle Delvallée, Dimitar Djilianov, Kwang-Soo Kim, Shigeru Nashimoto and Annie Paffen.

References

Aguettaz, P., Paffen, A., Delvallée, I., Van der Linde, P. and De Klerk, G.J. (1990) The development of dormancy in bulblets of *Lilium speciosum* generated *in vitro*. I. The effects of culture conditions. *Plant Cell Tissue and Organ Culture* 22, 167-172.

Al-Nimri, L.F. and Coolbaugh, R.C. (1990) Inhibition of abscisic acid biosynthesis in *Cercospora rosicola* by triarimol. *Journal of Plant Growth Regulation* 9, 221-225.

Bonner, J. (1965) *The Molecular Theory of Development*. Oxford University Press, London.

Borriss, H. (1940) Über die innere Vorgänge bei der Samenkeimung und ihre Beeinflussung durch Aussenfaktoren. *Jahrbuch Wissenschaftliche Botanie* 89, 254-339.

Borriss, H. and Arndt, M. (1956) Die Entwicklung isolierter *Agrostemma*-Embryonen in Abhängigkeit vom Reife- und Nachreifezustand der Samen. *Flora* 143, 492-498.

Cairns, A.L.P. and De Villiers, O.T. (1989) Effect of sucrose taken up by developing *Avena fatua* L. panicles on the dormancy and α-amylase synthesis of the seeds produced. *Weed Research* 29, 151-156.

Cornu, D. and Chaix, C. (1981) Multiplication par culture *in vitro* de merisiers adultes (*Prunus avium*). In: Boulay, M. (ed.) *Colloque International sur la Culture* In Vitro *des Essences Forestières*, Afocel, Nangis, pp. 71-79.

De Klerk, G.J. (1981) Degradation of early synthesized proteins in imbibed dormant and afterripened *Agrostemma githago* L. embryos. *Developmental Biology* 83, 183-187.

De Klerk, G.J. (1992) Hormonal control of dormancy and apical dominance in tissue-cultured plants. *Acta Botanica Neerlandica* 41, 443-451.

De Klerk, G.J., Ter Brugge, J., Smulders, R. and Benschop, M. (1990) Basic peroxidases and rooting in microcuttings of *Malus*. *Acta Horticulturae* 280, 29-36.

Delvallée, I., Paffen, A. and De Klerk, G.J. (1990) The development of dormancy in bulblets of *Lilium speciosum* generated *in vitro*. II. The effect of temperature. *Physiologia Plantarum* 80, 431-436.

Denissen, C.J.M., De Klerk, G.J., Albers, M.R.J., Ter Brugge, J. and Kunneman, B.P.A.M. (1992) Effect of accidental factors on rooting of *Malus*. *Agronomie* 12, 799-802.

Djilianov, D., Gerrits, M.M., Ivanova, A., Van Onckelen, H.A. and De Klerk, G.J. (1994) Levels of ABA and sensitivity during the development of dormancy in lily bulblets regenerated *in vitro*. *Physiologia Plantarum* 91, 639-644.

Gerrits, M., Kim, K.S. and De Klerk, G.J. (1992) Hormonal control of dormancy in bulblets of *Lilium speciosum* cultured *in vitro*. *Acta Horticulturae* 325, 521-527.

Kamerbeek, G.A., Beijersbergen, J.C.M. and Schenk, P.K. (1970) Dormancy in bulbs and corms. *Proceedings of 18th International Horticultural Congress*, Tel Aviv, Vol. 5, pp. 233-240.

Karssen, C.M. and Lacka, E. (1986) A revision of the hormone balance theory of seed dormancy: studies on gibberellin and/or abscisic acid-deficient mutants of *Arabidopsis thaliana*. In: Bopp, M. (ed.) *Plant Growth Substances 1985*. Springer Verlag, Berlin, Heidelberg, pp. 315-323.

Karssen, C.M., Brinkhorst-Van der Swan, D.L.C., Breekland, A.E. and Koornneef, M. (1983) Induction of dormancy during seed development by endogenous abscisic acid: studies on abscisic acid deficient genotypes of *Arabidopsis thaliana* L. Heynh. *Planta* 157, 158-165.

Kim, K.S., Davelaar, E. and De Klerk, G.J. (1994) Abscisic acid controls dormancy development and bulb formation in lily plantlets regenerated *in vitro*. *Physiologia Plantarum* 90, 59-64.

Koornneef, M., Hanhart, C.J., Hilhorst, H.W.M. and Karssen, C.M. (1989) *In vivo* inhibition of seed development and reserve protein accumulation in recombinants of abscisic acid biosynthesis and responsiveness mutants in *Arabidopsis thaliana*. *Plant Physiology* 90, 463-469.

Le Nard, M., Ducommun, C., Weber, G., Dorion, N. and Bigot, C. (1987) Observations sur la multiplication *in vitro* de la tulipe (*Tulipa gesneriana* L.) à partir de hampes florales prélevées chez des bulbes en cours de conservation. *Agronomie* 7, 321-329.

Le Page-Degivry, M.-T. and Garello, G. (1990) Involvement of endogenous abscisic acid in onset and release of *Helianthus annuus* embryo dormancy. *Plant Physiology* 92, 1164-1168.

Le Page-Degivry, M.-T. and Garello, G. (1992) *In situ* abscisic acid synthesis: a requirement for induction of embryo dormancy in *Helianthus annuus* embryo dormancy. *Plant Physiology* 98, 1386-1390.

McDonald, M.M. and Osborne, D.J. (1988) Synthesis of nucleic acids and protein in tuber buds of *Solanum tuberosum* during dormancy and early sprouting. *Physiologia Plantarum* 73, 392-400.

Monette, P.L. (1986) Cold storage of kiwifruit shoot tips *in vitro*. *HortScience* 21, 1203-1205.

Morris, C.F., Moffatt, J.M., Sears, R.G. and Paulsen, G.M. (1989) Seed dormancy and responses of caryopses, embryos and calli to abscisic acid in wheat. *Plant Physiology* 90, 643-647.

Morris, C.F., Anderberg, R.J., Goldmark, P.J. and Walker-Simmons, M.K. (1991) Molecular cloning and expression of abscisic acid-responsive genes in embryos of dormant wheat seeds. *Plant Physiology* 95, 814-821.

Murashige, T. and Skoog, F. (1962) A revised medium for rapid growth and bioassays with tobacco tissue cultures. *Physiologia Plantarum* 15, 473-497.

Niimi, Y., Endo, Y. and Arisaka, E. (1988) Effects of chilling- and GA_3-treatments on breaking of dormancy in *Lilium rubellum* Baker bulblets cultured *in vitro*. *Journal of the Japanese Society for Horticultural Science* 57, 250-257.

Ohkawa, K. (1979) Effects of gibberellins and benzyladenine on dormancy and flowering of *Lilium speciosum*. *Scientia Horticulturae* 10, 255-260.

Orlikowska, T. (1992) Effect of *in vitro* storage at 4°C on survival and proliferation of two apple rootstocks. *Plant Cell Tissue and Organ Culture* 31, 1-7.

Smulders, M.J.M., Croes, A.F. and Wullems, G.J. (1988) Polar transport of 1-naphthaleneacetic acid determines the distribution of flower buds on explants of tobacco. *Plant Physiology* 88, 752-756.

Stimart, D.P. and Ascher, P.D. (1982) Foliar emergence from bulblets of *Lilium longiflorum* Thunb. as related to *in vitro* generation temperatures. *Journal of the American Society for Horticultural Science* 106, 446-450.

Stimart, D.P., Ascher, P.D. and Wilkins, H.F. (1982) Overcoming dormancy in *Lilium longiflorum* bulblets produced in tissue culture. *Journal of the American Society for Horticultural Science* 107, 1004-1007.

Takayama, S. and Misawa, M. (1980) Differentiation in *Lilium* bulbscales grown *in vitro*: effects of activated charcoal, physiological age of bulbs and sucrose concentration on differentiation and scale leaf formation *in vitro*. *Physiologia Plantarum* 48, 121-125.

Van Aartrijk, J. and Blom-Barnhoorn, G.J. (1984) Adventitious bud formation from bulb-scale explants of *Lilium speciosum* Thunb. *in vitro*: interacting effects of NAA, TIBA, wounding and temperature. *Journal of Plant Physiology* 116, 409-416.

Vegis, A. (1964) Dormancy in higher plants. *Annual Review of Plant Physiology* 15, 185-224.

Walker-Simmons, M. (1987) ABA levels and sensitivity in developing wheat embryos of sprouting resistant and susceptible cultivars. *Plant Physiology* 84, 61-66.

Young, L.C.T., Winneberger, J.T. and Bennett, J.P. (1974) Growth of resting buds. *Journal of the American Society of Horticultural Science* 99, 146-149.

Zeevaart, J.A.D. and Creelman, R.A. (1988) Metabolism and physiology of abscisic acid. *Annual Review of Plant Physiology and Plant Molecular Biology* 39, 439-473.

8 Dormancy in Tuberous Organs: Problems and Perspectives

Jeffrey C. Suttle
US Department of Agriculture, Northern Crop Science Laboratory, PO Box 5677, State University Station, Fargo, ND 58105-5677, USA

Introduction

Used primarily as a source of carbohydrates for both human consumption and industrial use, tuberous crops comprise a small, but essential, element of agriculture in both developed and developing economies worldwide. In many areas, tubers are the principal source of dietary calories for low-income households. As with other crops such as cereal grains, a significant portion of each harvest is stored for year-round availability for producers and consumers and for use as vegetative propagules for the succeeding year's crop. In many cases, the percentage of crop stored far exceeds that used immediately following harvest. For example, more than 70% (or roughly 12 million metric tons) of the total US potato crop is stored annually to meet the demands of processors and consumers. However, unlike cereals, oilseeds and pulses, the agronomically useful portion of tuber-bearing crops is stored in a fully hydrated and highly perishable form. As a result, postharvest losses from both physiological and pathological processes can result in the loss of significant portions of the harvested yield.

In most cases, the harvested and stored tuber is a perennating organ whose biological function is to ensure the survival of the crop during times of environmental extremes. The phenomenon of dormancy is one of many developmental adaptations acquired to facilitate this survival function. At harvest, most tuberous crops are in a state of dormancy. Depending on the species, this dormancy is lost during the course of storage. The onset of sprouting that heralds the end of dormancy is accompanied by many physiological changes, most notable of which are increased water loss and a dramatic change in tuber composition due to storage reserve (i.e. carbohydrate) mobilization. These, together with other changes, result in a loss of

Table 8.1. Economically important tuber and root crops. (From FAO, 1993.)

Common name	Scientific name	Storage organ	World production (10⁶ MT)
Potato	*Solanum tuberosum*	Tuber	279.5
Cassava	*Manihot esculenta*	Root	153.1
Sweet potato	*Ipomoea batatas*	Root	128.4
Yams	*Dioscorea* spp.	Tuber	28.0
Taro	*Colocasia esculenta*	Corm	5.7

both nutritional and processing quality. In most cases, sprouting effectively ends the useful storage life of the tuber.

Although numerous species of plants are cultivated for their tubers or roots, five species represent the most agriculturally important crops (Table 8.1). Of these, this review will focus primarily on two, potato (*Solanum tuberosum* L.) and yam (*Dioscorea* spp.). The tuber of cassava (*Manihot esculenta*) is an enlarged root that is not a vegetative propagule (O'Hair, 1990). The tuber of sweet potato (*Ipomoea batatas*) is also an enlarged root but, in addition, serves as a vegetative propagule. Sweet potato exhibits no dormancy and must be put through a specific curing process prior to storage (Dempsey *et al.*, 1970). The edible organ of taro (*Colocasia esculenta*) is a corm that exhibits a protracted dormant period (O'Hair and Asokan, 1986). Due to the scarcity of information on this crop, it will not be discussed further.

General Dormancy Characteristics

In both potato and yam, the length or depth of dormancy is under genetic control, and both cultivar- and species-dependent dormant periods have been noted (Passam, 1982; Burton, 1989). In potatoes, genetic studies have indicated that dormancy exhibits a complex inheritance pattern and is controlled by many loci.

Temperature exerts a quantitative effect on potato and yam dormancy. Dormancy is extended as the storage temperature is reduced. This relationship holds until a low temperature threshold is reached (*c.* 2 to 3° and 10 to 12°C for potatoes and yams, respectively). Temperatures below these are considered to be stressful to the tuber and can actually hasten sprouting (Burton *et al.*, 1992; Osagie, 1992). It is important to note that the loss of dormancy in both crops does not require a low-temperature treatment.

In both crops, dormancy can be terminated prematurely by a variety of physical and chemical treatments. The effective physical treatments include high or low temperatures, hyperoxia, hypoxia and wounding. A diverse array of chemicals also break dormancy. These include sulphydryl-containing compounds, ethylene chlorohydrin, bromoethane, gibberellins (GAs), cytokinins and ethylene (van Es and Hartmans, 1981; Passam, 1982). The sheer diversity of these dormancy breaking agents renders simple interpretations of their actions impossible. No doubt, many of these treatments affect identical or convergent biochemical processes that ultimately result in the premature termination of dormancy. Exactly how they achieve this effect is a matter for future research.

Cell Biology of Dormancy

Historically, dormancy has been thought to result from the deficiency of a primary metabolic pathway that leads to the scarcity of a critical metabolite necessary for growth (e.g. Rappaport and Wolf, 1969a). While it is clear that the rates of many cellular processes are reduced during dormancy, the causes as well as the ramifications of these reductions are uncertain. In an analogous paradigm, it has been proposed that dormancy results from a global repression of deoxyribonucleic acid (DNA) template availability (Tuan and Bonner, 1964).

Although intuitively attractive, it is unlikely that dormancy is regulated in such a coarse fashion. Dormant tubers, although incapable of sustained growth, are none the less metabolically active and exhibit easily measurable rates of many cellular processes (Burton, 1989; Osagie, 1992). During dormancy, potato tubers continue to synthesize DNA, ribonucleic acid (RNA) and protein (Macdonald and Osborne, 1988). In yam, inhibition of protein synthesis results in a failure of deep dormancy to develop (Okagami, 1978). Furthermore, far from being metabolically quiescent, dormant tubers readily respond to a variety of both abiotic and biotic external stimuli. For example, physical wounding of dormant potato tubers elicits a massive stimulation of a variety of cellular processes including respiration, ethylene synthesis, cell wall modifications and protein and nucleic acid synthesis (Kahl, 1978; Bostock and Stermer, 1989). Exposure of dormant potatoes to pathogenic races of *Phytophthora infestans* (the causal agent of late blight) or its presumed elicitor, arachidonic acid, results in large increases in sesquiterpene phytoalexin biosynthesis (Choi *et al.*, 1992). Interestingly, recent studies have indicated that this elicitation process involves many signal-transduction systems, including protein kinases, calcium and lipoxygenase (Choi and Bostock, 1994). Thus, it is clear that dormant tubers are metabolically quite active and retain the ability to respond to a variety of external signals by significant shifts in cellular biochemistry. Such responses would not be expected if the dormant organ was functioning under severe metabolic restrictions.

By definition, the dormant state is characterized by the lack of visible growth (Lang et al., 1985). In potatoes, even microscopic growth is absent during the dormant period (van Ittersum, 1992). At the cellular level, the onset of bud growth following the termination of dormancy is the result of two interrelated processes: cell division and cell elongation. Tissue sections prepared from dormant potato buds exhibit few, if any, mitoses, indicating very low rates of cell division (Lesham and Clowes, 1972). Cell division increases within 24 h of ethylene chlorohydrin application (Rappaport and Wolf, 1969b). From these data, it would appear that the onset of cell division occurs early in the sequence of events following the termination of dormancy and is therefore a good place to begin studies of the molecular changes that accompany and regulate the loss of dormancy.

The eukaryotic cell cycle is divided into four phases: G_1, S, G_2 and M (Francis, 1991). As judged by flow cytometry, nuclei isolated from actively growing potato sprouts fall primarily into two populations, G_1 (27%) and G_2 (40%) (Campbell et al., 1996). Nuclei isolated from dormant buds (eyes) exhibit a distinctly different phase pattern, 77% G_1 and 13% G_2. These data indicate that dormancy in potato is associated primarily with a G_1 arrest. Consistent with this interpretation is an extremely low rate of thymidine incorporation into DNA by dormant buds, which indicates little progression into the S phase (Campbell et al., 1996).

Recent studies have demonstrated that progression through the cell cycle is absolutely dependent on the synthesis and activities of a number of proteins (Nurse, 1993). Chief among these regulatory proteins is the activity of a family of c. 34 kDa kinases, collectively termed P-34 kinases (Nurse, 1993). One member of this family (cdc-2 kinase) is involved intimately in several key points of the cell cycle and is highly conserved (both structurally and functionally) across all eukaryotic phyla examined to date (John et al., 1993).

Taking advantage of the highly conserved nature of the P-34 kinase family by using polymerase chain reaction (PCR) technology, a 413 bp fragment was cloned from a potato complementary DNA (cDNA) library that was shown by sequence analysis to be homologous to other cloned cdc-2 genes (Campbell and Suttle, 1994). Northern blot analysis of RNA isolated from actively growing sprouts, dormant buds and non-dormant (but not growing) buds detected at least two transcripts in all tissues examined. Immunoblot analysis of soluble protein extracts prepared from these same tissues revealed at least four antigenically related proteins. Together these data suggest that the G_1 arrest observed in dormant potato meristems is not the result of a lack of cdc-2 gene transcription or translation. It is known that these cell division-associated kinases are regulated post-translationally by reversible phosphorylation and association with other regulatory proteins, collectively called cyclins (Jacobs, 1992; Nurse, 1993). The involvement of these mechanisms in the regulation of cdc-2 kinase activity and cell cycle progression during potato tuber dormancy is being actively investigated in this laboratory.

The cell biology of dormancy termination in yam is quite different from that in potato. In potatoes, the future meristems (eyes) are present at the inception of dormancy and, following the loss of dormancy, growth resumes from one (typically the apical) of the preformed meristems. At the inception of dormancy in yam, no organized meristematic structures exist (Wickham *et al.*, 1981). Instead, at the end of dormancy, sprouting loci are formed from an inner zone of meristematic tissue. The mechanisms that determine the spatial location of sprout loci formation in yam are unknown.

Hormonal Regulation of Dormancy in Potato

As with many aspects of plant development, plant hormones have been assigned a principal role in the regulation of tuber dormancy. Based principally on the pioneering studies by Hemberg (1985), dormancy was initially thought to be controlled by the level of a group of inhibitory substances, collectively termed the β-inhibitor complex. Later this paradigm was modified somewhat to include a role for growth promotive substances as well. The effects, activities and proposed role of the five main classes of plant hormones in potato tuber dormancy are summarized in Table 8.2. Exogenously applied abscisic acid (ABA) elicits a transient inhibition of sprout growth but is ineffective in long-term sprout suppression assays (van Es and Hartmans, 1969). In general, internal ABA levels decline during storage prior to sprouting and they decline still further after the resumption of sprout growth (Korableva *et al.*, 1980; Suttle, 1995).

Table 8.2. Hormone activity during potato tuber dormancy.

	Exogenous effects on sprout growth	Internal levels		Method of quantification	Proposed role
		Dormancy	Sprouting		
Abscisic acid	Inhibitory	+/−[a]	−	B, GC, I[b]	Inhibitor
Auxin	Inhibitory/ promotive	+/0/−	+/−	B, H	Unknown
Cytokinins	Promotive	+	+	B, I	Promoter
Ethylene	Inhibitory/ promotive	?	+	GC	Unknown
Gibberellins	Promotive	−/+	+	B	Promoter

[a] +, Internal levels increase; −, internal levels decrease; 0, no change.
[b] B, bioassay; GC, gas chromatography; H, HPLC; I, immunoassay.

Exogenous auxins elicit a biphasic sprout growth response. At low concentrations, auxins slightly promote the growth of non-dormant sprouts, whereas at higher concentrations they inhibit sprout growth (Hemberg, 1985). This inhibition is of sufficient magnitude for the methyl ester of α-naphthalene acetic acid to have been periodically evaluated as a commercial sprout suppressant for bulk potato storage (van Es and Hartmans, 1981). Early studies relying on bioassays failed to find any consistent changes in endogenous auxin levels during storage (e.g. Hemberg, 1985). Recent studies have found no changes in free indoleacetic acid (IAA) levels during storage and only slight declines concomitant with sprouting (Sukhova et al., 1993).

The application of cytokinins to dormant tubers can result in the premature termination of dormancy (Hemberg, 1970). Cytokinins appear to be most effective when applied near the end of the natural dormant period (Turnbull and Hanke, 1985a). Earlier treatment produces erratic results. Endogenous cytokinin levels (principally zeatin riboside and isopentenyladenosine) are low at harvest, remain low during storage and then increase either prior to (Turnbull and Hanke, 1985b) or concomitant (Sukhova et al., 1993) with sprouting.

Depending on the duration of treatment, exogenous ethylene can either break dormancy or inhibit sprout growth (Rylski et al., 1974). Short-term treatment (i.e. ≤ 72 h) breaks dormancy and promotes sprouting, while extended or continuous treatment inhibits sprout growth. Interestingly, preharvest application of the ethylene-releasing compound, Ethrel, has been reported to extend the dormant period of stored potatoes (Sukhova et al., 1993). Dormant tubers produce extremely small amounts of ethylene (Okazawa, 1974). The rate of ethylene production increases following the initiation of sprout growth.

In general, dormancy is broken rapidly by GA_3 treatment (van Ittersum, 1992). In fact, GA treatment is widely used to promote sprouting of dormant seed tubers in situations requiring rapid turnaround time (i.e. seed multiplication and certification trials). As judged by bioassay, endogenous GAs either decline (Smith and Rappaport, 1961) or increase (Bialek, 1974) during storage prior to sprouting. However, endogenous GA levels increase substantially following the initiation of sprouting (Smith and Rappaport, 1961). Recent studies using GC-MS have identified GA_{20} and GA_1 in sprouts (Jones et al., 1988) and exogenous GA_{12} was metabolized in shoot apices to a variety of other gibberellins, including GA_1, GA_8, GA_{19}, GA_{20}, GA_{29}, GA_{44}, GA_{51} and GA_{53} (van den Berg et al., 1995). These results suggest that the early 13-hydroxylation pathway is the predominant route of GA synthesis and metabolism in potatoes. In other species possessing this pathway, GA_1 is thought to be the biologically active gibberellin (Phinney and Spray, 1982). The effects of storage and sprouting on endogenous GA_1 levels are unknown.

Taken together, these data have been incorporated into a tentative scheme for the hormonal regulation of tuber bud dormancy. As with other similar paradigms, the depth of dormancy or the onset of sprouting is thought to result from the interaction of growth-inhibiting and promoting substances. ABA is thought to be the principal inhibitor, while GAs and cytokinins are the main promoters. The roles of auxins and ethylene in tuber dormancy are enigmatic. Apart from inconsistencies in the data, there are many problems with interpreting this literature. In many instances, 'quantitative' analysis of hormones has been conducted solely using bioassays. Much is dependent on interpretation of application (pharmacological) studies, often using synthetic compounds rather than endogenous hormones. Finally, regardless of methodology, all studies to date have been correlative in nature. No attempts have been made to manipulate hormone levels and determine the resultant effects.

In an attempt to address these deficiencies, we have modified an *in vitro* tuberization system for use in physiological studies (Suttle and Hultstrand, 1994). The use of this system permits the manipulation of the developing tubers in a manner that previously was not possible. In particular, this system permits the feeding of selected inhibitors and/or precursors of a given pathway of hormone synthesis or action. For example, tubers generated *in vitro* in the presence of fluridone contained only traces of endogenous ABA and were essentially non-dormant. Morphologically, fluridone-treated tubers were identical to untreated controls. Micromolar concentrations of exogenous ABA restored internal ABA levels to control values and inhibited premature sprouting. These are the first data published that unequivocally establish a direct role for any hormone in the process of potato tuber dormancy.

Hormonal Regulation of Dormancy in Yam

The hormonal basis for dormancy in yam is even less clear than in potato. Endogenous levels of ABA are highest in dormant tubers or bulbils and decline as dormancy is broken (Hasegawa and Hashimoto, 1973). Exogenous ABA is ineffective in prolonging dormancy (Wickham *et al.*, 1984). However, in addition to ABA, a group of neutral growth inhibitors (batatasins) have been identified (Hashimoto *et al.*, 1972). The endogenous levels of these inhibitors correlate well with the depth of dormancy, and application of these inhibitors can extend yam dormancy (Ireland and Passam, 1984). However, the most remarkable difference between yams and other tuberous species concerns the response to GAs. Application of GA_3 or other bioactive GAs to dormant yam bulbils or tubers prolongs the dormant state (Okagami and Nagao, 1971). GA_4, which is endogenous to yam, is particularly active in this regard (Tanno *et al.*, 1992). In contrast, application of GA biosynthesis inhibitors promotes premature sprouting (Tanno *et al.*, 1992). These data suggest a novel role for gibberellins in this genus.

Lastly, the identification of the batatasins as possible contributors to dormancy regulation in yam underscores the need to continue searching for additional dormancy-regulating substances in other crops. For example, chromatography of acidic extracts of dormant potatoes reveals at least three zones of inhibitory activity, only one of which cochromatographs with ABA (Holst, 1971). Further, authentic ABA is only weakly active in the bioassays used to detect these inhibitors. Thus, it is clear that there are additional endogenous dormancy-regulating substances yet to be characterized and evaluated.

Conclusions and Prospects

The basic biology of tuber dormancy remains an enigma. Research to date has identified many physiological and biochemical alterations that characterize the dormant state. Which, if any, of these processes actually regulates entry or exit from dormancy? The problem of cause and effect continues to undermine the search for regulatory mechanisms. Being subterranean organs, tubers also present some interesting difficulties in the elucidation of their fundamental biology, including seasonal availability, non-uniform physiological age, inaccessibility and inability to modify internal physiology *in situ* by feeding studies. Many of these difficulties can be alleviated or eliminated altogether through the use of model systems. These include the *in vitro* microtubers of potato (described above) and aerial bulbils of yam, which appear to mimic the dormancy behaviour of the underground tuber (Okagami, 1986).

Furthermore, little attention has been given to the use of mutants in tuber dormancy studies. Being vegetatively propagated, true seeds of tuberous crops often exhibit a high degree of heterozygosity and resultant phenotypic variability (Dodds and Paxman, 1962). Thus, the use of physiologically defined mutants may be of great value in unravelling the biological intricacies of dormancy regulation. Technological improvements, such as hormone immunoanalysis, molecular cloning and reverse genetics, will also be of great value in future dormancy studies. In short, prospects for an improved understanding of tuber dormancy have never been better. New knowledge of this important developmental stage will undoubtedly lead to improved post-harvest storage capabilities which will, in turn, benefit producers and consumers alike.

References

Bialek, K. (1974) A preliminary study of activity of gibberellin-like substances in potato tubers. *Zeitschrift für Pflanzenphysiologie* 71, 370-372.

Bostock, R.M. and Stermer, B.A. (1989) Perspectives on wound healing in resistance to pathogens. *Annual Review of Phytopathology* 27, 343-371.

Burton, W.G. (1989) *The Potato*, 3rd edn. John Wiley & Sons, New York.

Burton, W.G., van Es, A. and Hartmans, K.J. (1992) The physics and physiology of storage. In: Harris, P. (ed.) *The Potato Crop*. Chapman and Hall, London, pp. 608-727.

Campbell, M.A., Suttle, J.C. and Sell, T.W. (1996) Changes in cell cycle status and expression of p34^{cdc2} kinase during potato tuber meristem dormancy. *Physiologium Plantarum* (in press).

Choi, D. and Bostock, R.M. (1994) Involvement of *de novo* protein synthesis, protein kinase, extracellular Ca^{2+}, and lipoxygenase in arachidonic acid induction of 3-hydroxy-3-methylglutaryl coenzyme A reductase genes and isoprenoid accumulation in potato (*Solanum tuberosum* L.). *Plant Physiology* 104, 1237-1244.

Choi, D., Ward, B.L., and Bostock, R.M. (1992) Differential induction and suppression of potato 3-hydroxy-3-methylglutaryl coenzyme A reductase genes in response to *Phytophthora infestans* and to its elicitor arachidonic acid. *Plant Cell* 4, 1333-1344.

Dempsey, A.H., Kushman, L.J. and Love, J.E. (1970) Storage. In: *Thirty Years of Cooperative Research 1939-1969*. Southern Cooperative Service Bulletin 159. Louisiana State University, Baton Rouge, pp. 36-38.

Dodds, K.S. and Paxman, G.J. (1962) The genetic system of cultivated diploid potatoes. *Evolution* 16, 154-167.

FAO (United Nations Food and Agriculture Organization) (1993) *World Crop Production Statistics 1992*. FAO, Rome.

Francis, D. (1991) The cell cycle in plant development: Tansley Review No. 38. *New Phytologist* 122, 1-20.

Hasegawa, K. and Hashimoto, T. (1973) Quantitative changes of batatasins and abscisic acid in relation to the development of dormancy in yam bulbils. *Plant and Cell Physiology* 14, 369-377.

Hashimoto, T., Hasegawa, K. and Kawarada, A. (1972) Batatasins: new dormancy-inducing substances of yam bulbils. *Planta* 108, 369-374.

Hemberg, T. (1970) The action of some cytokinins on the rest-period and the content of acid growth-inhibiting substances in potato. *Physiologia Plantarum* 23, 850-858.

Hemberg, T. (1985) Potato rest. In: Li, P.H. (ed.) *Potato Physiology*. Academic Press, New York, pp. 353-388.

Holst, U.B. (1971) Some properties of inhibitor β from *Solanum tuberosum* compared to abscisic acid. *Physiologia Plantarum* 24, 392-396.

Ireland, C.R. and Passam, H.C. (1984) The level and distribution of phenolic plant growth inhibitors in yam tubers during dormancy. *New Phytologist* 97, 233-242.

Jacobs, T. (1992) Control of the cell cycle. *Developmental Biology* 153, 1-15.

John, P.C.L., Zhang, K. and Dong, C. (1993) A p34^{cdc2}-based cell cycle: its significance in monocotyledonous, dicotyledonous and unicellular plants. In: Ormrod, J.C. and Francis, D. (eds.) *Molecular and Cell Biology of the Plant Cell Cycle*. Kluwer Academic, Dordrecht, pp. 9-34.

Jones, M.G., Horgan, R. and Hall, M.A. (1988) Endogenous gibberellins in the potato, *Solanum tuberosum*. *Phytochemistry* 27, 7-10.

Kahl, G. (1978) Induction and degradation of enzymes in aging plant storage tissues. In: Kahl, G. (ed.) *Biochemistry of Wounded Plant Tissues*. De Gruyter, Berlin, pp. 347-390.

Korableva, N.P., Kararaeva, K.A. and Metlitskii, L.V. (1980) Changes of abscisic acid content in potato tuber tissues in the period of deep dormancy and during germination. *Fiziologiya Rastenii* (English translation) 27, 441-446.

Lang, G.A., Early, J.D., Arroyave, N.J., Darnell, R.L., Martin, G.C. and Stutte, G.W. (1985) Dormancy: toward a reduced, universal terminology. *HortScience* 20, 809-812.

Lesham, B. and Clowes, F.A.L. (1972) Rates of mitosis in shoot apices of potatoes at the beginning and end of dormancy. *Annals of Botany* 36, 687-691.

Macdonald, M.M. and Osborne, D.J. (1988) Synthesis of nucleic acids and protein in tuber buds of *Solanum tuberosum* during dormancy and early sprouting. *Physiologia Plantarum* 73, 392-400.

Nurse, P. (1993) Cell cycle control. *Philosophical Transactions of the Royal Society of London, Section B* 341, 449-454.

O'Hair, S.K. (1990) Tropical root and tuber crops. *Horticultural Reviews* 12, 157-196.

O'Hair, S.K. and Asokan, M.P. (1986) Edible aroids: botany and horticulture. *Horticultural Reviews* 8, 43-99.

Okagami, N. (1978) Dormancy in *Dioscorea*: sprouting promotion by inhibitors of protein synthesis in bulbils and rhizomes. *Plant and Cell Physiology* 19, 221-227.

Okagami, N. (1986) Dormancy in *Dioscorea*: different temperature adaptation of seeds, bulbils and subterranean organs in relation to north-south distribution. *Botanical Magazine, Tokyo* 99, 15-27.

Okagami, N. and Nagao, M. (1971) Gibberellin-induced dormancy in bulbils of *Dioscorea*. *Planta* 101, 91-94.

Okazawa, Y. (1974) A relation between ethylene evolution and sprouting of potato tuber. *Journal of the Faculty of Agriculture, Hokkaido University, Sapporo* 57, 443-454.

Osagie, A.U. (1992) *The Yam Tuber in Storage*. Postharvest Research Unit, University of Benin, Benin City.

Passam, H.C. (1982) Dormancy of yams in relation to storage. In: Miege, J. and Lyonga, S.N. (eds) *Yams - Ignames*. Clarendon Press, Oxford, pp. 285-293.

Phinney, B.O. and Spray, C.R. (1982) Chemical genetics and the gibberellin pathway in *Zea mays* L. In: Wareing, P.F. (ed.) *Plant Growth Substances* 1982. Academic Press, London, pp. 101-110.

Rappaport, L. and Wolf, N. (1969a) The problem of dormancy in potato tubers. In: *Dormancy and Survival. Symposium, Society for Experimental Biology* 23, 219-240.

Rappaport, L., and Wolf, N. (1969b) Regulation of bud rest in tubers of potato *Solanum tuberosum* L. III. Nucleic acid synthesis induced by bud excision and ethylene chlorohydrin. In: *International Symposium Plant Growth Substances*, Calcutta, pp. 79-88.

Rylski, I., Rappaport, L. and Pratt, H.K. (1974) Dual effects of ethylene on potato dormancy and sprout growth. *Plant Physiology* 53, 658-662.

Smith, O.E. and Rappaport, L. (1961) Endogenous gibberellins in resting and sprouting potato tubers. *Advances in Chemistry Series* 28, 42-48.

Sukhova, L.S., Machackova, I., Eder, J., Bibik, N.D. and Kovableva, N.P. (1993) Changes in levels of free IAA and cytokinins in potato tubers during dormancy and sprouting. *Biologia Plantarum* 35, 387-391.

Suttle, J.C. (1995) Postharvest changes in ABA levels and ABA metabolism in relation to dormancy in potato tubers. *Physiologia Plantarum* 95, 233-240.

Suttle, J.C. and Hultstrand, J.F. (1994) Role of endogenous abscisic acid in potato microtuber dormancy. *Plant Physiology* 105, 891-896.

Tanno, N., Yokota, T., Abe, M. and Okagami, N. (1992) Gibberellins induce the dormancy of bulbils of *Dioscorea opposita* Thunb. *Proceedings of the Plant Growth Regulator Society of America* 19, 187-192.

Tuan, D.Y.H. and Bonner, J. (1964) Dormancy associated with repression of genetic activity. *Plant Physiology* 39, 768-772.

Turnbull, C.G.N. and Hanke, D.E. (1985a) The control of bud dormancy in potato tubers: evidence for the primary role of cytokinins and a seasonal pattern of changing sensitivity to cytokinin. *Planta* 165, 359-365.

Turnbull, C.G.N and Hanke, D.E. (1985b) The control of bud dormancy in potato tubers: measurement of the seasonal pattern of changing concentrations of zeatin-cytokinins. *Planta* 165, 366-376.

van den Berg, J.H., Davies, P.J., Ewing, E.E. and Halinska, A. (1995) Metabolism of gibberellin A_{12} and A_{12} aldehyde and the identification of endogenous gibberellins in potato (*Solanum tuberosum* ssp. *andigena*) shoots. *Journal of Plant Physiology* 146, 459-466.

van Es, A. and Hartmans, K.J. (1969) The influence of abscisin II and gibberellic acid on the sprouting of excised potato buds. *European Potato Journal* 12, 59-63.

van Es, A. and Hartmans, K.J. (1981) Dormancy, sprouting and sprout inhibition. In: Rastovski, A. and van Es, A. (eds) *Storage of Potatoes*. Pudoc, Wageningen, pp. 114-132.

van Ittersum, M.K. (1992) Dormancy and growth vigour of seed potatoes. PhD thesis, Wageningen Agricultural University.

Wickham, L.D., Wilson, L.A. and Passam, H.C. (1981) Tuber germination and early growth in four edible *Dioscorea* species. *Annals of Botany* 47, 87-95.

Wickham, L.D., Passam, H.C. and Wilson, L.A. (1984) Dormancy responses to post-harvest application of growth regulators in *Dioscorea* species. *Journal of Agricultural Science, Cambridge* 102, 427-432.

III Physiology/Temperature, Light, Stress

9 A Physiological Comparison of Vernalization and Dormancy Chilling Requirement

JAMES D. METZGER
Department of Horticulture and Crop Science, 2001 Fyffe Court, Ohio State University, Columbus, OH 43210-1096, USA

Introduction

Flower initiation represents a radical change in the organization of shoot apical meristems from leaf production on a vegetative axis to flower production in an inflorescence. This change in developmental state is signalled by either internal or environmental cues. Many temperate plant species use environmental cues to initiate reproductive development and ensure completion of seed production before the onset of periods (e.g. winter) that are unfavourable for growth. The most common cues are day length (photoperiodism) and low, non-freezing temperatures (usually 0–10°C). This latter process of cold induction of flowering is called vernalization. Temperatures below freezing are generally not effective in flower induction, indicating that vernalization is a biological process. Species that exhibit requirements for vernalization are represented in many families (Table 9.1).

Vernalization serves as a seasonal timing mechanism that allows plants to fill specific niches. The typical life cycle of a plant with a vernalization requirement begins with seed germination in the late summer or early autumn and subsequently develops overwintering vegetative rosettes. Reproductive development is initiated the following spring, with seed production completed by early summer. For some cold-requiring species the entire vernalization requirement may be met in the autumn before the onset of subfreezing temperatures, while in others with longer requirements vernalization is not completed until temperatures rise above freezing early the following spring.

Table 9.1. Examples of vernalization-requiring species.

Species (common name)	Family
Althaea rosea (hollyhocks)	Malvaceae
Thlaspi arvense (field pennycress)	Brassicaceae
Brassica oleracea (Brussels sprouts)	Brassicaceae
Hyoscyamus niger (black henbane)	Solanaceae
Triticum aestivum (winter wheat)	Poaceae
Beta vulgaris (beet)	Chenopodiaceae
Daucus carota (carrot)	Apiaceae
Apium graveolens (celery)	Apiaceae
Dendranthema grandiflora (chrysanthemum)	Asteraceae
Digitalis purpurea (foxglove)	Scrophulariaceae
Geum urbanum (geum)	Roseaceae

Species with vernalization requirements can be separated into two categories: winter annuals and biennials. Winter annuals have a facultative cold requirement, i.e. they will ultimately flower, albeit greatly delayed, at warm temperatures. Moreover, winter annuals are responsive to vernalizing temperatures at all stages of development, including imbibed seeds (seed vernalization). Biennials, on the other hand, have an obligate cold requirement that requires one full season of vegetative growth and exhibits a juvenile period during which plants are unable to respond to vernalizing temperatures.

Many cold-requiring plants also exhibit photoperiodic requirements in addition to vernalization. Apparently, vernalization functions to make plants sensitive to day length, since the cold treatment must occur prior to the inductive photoperiodic treatment (Lang, 1965). In general, plants with dual vernalization–photoperiodic requirements can be classified as long-day plants, although certain cold-requiring cultivars of chrysanthemum (*Dendranthema grandiflora*) also have a dual short-day requirement.

Vernalization and the chilling requirements for ending bud dormancy of deciduous temperate zone trees have similar functions in that both enable plants to sense a change of seasonal climatic conditions, i.e. determine the end of winter and the onset of a new period that is permissive of growth. In both processes, the duration of the period of low, non-freezing temperatures prior to and following winter serves as a mechanism for time measurement. This suggests that the fundamental molecular and biochemical basis for both vernalization and cold-induced loss of bud dormancy may be connected. In this chapter, the physiological characteristics of vernalization will be examined in more detail, and a comparison with fulfilling the chilling requirements of buds will be made.

Physiological Characteristics of Vernalization

Development of useful mechanistic models for the biochemical and molecular basis of any developmental process depends on the physiological criteria by which the validity of the model can be assessed. There are a number of physiological characteristics of vernalization that are common to diverse species, suggesting that the fundamental basis underlying vernalization is similar. In the following discussion, physiological characteristics exhibited by different species will be examined, with most of the examples being taken from work conducted on *Thlaspi arvense* L. (field pennycress), a cruciferous winter annual weed prevalent in cultivated fields of the Great Plains of North America.

In general, the temperature optimum of vernalization is relatively broad

In *Thlaspi*, a 4-week vernalization treatment at any temperature between 0 and 10°C results in similar levels of floral induction (Table 9.2). At 15°C, flowering is delayed by only 1 week. As in other winter annuals with a facultative vernalization requirement, plants grown continuously at 21°C eventually flowered, albeit greatly delayed. The broad range of vernalizing temperatures is important, since, in nature, daily temperatures can fluctuate by 10°C or more; a narrow temperature optimum or sharp temperature maximum reduces the probability that maximum levels of flower induction will be achieved.

Vernalization is a quantitative process

In other words, up to a point, the longer the vernalization treatment, the greater the amount of floral induction. An example of this is shown in Table

Table 9.2. Optimum vernalizing temperatures in *Thlaspi arvense*. Plants were grown at 21°C for 6 weeks and then transferred to a growth chamber at various temperatures for 4 weeks. Plants were then returned to 21°C. Values represent the time for open flowers to appear after the end of the vernalization treatment.

Temperature (°C)	Time to open flowers (weeks)
0	14
2	14
4	14
7	14
10	14
15	15
21	> 24

Table 9.3. Quantitative aspects of vernalization. Plants of *Thlaspi arvense* were grown at 21°C prior to vernalization treatments in a growth chamber at 6°C for various lengths of time and then returned to 21°C. All vernalization treatments ended on the same day.

Duration of vernalization treatment (weeks)	Time to open flowers (weeks)
0	24
1	18
2	15
3	14
4	13
6	13

9.3. In this experiment, *Thlaspi* plants were grown at 21°C prior to 6°C vernalization treatments of various durations, and then returned to 21°C. Longer durations of vernalizing temperatures resulted in more rapid flower development and hence seed production. This reduces the risk that a random adverse environmental change could prevent production of viable seed (Hazebroek and Metzger, 1990).

The time at vernalizing temperatures required for maximum induction is relatively long

In contrast to many photoperiodic species, in which as few as one inductive photoperiod is sufficient for full flower induction (Zeevaart, 1976), most cold-requiring species need at least 4 or more weeks of vernalization to achieve maximum induction (Table 9.4). The ecological significance of the relatively long requirement for plants to be subjected to vernalizing temperatures probably lies in a potential mechanism for increasing the probability that flower formation/development occurs after the coldest periods of winter are over.

Table 9.4. Vernalization requirements for various species.

Species	Time required for maximum vernalization (weeks)
Thlaspi arvense (CR_1)	4
Arabidopsis thaliana (ecotype Pitztal)	3–4
Lunaria annua (honesty)	8
Beta vulgaris	4–8
Lilium longifolium (Easter lily)	6
Triticum aestivum (winter wheat)	6–12

Table 9.5. Inverse relationship between flowering times and growth temperature in *Thlaspi arvense*.

Temperature (°C)	Time to appearance (weeks)		
	Flower primordia	Flower buds	Open flowers
15	9	12	14
18	12	14	16
21	19	21	24
28	> 36	–	–

Vernalization only alters the rate of progress to flowering

The mechanism by which vernalization acts is often thought of as a simple switch or trigger. However, the physiological data from experiments on *Thlaspi* do not support such a model. Plants grown at different temperatures show an inverse relationship between flowering times and temperature. In other words, the higher the temperature at which the plants are grown, the more flowering is delayed (Table 9.5). *Thlaspi* plants grown at 28°C never flowered, suggesting that they acquire an obligate cold requirement. Consistent with this is the fact that a 4-week vernalization treatment of these plants resulted in normal flowering (data not shown). In total, these results indicate that, because the plants eventually flower at most non-vernalizing temperatures, there is always a finite rate of progress towards flowering, but this rate becomes much slower as the temperature rises. Ultimately, there is a temperature at which progress towards flowering ceases and the plants essentially exhibit an obligate cold requirement. Thus, one of the characteristics that distinguishes winter annuals and biennials, namely facultative vs. obligate cold requirement, may in fact be due to relative differences in the temperature at which progress toward flowering ceases.

Vernalization is a cell-autonomous process

The site of perception of vernalizing temperatures is generally considered to be the meristematic regions of the shoot tip (Curtis and Chang, 1930; Chroboczek, 1934; Purvis, 1940; Schwabe, 1954; Metzger, 1988). As a general rule, there does not appear to be a graft-transmissible signal induced by vernalization that is analogous to the floral stimulus in photoperiodic plants (Schwabe, 1954; Zeevaart, 1976; Metzger, 1988). There are exceptions to this generalization, however. Most of these plants have dual obligate photoperiodic requirements, and it appears that vernalization induces the capacity of leaves to perceive photoperiod and/or generate the floral stimulus (Lang, 1965; Crosthwaite and Jenkins, 1993). Nevertheless, there does not appear to be cell-to-cell transmission of the vernalization-induced state, i.e. vernalization is a cell-autonomous process. In other words, for a cell to be induced, it must be directly subjected to vernalizing temperatures. As a result of this property, there is a direct mitotic transmission of the induced state because the floral structures are ultimately derived from vernalized cells in the shoot apical meristem. Thus, there is a mitotic memory of the induced state which is stable through many cell divisions.

For successful flower induction, cell division must occur during the vernalization period

The fact that the shoot apical meristem appears to be the site of perception of vernalizing temperatures indicates an association between cell division and sensitivity to vernalization. That cell division during vernalization is probably required for flower induction has been demonstrated by vernalization temperature 'induction' of tissues other than the shoot apical meristem, e.g. leaves (Wellensiek, 1964; Metzger, 1988) and roots (Metzger *et al.*, 1992; Burn *et al.*, 1993). For flower induction to occur in either organ, the presence of dividing cells during the vernalization treatment is required. Moreover, the cell type is irrelevant. Cells from the growing regions of the organs and dividing callus cells present during the regeneration process are inducible by vernalization (Wellensiek, 1964; Metzger *et al.*, 1992; Burn *et al.*, 1993). These observations are also consistent with the observation that cold-requiring plants lack a mobile flowering stimulus. It is clear that the induced state is a cellular property that is highly stable, even through differentiation and many cell divisions.

Vernalization is an epigenetic process

In all cold-requiring plants, the developmental pattern is reset in the next generation and the progeny of vernalized plants must also be vernalized in order to flower. Thus, cold-induced flowering is an epigenetic process, restricted to a single generation.

Molecular Mechanism of Vernalization

It is likely that vernalization results in a substantial alteration in the pattern of gene expression, which is responsible for flower induction, but the molecular mechanism by which this is accomplished is not understood. However, a molecular model has been proposed recently that accounts for all of the above physiological properties (Burn *et al.*, 1993). The central feature of this model is the role of deoxyribonucleic acid (DNA) methylation in the control of gene expression. In both plants and animals, there is increasing evidence that the pattern of DNA methylation is an overriding control mechanism of many epigenetic processes. Cytosines of DNA can be methylated at the C-5 position, which can alter the transcriptional activity of genes if the methylated cytosines lie within the promoter (Holliday, 1990). The lack of transcriptional activity of genes is often associated with the presence of 5-methylcytosines.

The pattern of cytosine methylation is propagated by the activity of a methyltransferase enzyme (the so-called 'maintenance' methylase), which prefers as a substrate a hemimethylated double-stranded sequence of DNA that occurs as a result of DNA replication. In this way, methylation patterns are maintained with high fidelity through each cycle of DNA replication and are transmitted to both daughter cells resulting from cell division (Holliday, 1990). Although the maintenance methylase faithfully replicates methylation patterns through cell division, the fidelity is not complete. These errors in cytosine methylation in hemimethylated DNA could lead to a cell lineage in which previously silent genes become expressed. Thus, highly stable alterations in the pattern of gene expression can occur as a result of cell division. The pattern of DNA methylation is reset in the progeny, probably during meiosis or shortly after fertilization. In total, epigenetic change in patterns of gene expression through demethylation of DNA shares many of the physiological characteristics of vernalization: both are cell autonomous processes that require cell division, are highly stable and are reset in the progeny.

On the basis of these similarities, it was proposed that, in vegetative plants, one or more key genes for flower initiation is not expressed because of the presence of methylated cytosines in the promoters. According to this model, vernalization results in the the demethylation and subsequent activation of the key flowering genes. One prediction of this model is that fidelity of methylation by the maintainence methylase is related to temperature. As the temperature declines, the number of errors in the methylation of hemimethylated DNA sequences increases, as the temperature approaches the optimum temperature range for vernalization. This provides a possible biochemical mechanism for the observations that: (i) vernalization only alters the rate of progress toward flowering; (ii) many species exhibit a facultative cold requirement; and (iii) an absolute cold requirement is apparently induced in plants grown at high temperatures (Table 9.5).

Two lines of correlative evidence support the methylation/demethylation hypothesis of vernalization. First, treatment with 5-azacytidine, a potent demethylating drug, promoted flowering at nonvernalizing temperatures in cold-requiring ecotypes of *Thlaspi* and *Arabidopsis thaliana*. The activation of gene transcription by 5-azacytidine treatment has been correlated with demethylation in both plants and animals. Second, vernalizing temperatures also result in extensive demethylation of plant DNA (Burn *et al.*, 1993). Definitive proof of this hypothesis, however, must await identification and molecular analysis of the primary gene(s) that responds to vernalizing temperatures by initiating flowering.

Vernalization and Bud Dormancy: Common Mechanisms?

The chilling requirement for removing dormancy in buds of temperate-zone deciduous trees is reminiscent of vernalization. This begs the question of common mechanisms for the two processes. Indeed, there are physiological similarities between cold-induced dormancy loss and flower induction by vernalization. First of all, both processes share similar broad temperature optima between 0 and 10°C (Powell, 1987). Second, the site of perception of cold apparently resides in the meristematic regions of the bud (Swartz et al., 1984), suggesting a cell division requirement. Third, cold-induced loss of dormancy is a non-transmissible, cell-autonomous process (Salisbury and Ross, 1969). Fourth, the time necessary to meet the chilling requirement for dormancy loss is similar to that necessary to achieve maximum flower induction (Powell, 1987). Thus, from a physiological perspective, vernalization and chilling-induced loss of bud dormancy are very similar. However, it remains to be seen if bud dormancy shares other characteristics, such as the cell division requirement and the epigenetic stability of the vernalization-induced state. If future work bears this out, it may very well be that both processes have common molecular mechanisms.

References

Burn, J.E., Bagnall, D.J., Metzger, J.D., Dennis, E.S. and Peacock, W.J. (1993) DNA methylation, vernalization, and the initiation of flowering. *Proceedings of the National Academy of Sciences, USA* 90, 287-291.

Chroboczek, E. (1934) A study of some ecological factors influencing seed-stalk development in beets (*Beta vulgaris* L.). *Cornell Agricultural Experiment Station Memoirs* 154, 1-84.

Crosthwaite, S.K. and Jenkins, G.I. (1993) The role of leaves in the perception of vernalizing temperatures in sugar beet. *Annals of Botany* 69, 123-127.

Curtis, O.F. and Chang, C.K. (1930) The relative effectiveness of the temperature of the crown as contrasted with that of the rest of the plant upon flowering of celery plants. *American Journal of Botany* 17, 1047-1048.

Hazebroek, J.P. and Metzger, J.D. (1990) Seasonal pattern of seedling emergence, survival, and reproductive behavior in *Thlaspi arvense* (Cruciferae). *American Journal of Botany* 77, 954-962.

Holliday, R. (1990) Mechanisms for the control of gene activity during development. *Biological Reviews* 65, 431-471.

Lang, A. (1965) Physiology of flower initiation. In: Ruhland, H. (ed.) *Encyclopedia of Plant Physiology*. Springer-Verlag, Berlin, pp. 1380-1536.

Metzger, J.D. (1988) Localization of the site of perception of thermoinductive temperatures in *Thlaspi arvense* L. *Plant Physiology* 88, 424-428.

Metzger, J.D., Dennis, E.S. and Peacock, W.J. (1992) Tissue specificity of thermoinductive processes: *Arabidopsis* roots respond to vernalization. *Plant Physiology* S99, 52.

Powell, L.E. (1987) The hormonal control of bud and seed dormancy in woody plants. In: Davies, P.J. (ed.) *Plant Hormones and Their Role in Plant Growth and Development*, Martinus Nijhoff Publishers, Boston, pp. 539-552.

Purvis, O.N. (1940) Vernalization of fragments of embryo tissues. *Nature* 145, 462.

Salisbury, F.B. and Ross, C. (1969) *Plant Physiology*. Wadsworth, Belmont, California.

Schwabe, W.W. (1954) Factors controlling flowering in the chrysanthemum. *Journal of Experimental Botany* 5, 389-400.

Swartz, H.J., Geyer, A.S., Powell, L.E. and Lin, S.-H.C. (1984) The role of bud scales in the dormancy of apples. *Journal of the American Society of Horticultural Science* 109, 745-749.

Wellensiek, S.J. (1964) Dividing cells as the prerequisite for vernalization. *Plant Physiology* 39, 832-835.

Zeevaart, J.A.D. (1976) Physiology of flower formation. *Annual Review of Plant Physiology* 27, 321-348.

10 Dormancy Breakage By Chilling: Phytochrome, Calcium and Calmodulin

JAMES D. ROSS
Department of Botany, University of Reading, Whiteknights, Reading, Berkshire RG6 2AS, UK

Introduction

The phenomenon of seed dormancy alleviation is second only to the onset of flowering in its obviousness and drama, yet despite the endeavours of many years, the keys to both enigmas remain elusive. At various times the signal route from environmental perception to the onset of rapid embryo growth has been attributed to the loss of an inhibitor and/or the increase in growth promoters, change in sensitivity to growth substances, increased chemical energy from respiration, availability of biochemical precursors, changes in cell wall rigidity and alterations to membrane characteristics and properties.

The history of science demonstrates that, when evidence is found supporting more than one hypothesis, it is probable that all are correct, at least partially. It would appear that a process as intricate as the alleviation of seed dormancy involves many parallel and concomitant pathways. Activation of just one or even a few does not successfully complete the process; indeed, it may often result in the initiation of a further and often deeper secondary dormancy. Somehow the plethora of metabolic pathways must be regulated in a coordinated and balanced manner.

For many species the cold period of winter is the environmental stimulus which breaks dormancy and leads to germination when temperatures rise in the spring. One such species is the hazelnut, *Corylus avellana* L., which has been utilized as a model system for 30 years. Held in the imbibed state at 5°C for about 30 days, the hazel seed will germinate readily when temperatures rise. Originally this was suggested to be the result of a decrease in inhibitor concentration and an increase in gibberellins during the chilling treatment (Frankland and Wareing, 1962); later it was shown that the *de novo* increase in gibberellin biosynthesis occurred after the shift to higher temperature and at the same time as radicle growth (Ross and Bradbeer, 1968, 1971) and therefore is probably more involved with growth after the onset of germination. The inhibitor, abscisic acid (ABA), did not necessarily decline in concentration during chilling; however, the embryo lost its sensitivity to physiologically normal concentrations (Williams *et al.*, 1973), which suggests that ABA was more likely to be functioning in the suppression of vivipary than in postshedding dormancy.

As the fashion for hormonal explanations began to fade, subsequent studies focused on the metabolic and ultrastructural changes that occur during cold treatment. Similar work proceeded with various forestry species as well as fruit crops, such as apple, and has been reviewed to some extent elsewhere (Ross, 1984). One such area of investigation was the possible role of the cytosolic pentose phosphate pathway, proposed by Roberts and Smith (1977) and demonstrated to increase in activity in seeds of sour cherry (La Croix and Jaswal, 1967) and hazel (Gosling and Ross, 1980). This pathway has as its primary functions the production of reduced nicotinamide adenine dinucleotide phosphate (NADP) necessary for biosynthetic reactions, ribose-5-P for nucleotide production and erythrose-4-P, an essential feedstock in the production of the three aromatic amino acids and a number of secondary metabolites. The initial enzyme of the pathway, glucose-6-phosphate dehydrogenase, increases progressively in the hazel embryonic axis during chilling (Gosling and Ross, 1980). However, regulation of the pathway in other plant organs is thought to be by the cytosolic availability of the cofactor $NADP^+$ and the $NADPH/NADP^+$ ratio.

The first calcium/calmodulin-dependent enzyme to be isolated from plant tissue was nicotinamide adenine dinucleotide (NAD) kinase (Muto and Miyachi, 1977), although it was found later that not all NAD kinase isozymes showed such dependency (Simon *et al.*, 1984). Calcium as a secondary messenger has been implicated in a number of physiological events but relatively little evidence has come from seeds. Cocucci (1984) and Cocucci and Negrini (1988) reported that calmodulin levels in radish seeds were very low initially and rose considerably during germination; inhibitory treatments, including ABA application, suppressed this rise. The germination of embryonic axes of *Cicer arietinum* has also been reported to be accompanied by increases in calmodulin concentrations when measured by radioimmunoassay (Hernandez-Nistal *et al.*, 1989). Two calcium-dependent protein kinases from wheat embryos were reported to be inhibited by a variety of calmodulin antagonists (Polya and Micucci, 1985). These authors, however, also showed that, while some calmodulin inhibitors acted against these enzymes, other compounds with similar properties did not, although they did have effects on plant developmental processes. The breakdown of phospholipid by microsomal membranes in *Phaseolus vulgaris* cotyledons has been shown to be regulated by calcium and calmodulin (Paliyath and Thompson, 1987). Investigation of chill-induced injury of whole plants has suggested that intracellular calcium concentrations can rise due to cold inactivation of a calcium-transport adenosine triphosphatase (ATPase) (Minorsky, 1985). Roberts and Harmon (1992) have reviewed the regulatory activities of calmodulin in plants.

The modulation of calcium/calmodulin-dependent phenomena by light acting through the phytochrome system is well documented (Haupt and Weiseneel, 1976; Dieter and Marmé, 1986; Roux *et al.*, 1986). Although the hazel seed is encased within a hard woody pericarp and is therefore not normally exposed to light in its mature state, it has been reported that light stimulates metabolism and germination when dissected half-seeds are used (Shannon *et al.*, 1983). Irradiation by red light enhances protein synthesis in excised embryonic axes (Ratanakosum, 1986). We have therefore attempted to assay the calcium/calmodulin-dependent NAD kinase activity during cold-treatment dormancy breakage, to investigate any concurrent changes in calmodulin and to examine the effect of light on these processes.

Material and Methods

Plant material

Full details of materials and methods are found in Smith (1990). Hazelnuts were obtained commercially, cupules removed and the nuts air-dried and stored in sealed cans in a refrigerator. Complete fruits were placed directly in trays of moist vermiculite at either 5°C or 20°C for the appropriate periods, with additional watering as required. For each assay 100 embryonic axes were removed and placed directly into the appropriate extraction buffer. The samples were then blotted dry and weighed prior to being extracted first with a pestle and mortar and then with an all-glass homogenizer.

Assays

Maximum catalytic activity of NAD kinase was measured using the method of Muto and Miyachi (1977). NADP and NAD levels were quantified using the enzymatic coupling method of Matsumura and Miyachi (1980). Purification of calcium/calmodulin-dependent and independent NAD kinases was carried out by calmodulin-affinity chromatography modified from Dieter and Marmé (1986) and hydrophobic interaction chromatography as by Battey and Venis (1988). Calmodulin was extracted in cold acetone and purified by FPLC prior to quantification by measured enhancement of NAD kinase activity (Smith, 1990).

Results

The nicotinamide nucleotides were extracted from the original dry seeds and then throughout the chilling and control (warm) treatments of the imbibed seeds. The ratio of (NADP$^+$ + NADPH)/(NAD$^+$ + NADH) increased rapidly in the chilled seeds during the first week of the treatment (Fig. 10.1). Preliminary experiments showed that the dormant axes excised from dry seed contained NAD kinase activity which increased 3.5-fold in response to calcium and calmodulin and was inhibited by 85% when assayed with TFP, a calmodulin antagonist. Thus, there are two forms of NAD kinase, one dependent and the other independent on calcium/calmodulin. Total levels of NAD kinase rose rapidly during the first 2 weeks of chilling, while control axes, imbibed at 20°C, lost activity (Fig. 10.2).

In the dry dormant seed, NAD kinase activity was largely (85%) attributable to the calcium/calmodulin-dependent form. During the early stages of the chilling treatment, the increase in activity was entirely due to this form (Fig. 10.3). In the later stages during chilling, the dependent form decreased in activity while the independent form increased to become the majority constituent. A different pattern of activities was evident in the warm-incubated seeds, which did not lose dormancy (Fig. 10.4).

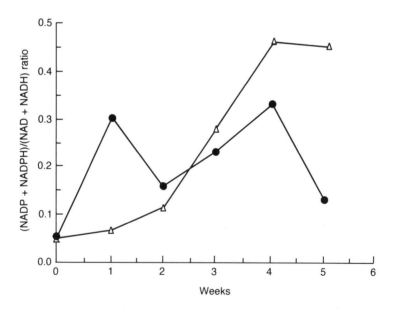

Fig. 10.1. (NADP + NADPH)/(NAD + NADH) ratio in hazel embryonic axes during stratification at 5°C (●) and warm incubation at 20°C (Δ). (From Smith, 1990.)

Protein was extracted with cold acetone from defatted extracts of embryonic axes and subjected to phenyl-Sepharose chromatography. The calmodulin content was assayed by using the NAD kinase reaction standardized with bovine calmodulin. The content in the dry dormant tissue was found to be equivalent to 23 μg per 100 axes, or 60 μg g^{-1} fw. This concentration rose in the early stages of chilling before declining, while controls held at 20°C did not show the early increase (Fig. 10.5). Unchilled seeds excised from the pericarp and with the opaque testa removed, imbibed at 20°C and irradiated with continuous red light (10 μmol m^{-2} s^{-1}) germinated rapidly, while those kept in continuous darkness did not germinate during the experimental period (Fig. 10.6).

The presence of phytochrome in the embryonic axes was demonstrated by measuring the photoreversibility of the P_r and P_{fr} forms in excised axes packed into a cuvette and alternately irradiated with red or far-red pulses. These measurements suggest that between 5 and 10% of the phytochrome exists as P_{fr} after axes from dormant seeds have been partially imbibed in darkness for 30 min at room temperature.

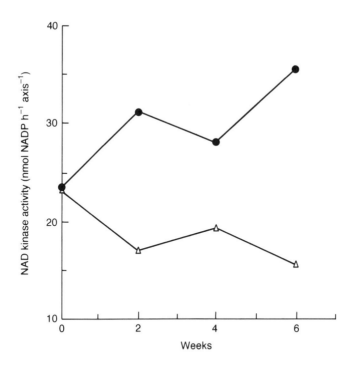

Fig. 10.2. Activity of total NAD kinase in embryonic axes stratified at 5°C (●) or warm-incubated at 20°C (Δ). Activity measured in the presence of 5 mM Ca^{2+} and 5 μg ml^{-1} calmodulin. Values are the mean of three replicates. (From Smith, 1990.)

To examine the possibility of light altering the NADP/NAD ratio through the action of the calcium/calmodulin-dependent NAD kinase system, dry dormant seeds were removed from the pericarp and allowed to imbibe for 4 h (to aid dissection). For each light treatment, 50 excised axes (c. 200 mg fw) were imbibed for 30 min, followed by a 30 min incubation period in buffer, sucrose and the calcium ionophore A23187. Either 0.1 mM $CaCl_2$ or 0.1 mM EGTA was added to the incubation buffer as appropriate. The Petri dishes containing the axes and test solutions were given light treatments as follows: (i) red-light treatment – initial imbibition under normal laboratory light (cool white fluorescent) and the incubation period under red light (12 μmol m^{-2} s^{-1}); (ii) dark treatment – initial imbibition under far-red light (1 μmol m^{-2} s^{-1}) and the incubation period in darkness. At the end of the treatment periods, samples were frozen in liquid nitrogen and stored in an ultralow freezer until the NADP/NAD ratio could be assayed. The results show that each treatment changed the NADP/NAD ratio (Fig. 10.7), with the greatest change being between the dark minus calcium (0.20) and the red plus calcium (1.75) treatments. The ratio for the latter treatment was substantially higher than that in 5-week-chilled axes.

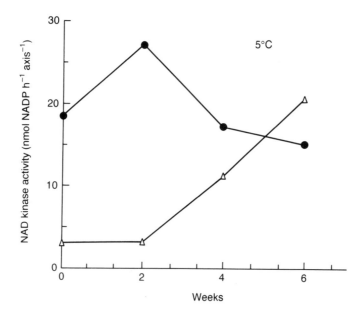

Fig. 10.3. Activity of calcium/calmodulin-dependent and independent NAD kinase in stratified embryonic axes. Dependent activity (●) measured in samples purified by calmodulin–agarose in the presence of calcium (5 mM) and calmodulin (5 μg ml^{-1}). Independent activity (Δ) purified on phenyl-Sepharose in the presence of EGTA (2 mM). Values are the mean of three replicates. (From Smith, 1990.)

Discussion

We have established that the dormant embryonic axis of hazel contains two forms of NAD kinase which, on purification, was shown to be 80% constituted of the calcium/calmodulin-dependent isozyme. The normal dormancy breaking treatment of chilling induced an increase in the total activity of NAD kinase not evident in warm-incubated seeds which remained dormant. The initial rise in activity was of the dependent isozyme whereas the later increase was of the independent form. Calmodulin concentrations in the axes also increased during the first 2 weeks of chilling. One consequence of these two increases may be the observed change in the (NADP + NADPH)/(NAD + NADH) ratio during the first phase of dormancy loss. The phosphorylation of NAD would in turn enable the activation of the NADP-dependent pentose phosphate pathway.

The fact that a seed usually thought to be light-insensitive has a high phytochrome content in its embryonic axes and that its dormancy can be broken by light when removed from its masking pericarp and testa must focus attention on whether the cold-stimulated and light-stimulated dormancy breakage mechanisms are indeed one and the same. It is entirely

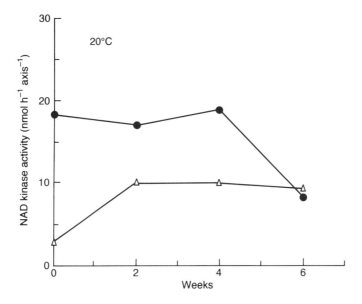

Fig. 10.4. Activity of calcium/calmodulin dependent and independent NAD kinase in axes of warm (20°C)-incubated seed. Dependent activity (●) measured in samples purified by calmodulin – agarose in the presence of calcium (5 mM) and calmodulin (5 µg ml^{-1}). Independent activity (Δ) purified on phenyl-Sepharose in the presence of EGTA (2 mM). Values are the mean of three replicates. (From Smith, 1990.)

probable that either P_{fr}, or its precursor, *meta*-Rb, formed by phytochrome cycling while the developing fruit was translucent and able to absorb light and then trapped on seed drying, may be the controlling principle in cold sensitive seeds such as hazel (see discussion on phytochrome transformations in seeds in Bewley and Black, 1985). Using the data of Taylorson and Hendricks (1969) for the half-life of P_{fr} at various temperatures in *Amaranthus retroflexus*, it can be estimated that the P_{fr} half-life at 5°C could be as long as 2000 h. This P_{fr} would be lost at 20°C by the temperature-sensitive dark conversion process.

In some species, such as *Betula papyrifera*, seeds may be stimulated to germinate either by red light or by prechilling (Bevington and Hoyle, 1981). A preirradiation by red light prior to chilling was even more effective, but the effect was diminished if chilling temperature was increased. There exists a formidable literature on light-stimulated germination of seeds, in particular lettuce, in which P_{fr} is necessary for successful initiation of the growth response. There is also a correlation between prechilling and increased sensitivity to P_{fr} in lettuce (Van Der Woude and Toole, 1980). However, only recently have results been reported on the effect of red light on the activity of

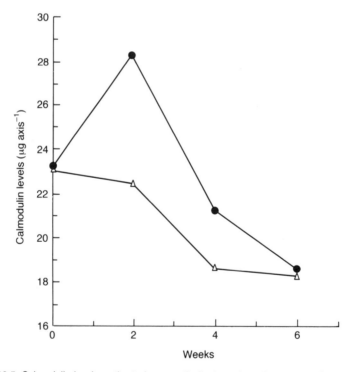

Fig. 10.5. Calmodulin levels, estimated enzymatically, in embryonic axes stratified at 5°C (●) or warm-incubated at 20°C (△). (From Smith, 1990.)

the calcium/calmodulin-dependent isozyme of NAD kinase and the subsequent changes in cofactor levels in lettuce seeds. Zhang *et al.* (1994b) have demonstrated that in half-seeds the calcium/calmodulin-dependent isozyme increased in activity following red-light irradiation compared with levels after far-red treatment. They further showed that the ratio of the phosphorylated to non-phosphorylated cofactor was strikingly and rapidly increased after irradiation (comparing red with far-red treatment), which would be consistent with *in vivo* activation of the NAD kinase.

The same group has also been able to show that, in gibberellin-treated lettuce seeds held dormant by far-red irradiation, there is a similar rise in NAD kinase-specific activity and that, once again, NADP(H)/NAD(H) ratios were higher in the gibberellin-treated seeds (Zhang *et al.*, 1994a). Thus, there is an essentially similar set of linked events in the red-light-stimulated and gibberellin-treated lettuce seeds compared with the chilled hazel seeds. Unfortunately, as yet there is no report of the calmodulin levels in lettuce.

Other model systems, such as red-light-induced fern spore germination (Wayne and Hepler, 1985; Iino *et al.*, 1989), also implicate calcium fluxes in the transduction of the phytochrome signal. Similarly, Kansara *et al.* (1989) showed NAD kinase activity to be photoregulated via phytochrome in buds of

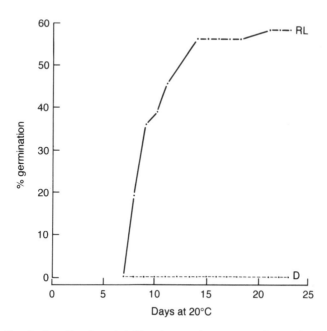

Fig. 10.6. Germination of hazel seeds (with pericarp and testa removed) at 20°C without previous chilling. Incubated under red light (10 μmol m^{-2} s^{-1}) or in darkness.

pea seedlings. Many other examples of calcium and calmodulin being involved in the regulation of metabolism and development from a wide range of species and cell types have been reviewed in Marmé (1989). Perhaps a greater understanding of such orchestration will come as advances in visualizing the changes in calcium and calmodulin location become more refined, and such changes in seeds can be investigated in the manner used by Tirlapur and Cresti (1992) with germinating tobacco pollen. However, it is clear that the best tool available for the future will be the widening variety and availability of *Arabidopsis* mutants (Karssen *et al.*, 1983). Johnson *et al.* (1994) have shown that results from such mutants suggest that there are two separate light responses controlling germination, regulated by different forms of phytochrome. From such cumulative evidence as reviewed and presented here, it is possible to hypothesize that the parallel metabolic and physiological processes which ensure successful dormancy loss and germination may be directed by phytochrome and mediated through the calcium-calmodulin regulating system.

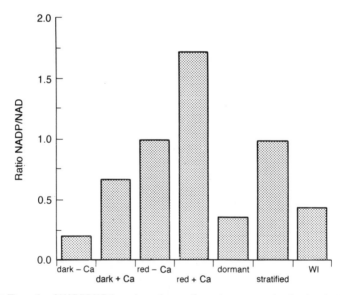

Fig. 10.7. The ratio of NADP/NAD in embryonic axes from dormant seeds after incubation under different light regimes and in the presence or absence of calcium (Ca). Samples with calcium were incubated with 0.1 mM $CaCl_2$ and those minus calcium with 0.1 mM EGTA. The calcium specific ionophore A23187 was present in the incubation medium. Values obtained from dry dormant, 5-week chilled and 5-week warm-imbibed (WI) axes are included for comparison. (From Smith, 1990.)

References

Battey, N.H. and Venis, M.A. (1988) Separation of calmodulin from calcium-activated protein kinase using calcium-dependent hydrophobic interaction chromatography. *Analytical Biochemistry* 170, 116-122.

Bevington, J.M. and Hoyle, M.C. (1981) Phytochrome action during prechilling induced germination of *Betula papyrifera* Marsh. *Plant Physiology* 67, 705-710.

Bewley, J.D. and Black, M. (1985) *Seeds: Physiology of Development and Germination*. Plenum Press, New York, London.

Cocucci, M. (1984) Increase in calmodulin level in the early phases of radish seed (*Raphanus sativus*) germination. *Plant, Cell and Environment* 7, 215-221.

Cocucci, M. and Negrini, N. (1988) Changes in the levels of calmodulin and of a calmodulin inhibitor in the early phases of radish (*Raphanus sativus* L.) seed germination. *Plant Physiology* 88, 910-914.

Dieter, P. and Marmé, D. (1986) NAD kinase in corn: regulation by far red light is mediated by Ca^{2+} and calmodulin. *Plant Cell Physiology* 27, 1327-1333.

Frankland, B. and Wareing, P.F. (1962) Changes in endogenous gibberellins in relation to chilling of dormant seeds. *Nature* 194, 313-314.

Gosling, P.G. and Ross, J.D. (1980) Pentose phosphate metabolism during dormancy breakage in *Corylus avellana* L. *Planta* 148, 362-366.

Haupt, W. and Weiseneel, M.H. (1976) Physiological evidence and some thoughts on localized responses, intracellular localization and action of phytochrome. In: Smith, H. (ed.) *Light and Plant Development*. Butterworths, Boston/London, pp. 63-74.

Hernandez-Nistal, J., Rodriguez, D. Nicolas, G. and Aldasoro, J.J. (1989) Abscisic acid and temperature modify the levels of calmodulin in embryonic axes of *Cicer arietinum*. *Physiologia Plantarum* 75, 255-260.

Iino, M., Endo, M. and Wada, M. (1989) The occurrence of a calcium-dependent period in the red light-induced G1 phase of germinating *Adiantum* spores. *Plant Physiology* 91, 610-616.

Johnson, E., Bradley, M., Harberd, N.P. and Whitelam, G.C. (1994) Photoresponses of light-grown *phyA* mutants of *Arabidopsis*. *Plant Physiology* 105, 141-149.

Kansara, M.S., Ramdas, J. and Srivistava, S.K. (1989) Phytochrome mediated photoregulation of NAD kinase in terminal buds of pea seedlings. *Journal of Plant Physiology* 134, 603-607.

Karssen, C.M., Brinkhorst-van der Swan, D.L.C., Breekland, A.E. and Koornneef, M. (1983) Induction of dormancy during seed development by endogenous abscisic acid: studies on abscisic acid deficient genotypes of *Arabidopsis thaliana* L. Heynh. *Planta* 157, 158-165.

La Croix, L.J. and Jaswal, A.S. (1967) Metabolic changes in after-ripening seed of *Prunus cerasus*. *Plant Physiology* 42, 479-480.

Marmé, D. (1989) The role of calcium and calmodulin in signal transduction. In: Boss, W.F. and Morre, D.J. (eds) *Second Messengers in Plant Growth and Development*. Alan R. Liss, New York, pp. 57-80.

Matsumura, H. and Miyachi, S. (1980) Cycling assay for nicotinamide adenine nucleotides. *Methods in Enzymology* 69, 465-470.

Minorsky, P.V. (1985) An heuristic hypothesis of chilling injury in plants: a role for calcium as the primary physiological transducer of injury. *Plant, Cell and Environment* 8, 75-94.

Muto, S. and Miyachi S. (1977) Properties of a protein activator of NAD kinase from plants. *Plant Physiology* 59, 55-60.

Paliyath, G. and Thompson, J.E. (1987) Calcium- and calmodulin-regulated breakdown of phospholipid by microsomal membranes from bean cotyledons. *Plant Physiology* 83, 63-68.

Polya, G.M. and Micucci, V. (1985) Interaction of wheat germ Ca^{2+}-dependent protein kinases with calmodulin antagonists and polyamines. *Plant Physiology* 79, 968-972.

Ratanakosum, K. (1986) An investigation into protein mobilization during dormancy breakage in hazel seeds (*Corylus avellana* L.). PhD thesis, University of Reading, UK.

Roberts, D.M. and Harmon, A.C. (1992) Calcium-modulated proteins: targets of intracellular calcium signals in higher plants. *Annual Review of Plant Physiology and Plant Molecular Biology* 43, 375-414.

Roberts, E.H. and Smith, R.D. (1977) Dormancy and the pentose phosphate pathway. In: Khan, A.A. (ed.) *The Physiology and Biochemistry of Seed Dormancy and Germination.* Elsevier/North Holland Biomedical Press, Amsterdam, pp. 385-412.

Ross, J.D. (1984) Metabolic aspects of dormancy. In: Murray, D.R. (ed.) *Seed Physiology,* Vol. 2, *Germination and Reserve Mobilization.* Academic Press, Sydney, pp. 45-75.

Ross, J.D. and Bradbeer, J.W. (1968) Concentrations of gibberellin in chilled hazel seeds. *Nature* 220, 85-86.

Ross, J.D. and Bradbeer, J.W. (1971) Studies in seed dormancy, V. The content of endogenous gibberellins in seeds of *Corylus avellana* L. *Planta* 100, 288-302.

Roux, S.J., Wayne, R.O. and Datta, N. (1986) Role of calcium ions in phytochrome responses: an update. *Physiologia Plantarum* 66, 344-348.

Shannon, P.R.M., Jeavons, R.A. and Jarvis, B.C. (1983) Light-sensitivity of hazel seed with respect to the breaking of dormancy. *Plant Cell Physiology* 24, 933-936.

Simon, P., Bonzon, M., Greppin, H. and Marmé, D. (1984) Subchloroplastic localization of NAD kinase activity: evidence for a CA^{2+}, calmodulin-dependent activity in the envelope and for a Ca^{2+}, calmodulin-independent activity in the stroma of chloroplasts. *FEBS Letters* 167, 332-338.

Smith, S.B. (1990) An investigation into the mechanism of dormancy breakage. PhD thesis, University of Reading, UK.

Taylorson, R.B. and Hendricks, S.B. (1969) Action of phytochrome during prechilling of *Amaranthus retroflexus* L. seeds. *Plant Physiology* 44, 821-825.

Tirlapur, U.K. and Cresti, M. (1992) Computer-assisted video image analysis of spatial variations in membrane-associated Ca^{2+} and calmodulin during pollen hydration, germination and tip growth in *Nicotiana tabacum* L. *Annals of Botany* 69, 503-508.

Van Der Woude, W.J. and Toole, V.K. (1980) Studies on the mechanism of enhancement of phytochrome-dependent lettuce seed germination by prechilling. *Plant Physiology* 66, 220-224.

Wayne, R. and Hepler, P.K. (1985) Red light stimulates an increase in intracellular calcium in the spores of *Onoclea sensibilis*. *Plant Physiology* 77, 8-11.

Williams, P.M., Ross, J.D. and Bradbeer, J.W. (1973) Studies in seed dormancy VII; the abscisic acid content of the seeds and fruits of *Corylus avellana* L. *Planta* 110, 303-310.

Zhang, Y., Ross, C.W. and Orr, G.L. (1994a) Enhanced activity of a calcium-calmodulin-dependent isozyme of NAD kinase caused by gibberellic acid in photodormant lettuce seeds. *Journal of Plant Physiology* 143, 687-692.

Zhang, Y., Ross, C.W. and Orr, G.L. (1994b) Enhanced activity of a calcium-calmodulin-dependent isozyme of NAD kinase and changes in nicotinamide coenzyme levels in lettuce half-seeds caused by red light. *Journal of Plant Physiology* 143, 693-698.

11 Conifer Bud Dormancy and Stress Resistance: A Forestry Perspective

FRANCINE J. BIGRAS
Natural Resources Canada, Canadian Forest Service – Quebec Region, 1055 du PEPS, PO Box 3800, Sainte-Foy, Quebec G1V 4C7, Canada

Introduction

Millions of conifer seedlings intended for reforestation are produced each year in nurseries in the northern temperate region. The period between onset of dormancy and bud break and the related phenomena of cold acclimatization, deacclimatization and stress resistance cause many difficulties for nurseries in temperate climates. Seedlings must reach a desired height and diameter in 1 or 2 years (depending on local criteria) to be outplanted. The resulting constraints imposed by increased fertilization, irrigation, high temperature and light during the growing period are often quite different from the natural autumn conditions favourable to seedling dormancy, cold acclimatization, and deacclimatization. Consequently, dormancy induction and release treatments must be used to accelerate or delay dormancy and bud break.

The increased demand for containerized seedlings, a production method that exposes the root system to temperature extremes, adds to the difficulties encountered during autumn, winter and spring. For both bare root and container production, it is important to determine the time of maximal resistance for lifting, overwintering and planting operations. To attain expected field performances, seedlings must establish rapidly. Therefore, the assessment of root system viability has always been an important assessment for foresters.

In physiology textbooks and thousands of research papers, cold acclimatization (cold hardiness) and dormancy are often treated separately, reflecting the current idea of distinct phenomena. However, these should be considered inseparable. Moreover, most papers related to cold acclimatization, dormancy and stress resistance refer separately to the aerial parts and the root system. However, as pointed out by Lavender (1991), the phenomenon of dormancy has rarely been examined from a whole plant perspective. Therefore, it is important that the study of cold hardiness, dormancy and stress resistance be examined in this perspective.

The objectives of this chapter are to present the relationships between growth, dormancy, cold hardiness and stress resistance of shoots and roots of conifer seedlings used for reforestation in northern temperate regions, including dormancy manipulations. First, shoot growth, bud dormancy and shoot stress resistance will be discussed. Second, these physiological characteristics will be examined for root systems. Finally, cultural manipulations associated with dormancy and frost resistance will be examined with an emphasis on short-day (SD) treatments and their impact on the growth cycle.

Shoot Growth and Dormancy

Fuchigami *et al.* (1982) developed a numerical procedure, called the degree growth stage (°GS) model, to quantify the annual development of temperate zone woody species. This model will be used to illustrate the growth cycle. In temperate climates, growth of pines (*Pinus* spp.) and spruces (*Picea* spp.) results from expansion of terminal buds after bud break (0°GS) on the main axis and its branches in a single flush (Kramer and Kozlowski, 1979). Lavender *et al.* (1973) suggested that air and soil temperatures, as well as photoperiod, interact to allow the earliest possible bud activity compatible with the risk of frost for any given year. Lateral buds begin to grow earlier than terminal buds (Sweet, 1965). For most species of pine and spruce, active shoot elongation is rapid and largely complete within a period of 30 to 90 days (Lavender, 1981). During shoot elongation, bud scales are initiated; at the end of shoot elongation, a period of formation of leaf primordia occurs (Glerum, 1982), which ends at dormancy (Owens and Molder, 1977) (Fig. 11.1).

After elongation of the terminal shoots is complete, new terminal buds form and expand. In temperate climates, under natural conditions of growth, bud formation occurs quite early during the season (Lyr and Hoffmann, 1967). Before the maturity induction point (90°GS, when plants become sensitive to decreasing photoperiod), growth cannot be inhibited by short photoperiods. As buds develop from 90°GS to 180°GS, shoot growth is reduced. Under controlled conditions, bud formation is more rapid under a short photoperiod (Dormling *et al.*, 1968), but bud formation may also happen earlier in natural environments when conditions (e.g. drought, excessive heat, nutrient deficiencies) are unfavourable to growth (i.e. summer quiescence) (Lavender, 1980). This early bud set can accelerate the growth cycle. Conversely, a long photoperiod, high fertility or irrigation can counterbalance the effect of decreasing photoperiod and can delay bud formation and the growth cycle.

The delay or acceleration of bud set and dormancy is an important characteristic of the growth cycle associated with environmental or cultural factors. For instance, first-year seedlings of most woody plants behave indeterminately, and growth is sometimes difficult to stop during nursery production. These aspects will be discussed in more detail in the section dealing with dormancy manipulation.

One important aspect of shoot dormancy is that it applies only to the apical meristem of shoots (i.e. the buds). Lavender *et al.* (1970) and Worrall

Vegetative Buds

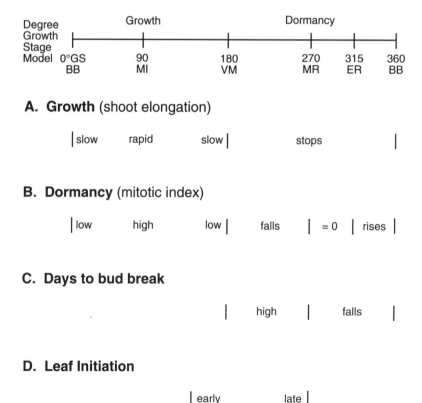

Fig. 11.1. Changes in (A) growth, (B) mitotic index, (C) days to bud break and (D) primordia formation in relation to °GS model. BB, bud break; MI, maturity induction; VM, vegetative maturity; MR, maximum rest; ER, end of rest ((C) from Burr, 1990; (D) from Glerum, 1982).

(1971) both reported that lateral cambia do not enter rest. Conifer root growth is also independent of bud dormancy (Zaerr and Lavender, 1974).

Based on Fuchigami's °GS model, Burr (1990) described the changes in shoot growth, days to bud break (DBB) and mitotic index (MI) of conifer seedlings in relation to bud dormancy (Fig. 11.1). Intensity of dormancy increases from 180° (the vegetative maturity point or onset of rest) to 270°GS (maximum rest) and is maintained internally (when DBB is high, the MI falls). A chilling temperature, ranging within very narrow limits around 5°C, is required to overcome rest under natural conditions. The length of time a bud must be exposed to these chilling temperatures to make the transition from rest to quiescence varies with species and possibly with ecotypes (Lavender, 1981). The chilling requirements for coastal Douglas fir (*Pseudotsuga menziesii*) (Lavender and Hermann, 1970), interior Douglas fir (var. *glauca*) (Wells, 1979), western hemlock (*Tsuga heterophylla*) (Nelson and Lavender, 1979) and for several species of spruces (Nienstaedt, 1966, 1967) have been established. Usually, temperatures below 5°C do not satisfy the chilling requirements.

As chilling requirements are met (between 270° and 315°GS), dormancy decreases (DBB decreases, MI approaches zero near 270°GS) (Fig. 11.1). The period between 315° and 360°GS corresponds to quiescence and the absence of shoot growth due to environmental conditions (MI rises). It lasts as long as environmental factors are unfavourable. After the quiescence period, shoot growth will resume.

Because of the practical implications of the relationships between dormancy and stress resistance (discussed below), many tests have been proposed to assess intensity of dormancy in conifer seedlings: bud break tests, dormancy release index, chilling sums, oscilloscope technique, dry-weight fraction, MI, hormone analysis and electrical resistance (Ritchie, 1984). So far, the bud break test remains the most reliable method to assess the intensity of dormancy in conifer seedlings.

Shoot Temperature Stress Resistance

Conifer seedlings that are mass-produced must be resistant to a wide range of stresses including low and high temperatures, drought and mechanical stress. In this section, stress resistance applies to the whole aerial part of the conifer seedling. The degree of resistance is highly variable according to the growth stage and follows a pattern related to the dormancy cycle. However, the resistance pattern differs according to the nature of the stress.

Cold hardiness occurs in stages on the entire aerial portion of the plant (Weiser, 1970). For northern conifers, the first stage of acclimatization is induced by the shortening of photoperiod in late summer (Zehnder and Lanphear, 1967; Bigras *et al.*, 1989a). Under controlled conditions, frost tolerance induction is generally greater under short (6 to 8 h) than under long photoperiods (van den Driessche, 1969; Jonsson *et al.*, 1981; Bigras and D'Aoust, 1993). During this first stage, cold hardiness increases from a few degrees below zero to between $-10°$ and $-20°C$ at the end of this stage. Most authors agree that growth cessation is a prerequisite for cold hardening at this stage (van den Driessche, 1970; Christersson, 1978). Reduction of photoperiod induces both growth cessation and cold hardening.

The second stage of hardening, during which cold tolerance increases greatly, is triggered by low temperatures (slightly above freezing to subfreezing) and is generally independent of photoperiod. A third stage of hardening, triggered by very low temperatures ($-30°$ to $-60°C$), occurs in some species having a high degree of frost tolerance. During the first two stages of hardening, shoot water content decreases and can be used as an indicator of frost hardiness during conifer seedling production (Hultén and Lindell, 1980; Calmé *et al.*, 1993). Growth cessation, hardening and onset of rest happen at about the same time that major biochemical changes occur, involving sugars, amino acids, proteins and growth regulators. Since these events occur almost simultaneously in conifers, it is difficult to associate each biochemical change with a given physiological process. A great deal of literature has been published on changes of biochemical constituents and physiological processes, and extensive reviews have been produced (Levitt, 1980; Sakai and Larcher, 1987). However, an understanding of the functional relationship between the biochemical and physiological processes is still lacking.

After chilling requirements are met and plants are exposed to warm temperatures, cold hardiness is progressively lost, and returns to its initial level; the change at this time is rapid and stages of dehardening have not been identified (Aronsson, 1975; Burr *et al.*, 1989; Ritchie, 1991). Temperature is the principal factor causing dehardening. However, some studies have suggested that photoperiod can act in dehardening (Greer and Stanley, 1985; Greer *et al.*, 1989; Hawkins, 1993). Methods for assessing cold hardiness of shoots have been the subject of extensive reviews (Timmis, 1976; Warrington and Rook, 1980; Ritchie, 1991) and will not be discussed here. Acclimatization and deacclimatization of conifer seedling shoots, in relation to the °GS model, is depicted in Fig. 11.2A.

Parker (1971) reported that from October through April, twigs of eastern red cedar (*Juniperus virginiana*) were more resistant to heat than those of Colorado spruce (*Picea pungens*). In *Juniperus*, percentage survival decreased in spring and increased in late summer and early autumn. Seasonal trends were more erratic in *Picea*. However, while some *Picea* leaves always survived the heat treatment in winter, none survived it in summer. Koppenaal and Colombo (1988) showed in black spruce (*Picea mariana*) seedlings that:

(i) heat damage to the current-year shoots was lower in bud-initiated seedlings than in actively growing seedlings; (ii) current-year shoot growth was more sensitive to heat stress than the lignified first-year shoots; and (iii) dormant seedlings were the most heat-tolerant. Several other authors have also reported a maximum heat tolerance occurring during winter in white spruce (*Picea glauca*) (Riedmüller-Scholm, 1974), Norway spruce (*Picea abies*) and *Abies alba* (Pisek et al., 1968), *Pinus cembra* (Schwarz, 1970), and Douglas fir and Englemann spruce (*Picea engelmannii*) (Burr et al., 1993), whereas the minimum tolerance was observed during spring and early summer. Based on these data, an annual pattern for high temperature resistance for temperate zone conifers is presented in Fig. 11.2B.

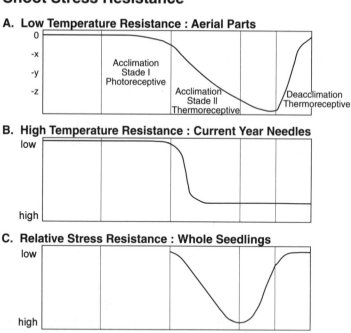

Fig. 11.2. Changes in (A) low temperature resistance for aerial parts, (B) high temperature resistance for current-year needles and (C) relative stress resistance of whole seedlings in relation to °GS model. Abbreviations as in Fig. 11.1 ((C) from Lavender, 1991).

Shoot Relative Stress Resistance

In forestry the term 'relative stress resistance', often called 'hardening off', refers to a general stress resistance of the whole seedling (resistance to nursery operations that result in mechanical stress or exposure to desiccation or herbicides, for example) (Lavender, 1991). This term was probably developed in relation to bare root production, for which a period favourable to lifting, storing and planting must be defined. However, even for container-produced seedlings, it remains necessary to define this general resistance in order to establish the right period for container extraction, storage or outside overwintering and planting. A relative stress resistance curve, based on published data for Douglas fir (Hermann, 1967; Lavender and Wareing, 1972), is depicted in Fig. 11.2C.

One important aspect in forestry applications is that the relative stress resistance curve correlates well with the success of plantation. For Douglas fir seedlings, the stress resistance curve showed that this species is most resistant to nursery harvest, storage and outplanting from December to February. That is, if seedlings are disturbed when their relative resistance to stress is low, their potential for survival and growth may be correspondingly low. If, on the other hand, reforestation procedures are conducted when resistance to stress is relatively high, seedling survival potential will be higher (Lavender, 1985). This relative stress resistance curve is inverse to mitotic activity in the cells of buds (Owens and Molder, 1973); when MI equals zero (i.e. during maximum rest (Fig. 11.1)), relative stress resistance is maximum.

From the examination of the freezing temperature, high temperature or relative stress resistances, it can be concluded that stress resistance increased following vegetative maturity, reached a maximum during dormancy and decreased to the initial value at bud break. However, patterns of stress resistance may differ between species, as shown for heat resistance by Burr *et al.* (1993), and between types of stress.

Root Growth and Dormancy

Growth, dormancy, cold tolerance and stress resistance of conifer root systems have received little experimental attention (Glerum, 1990). Romberger (1963) reviewed the information available since 1758 on seasonal and episodic root growth. Based on this review, four conclusions may be drawn: (i) root growth begins before bud break in spring and continues after shoot growth in the autumn as root requirements for optimal growing temperature are lower than shoot requirements; (ii) trees have two main periods of root growth, one in the spring and one in the autumn; (iii) the total growth during the spring is usually greater than in the autumn; and (iv) the occurrence of root growth in winter is uncertain.

Lyr and Hoffmann (1967), in their detailed review on root growth of trees, considered the cessation of root growth in midsummer to be a consequence of unfavourable environmental conditions (e.g.: drought or high temperature). They also reported that, in general, conifer root growth is more uniform throughout the whole vegetation period as compared with deciduous species. In areas with low winter temperatures, root growth usually stops in the autumn after shoot growth has ceased. Winter growth of roots is thought to be restricted to regions with mild winter temperatures and frost-free soils.

More recently, Johnson-Flanagan and Owens (1985) showed that root elongation of white spruce in the spring is followed by the production of root hairs and by another period of root elongation in the autumn (e.g. Fig. 11.3). This root growth pattern was also reported for other species of conifer seedlings (Ritchie and Dunlap, 1980). Ritchie and Tanaka (1990) have shown that root growth potential (RGP), the ability of the seedling to grow roots when placed in a highly favourable environment for root production, also shows a cyclical pattern (Fig. 11.3). They defined two periods of high RGP, one between 90° and 180°GS and the other between 270° and 360°GS,

Vegetative Buds

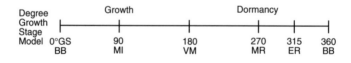

A. Root Growth

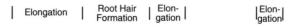

B. Root Growth Potential

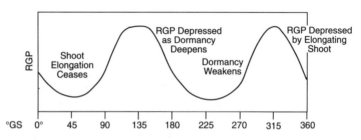

Fig. 11.3. Changes in (A) root growth and (B) root growth potential in relation to °GS model. Abbreviations as in Fig. 11.1 ((B) from Ritchie and Tanaka, 1990).

interrupted by two periods of low potential. They based their conclusion, in part, on the results obtained experimentally by Burr *et al.* (1989) on ponderosa pine (*Pinus ponderosa*), Douglas fir, and Engelmann spruce, in which cold tolerance of aerial parts, the number of days to 50% bud break, and root growth were assigned between 180° and 360°GS. Cannell *et al.* (1990) also established, for Sitka spruce (*Picea sitchensis*), a relationship between root growth potential, dormancy status (DBB) and shoot frost hardiness.

In comparison with the well-defined pattern of dormancy reported for the aerial part of the plant, root dormancy is much less understood and its existence remains highly controversial even if there exists some agreement that root growth stops in the autumn in areas where winters are cold. Romberger (1963), Lavender *et al.* (1970) and others have concluded that there is no evidence that roots enter rest. Lyr and Hoffmann (1967) also agreed that, despite the fact that trees in temperate latitudes have a period of rest in winter, there was no evidence for an internally controlled period of dormancy in roots. It is known that roots may enter quiescence as a result of either unfavourable soil moisture or unfavourable temperature conditions (Lavender, 1985).

Johnson-Flanagan and Owens (1985), studying the root system of white spruce in British Columbia, observed that, following a brief period of growth in the autumn, the roots became dormant early in December. In early winter, they noted a gradual transition from dormancy to quiescence. They suggested that this general trend in root growth was partly obscured by independent growth cycles in individual roots. Little progress has been made recently on the subject, and the question of root dormancy still remains fully open today. The conclusion of Lyr and Hoffmann (1967) that 'root rest and root growth resumption need more experimental data' remains true.

To orientate research on root dormancy and root quiescence, the °GS model developed for shoots could probably be applied to roots of conifer seedlings with some modifications (Fig. 11.4). However, thus far, only a few point events can be identified in the cycle, such as the beginning of root elongation, the end of root elongation and the onset of rest or quiescence. Experiments will be necessary to identify other point events to verify the cycle's validity. For example, it is not known if the chilling requirement event applies to the root system, whether photoperiod can influence root growth or whether root 'dormancy' exists. In terms of practical application, one must be cautious since 0°GS for roots does not correspond to the same calendar date as 0°GS for shoots.

Root Temperature Stress Resistance

As for other root processes, the low temperature resistance pattern of roots remains poorly understood. Some authors have identified distinct stages of

Vegetative Buds

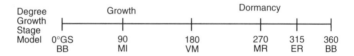

Root systems

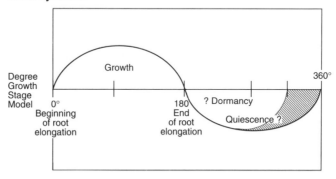

Fig. 11.4. Growth model proposed for root systems in relation to °GS model. Abbreviations as in Fig. 11.1.

acclimatization for roots (Bigras *et al.*, 1989a) similar to those of shoots. Others consider the acclimatization process to be a progressive phenomenon. Temperature is known to be the major factor controlling root cold hardiness (Simpson, 1993). Johnson and Havis (1977) found a marked influence of both photoperiod and temperature on the hardening of secondary mature roots of white spruce. A small effect of short photoperiods was also noted for the root hardiness of black spruce (Colombo, 1994), whereas others have found no effect of photoperiod for Scots pine (*Pinus sylvestris*) (Smit-Spinks *et al.*, 1985), Pfitzer juniper (*Juniperus chinensis* 'Pfitzerana') (Bigras *et al.*, 1989a) or black and white spruce (Bigras and D'Aoust, 1992). In Japanese yew (*Taxus cuspidata*), root tips do not acclimatize at all (Mityga and Lanphear, 1971). Since acclimatization of roots seems to be influenced mainly by low temperatures, root acclimatization occurs later than acclimatization of aerial parts (Bigras *et al.*, 1989a,c; Bigras and D'Aoust, 1992).

In spring, deacclimatization of roots occurs much earlier than that of aerial parts (Bauer *et al.*, 1971) and dehardening of roots is very rapid (Bigras and D'Aoust, 1992). Consequently, roots are at a maximal resistance for a shorter period of time than the aerial parts (Sakai and Larcher, 1987). Glerum (1982) reported that, in spring, the roots of bare root conifer seedlings were actively elongating and had lost their hardiness, while the tops were still very hardy; in fact, the tops did not deharden until 8 weeks later.

As with aerial parts, root growth must cease before hardening. Conifer roots do not become as hardy as shoots (Pellett, 1971; Studer *et al.*, 1978) and woody roots are more hardy than fine roots (Mityga and Lanphear, 1971; Studer *et al.*, 1978; Bigras *et al.*, 1989a,b,c). Based on these results, a model for the acclimatization and deacclimatization of conifer seedling root systems is presented in Fig. 11.5A. Various methods for measuring root cold hardiness have been assessed in Bigras and Calmé (1994) and will not be discussed here.

Experiments on the heat resistance of conifer seedling roots are rare. Hermann (1967) exposed roots of Douglas fir seedlings to a temperature of 32°C for 5 to 120 min in November, January and March. Seedlings lifted in November and March showed low tolerance to exposure (Fig. 11.5B).

Vegetative Buds

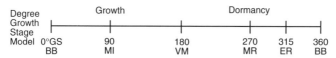

Root Stress Resistance

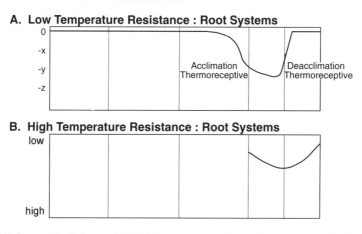

Fig. 11.5. Changes in (A) low and (B) high temperature resistance for root systems in relation to °GS model. Abbreviations as in Fig. 11.1 ((B) from Hermann, 1967).

Growth, Dormancy and Stress Resistance Relationships

The relationship between growth, bud dormancy and stress resistance (low and high temperatures) has been examined for many forest species (see Table 11.1). These relationships were examined to determine the time of maximal stress resistance and to find correlations between these physiological processes and measurable variables. As stress resistance, dormancy and cold hardiness are correlated for a given species, it has been suggested that the determination of one characteristic could possibly be used to determine the others. For example, results of cold hardiness tests can be readily obtained in 7 days or less, while tests for stress resistance or dormancy are much longer.

Based on the studies presented in Table 11.1 and on others reviewed above, some general conclusions can be drawn (see Fig. 11.6).

1. Root elongation and shoot apical MI continue after height growth ceases.
2. Correlated inhibition is characterized by high levels of MI associated with needle primordia initiation.
3. Cessation of needle primordia formation is correlated to increases in frost hardiness and the transition from a state of correlated inhibition to rest.
4. Plants cannot achieve full hardiness until growth has stopped.
5. Frost hardening occurs during the later stages of bud development while shoot apical MI is decreasing.
6. Buds do not become dormant (i.e. acquire a chilling requirement) until shoot MI has virtually ceased.
7. Onset of deep dormancy coincides with the time when cell division ceases in the shoot apices.
8. Frost hardening begins before the buds become dormant.
9. Cold hardening proceeds as the number of DBB begins to decline.
10. Shoots are most frost hardy in midwinter when they have intermediate levels of dormancy.
11. Changes in heat resistance do not coincide with changes in cold hardiness or bud dormancy.
12. Root frost hardiness is at a maximum resistance for a shorter period than the aerial parts.

Although little is known about the biochemical and physiological changes associated with bud dormancy and stress resistance, Berrie (1987) proposed a model defining conditions that impose and break bud dormancy. The model also incorporates biochemical and morphological changes in the plant. For forestry applications, a model that would integrate the morphology, physiology and biochemistry of the root system would be useful and practical. An understanding of the relationship between shoots and roots regarding dormancy and stress resistance would help the understanding of whole plant

Table 11.1. Variables measured during several studies on the relationship between growth, bud dormancy and stress resistance of conifer seedlings.

	Bud dormancy			Root growth		Shoot growth	Low temp resistance			Heat resistance			Other physiological measurements
	Bud phenology	DBB	MI	Per se	RGP		Aerial parts	Root	WS	Aerial parts	Root	WS	
Glerum, 1982	X	X				X	X						
Ritchie, 1986		X					X					X	Mechanical stress
Burr et al., 1989		X			X		X						
Cannell et al., 1990		X	X	X	X	X	X						Root and shoot DM, carbohydrates
Colombo, 1990	X	X					X						Shoot water content
Silim and Lavender, 1991	X				X		X						Damage by cold storage, physical stress
Burr et al., 1993		X					X			X			
Dormling, 1993						X	X						Frost drought, photoinhibition

DDB, days to bud break; MI, mitotic index; WS, whole seedlings; DM, dry mass.

physiology and could be used to predict the reactions of conifer seedlings to dormancy manipulations.

Dormancy Manipulations

The production of conifer seedlings in temperate regions is strongly dependent on manipulations to induce or to release dormancy. These manipulations use the properties of the natural growth cycle of the plant. Dormancy induction treatments are used to hasten cessation of growth, induce dormancy and increase frost resistance, while dormancy release treatments are used to accelerate bud break and growth. In northern forestry, dormancy release treatments are rarely used during conifer seedling production because hastening bud break could result in an increase in spring frost damage. Thus, the following discussion will be limited to dormancy induction.

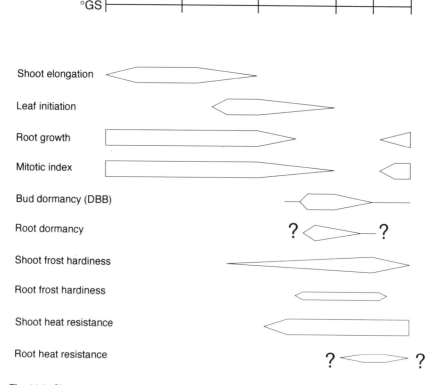

Fig. 11.6. Changes in the growth, dormancy and stress resistance for conifer seedlings of northern temperate zones in relation to °GS model.

In conifer seedling production, dormancy is usually induced by one or more of the following treatments: SD photoperiod, moisture stress or nutrient stress (Table 11.2). The application of SD treatment and its effect on frost resistance is particularly efficient for conifer seedlings (e.g. Sandvik, 1980). For that reason, the present discussion will be limited to this phenomenon. These treatments consist of an 8 : 16 h day : night for 2 to 4 weeks. Induction by SD of the frost-hardening process of aerial parts is well established for many conifer species, namely Scots pine and Norway spruce (Christersson, 1978), Douglas fir (van den Driessche, 1969), Monterey pine (*Pinus radiata*) (Greer and Warrington, 1982), black spruce (Colombo, 1986) and white spruce (Bigras and D'Aoust, 1990, 1992). However, Bigras and D'Aoust (1992, 1993) showed that SD treatments have little effect on root hardening, since root hardening is influenced mainly by temperature. Consequently, for conifer seedling production, SD has no utility for promoting root hardiness.

Table 11.2. Some examples of dormancy induction treatments applied to conifer seedlings.

Species	Treatment	References
Abies procera	Photoperiod	Tung and De Yoe, 1991
A. magnifica	Thermoperiod	
var. *shastensi*	Moisture regimes	
Picea mariana	Photoperiod	Bigras and D'Aoust, 1992, 1993
Picea glauca		
Pinus taeda	Ethephon	Gagnon and Johnson, 1988
Pinus elliottii		
Picea glauca	Moisture stress	Macey and Arnott, 1986
	Nutrient stress	
Picea abies	ABA	Heide, 1986
Picea pungens	Temperature	Young and Hanover, 1978
	Nutrient stress	
	Moisture stress	
Picea abies	Temperature	Dormling et al., 1968
	Photoperiod	
Picea mariana	Temperature	D'Aoust and Cameron, 1982
	Photoperiod	
	Moisture stress	
Abies alba	ABA	Little and Eidt, 1968
Picea glauca		
Tsuga heterophylla	Photoperiod	Major et al., 1994
	Moisture stress	
Pseudotsuga menziesii	Photoperiod	MacDonald and Owens, 1993
	Moisture stress	

It has been shown, moreover, that SD treatments cause an early bud break in white spruce (Bigras and D'Aoust, 1990, 1992), Norway spruce (Dormling et al., 1968), Douglas fir (Lavender, 1989), and black spruce (Bigras and D'Aoust, 1990, 1992), thus enhancing the risk of spring frost damage. Moreover, seedlings receiving SD treatments extend their growth period later the following autumn, thus delaying cold hardiness and increasing the probability of autumn frost damage (Odlum and Colombo, 1988).

After studying the effect of SD treatment on white spruce seedlings, Bigras and D'Aoust (1993) concluded that the whole growth cycle of the aerial part was influenced by SD treatments. They suggested that SD treatments accelerated the °GS cycle, whereas long-day (LD) treatment delayed the cycle. Furthermore, the interaction between photoperiod treatment and temperature, as proposed by Kobayashi et al. (1983), led Bigras and D'Aoust (1993) to present a model for dehardening of woody plants under different conditions of day length and temperature. However, this model is valid only for aerial parts, which are photoreceptive during the first stage of acclimatization; it does not apply to roots, which are thermoreceptive during acclimatization and deacclimatization.

Conclusions

In temperate climate forestry, an understanding of the relationship between dormancy, frost hardening and resistance to other stresses is crucial to assess the effect of climatic extremes on the proper selection and efficient production of seedlings for reforestation. Information developed on the functional parts of seedlings (i.e. buds, shoots, roots) must be integrated into the perspective of the whole plant. Treatments to manipulate dormancy in conifer seedling production systems should take the whole cycle of dormancy and stress resistance into account to prevent inadvertent enhancement of other risks, such as spring frost damage. The study of dormancy and stress resistance should also be considered in the context of potential future effects of global changes, especially in relation to the selection of genetically improved plants for future reforestation.

References

Aronsson, A. (1975) Influence of photo- and thermoperiod on the initial stages of frost hardening and dehardening of phytotron-grown seedlings of Scots pine and Norway spruce. *Studia Forestalia Suecica* 128, 1-20.

Bauer, H., Harrasser, J., Bendetta, G. and Larcher, W. (1971) Jahresgang der Temperaturresistenz Junger Holzplanzen im Zusammenhang mit Ihrer Jahreszeilichen Entwicklung. *Berichte der Deutschen Botanischen Gesellschaft* 84, 561-570.

Berrie, A.M.M. (1987) Bud dormancy. In: Newman, D. and Wilson, K.G. (eds) *Models in Plant Physiology and Biochemistry*, Vol. II. CRC Press, Boca Raton, Florida, pp. 131-133.

Bigras, F.J. and Calmé, S. (1994) Viability tests for estimating root cold tolerance of black spruce seedlings. *Canadian Journal of Forest Research* 24, 1039-1048.

Bigras, F.J. and D'Aoust, A.L. (1990) Short day treatments cause early dehardening in conifer seedlings. *HortScience* 25, 1090 (abstract).

Bigras, F.J. and D'Aoust, A.L. (1992) Hardening and dehardening of shoots and roots of containerized black spruce and white spruce seedlings under short and long days. *Canadian Journal of Forest Research* 22, 388-396.

Bigras, F.J. and D'Aoust, A.L. (1993) Influence of photoperiod on shoot and root frost tolerance and bud phenology of white spruce seedlings (*Picea glauca*). *Canadian Journal of Forest Research* 23, 219-228.

Bigras, F.J., Paquin, R., Rioux, J.A. and Therrien, H.P. (1989a) Influence de la photopériode et de la température sur l'évolution de la tolérance au gel, de la croissance et de la teneur en eau, sucres, amidon et proline des rameaux et des racines de genévrier (*Juniperus chinensis* L. '*Pfitzerana*'). *Canadian Journal of Plant Science* 69, 305-316.

Bigras, F.J., Rioux, J.A., Paquin, R. and Therrien, H.P. (1989b) Action des fertilisations tardives sur le genévrier (*Juniperus chinensis* '*Pfitzerana*' *Aurea*) cultivé en contenants. *Canadian Journal of Plant Science* 69, 967-977.

Bigras, F.J., Rioux, J.A., Paquin, R. and Therrien, H.P. (1989c) Influence de la prolongation de la fertilisation à l'automne sur la tolérance au gel et sur la croissance printanière de *Juniperus chinensis* '*Pfitzerana*' cultivé en contenants. *Phytoprotection* 70, 75-84.

Burr, K.E. (1990) The target seedling concepts: bud dormancy and cold-hardiness: Target Seedling Symposium. *Proceedings of the Combined Meeting of the Western Forest Nursery Associations* (August 13-17, 1990). Roseburg, Oregon, Fort Collins, Colorado, United States Department of Agriculture, Forest Service, Rocky Mountain Forest and Range Experimental Station, General Technical Report RM-200, pp. 79-90.

Burr, K.E., Wallner, S.J. and Tinus, R.W. (1993) Heat tolerance, cold hardening, and bud dormancy relationship in seedlings of selected conifers. *Journal of the American Society for Horticultural Science* 118, 840-844.

Burr, K.E., Tinus, R.W., Wallner, S.J. and King, R.M. (1989) Relationships among cold hardiness, root growth potential and bud dormancy in three conifers. *Tree Physiology* 5, 291-306.

Calmé, S., Margolis, H.A. and Bigras, F.J. (1993) Influence of cultural practices on the relationship between frost tolerance and water content of containerized black spruce, white spruce, and jack pine seedlings. *Canadian Journal of Forest Research* 23, 503-511.

Cannell, M.G.R., Tabbush, P.M., Deans, J.D., Hollingsworth, M.K., Sheppard, L.J., Philipson, J.J. and Murray, M.B. (1990) Sitka spruce and Douglas fir seedlings in the nursery and in cold storage: root growth potential, carbohydrate content, dormancy, frost hardiness and mitotic index. *Forestry (Oxford)* 63, 9-27.

Christersson, L. (1978) The influence of photoperiod and temperature on the development of frost hardiness in seedlings of *Pinus sylvestris* and *Picea abies*. *Physiologia Plantarum* 44, 288-294.

Colombo, S.J. (1986) Second-year shoot development in black spruce *Picea mariana* (Mill.) B.S.P. container seedlings. *Canadian Journal of Forest Research* 16, 68-73.

Colombo, S.J. (1990) Bud dormancy status, frost hardiness, shoot moisture content, and readiness of black spruce container seedlings for frozen storage. *Journal of the American Society for Horticultural Science* 115, 302-307.

Colombo, S.J. (1994) Timing of cold temperature exposure affects root and shoot frost hardiness of *Picea mariana* container seedlings. *Scandinavian Journal of Forest Research* 9, 52-59

D'Aoust, A.L. and Cameron, S.I. (1982) The effect of dormancy induction, low temperatures and moisture stress on cold hardening of containerized black spruce seedlings. In: Scarratt, J.B., Glerum, C. and Plexman, C.A. (eds) *Proceedings of the Canadian Containerized Tree Seedling Symposium* (September 14-16, 1981). Toronto, Ontario. Department of the Environment, Canadian Forestry Service, Great Lakes Forest Research Centre, Sault Ste. Marie, Ontario, COJFRC Symposium Proceedings O-P-10, pp. 153-161.

Dormling, I. (1993) Bud dormancy, frost hardiness, and frost drought in seedlings of *Pinus sylvestris* and *Picea abies*. In: Li, P.H. and Christersson, L. (eds) *Advances in Plant Cold Hardiness*. CRC Press, Boca Raton, Florida, pp. 285-298.

Dormling, I., Gustafsson, A. and von Wettstein, D. (1968) The experimental control of the life cycle in *Picea abies* (L.) Karst. I. Some basic experiments on the vegetative cycle. *Silvae Genetica* 17, 44-64.

Fuchigami, L.H., Weiser, C.J., Kobayashi, K., Timmis, R. and Gusta, L.V. (1982) A degree growth stage (°GS) model and cold acclimation in temperate woody plants. In: Li, P.H. and Sakai, A. (eds) *Plant Cold Hardiness and Freezing Stress*. Academic Press, New York, pp. 93-116.

Gagnon, K.G. and Johnson, J.D. (1988) Bud development and dormancy in slash and loblolly pine. II. Effects of ethephon applications. *New Forests* 2, 269-274.

Glerum, C. (1982) Frost hardiness and dormancy in conifers. *Proceedings of the Northeastern Area Nurserymen's Conference* (July 25-29, 1982). Halifax, Nova Scotia, pp. 37-46.

Glerum, C. (1990) The status of frost hardiness research in North American forestry. *Proceedings of the International Union of Forest Research Organizations, XIX World Congress* (August 5-11, 1982). Montreal, Quebec, pp. 88-92.

Greer, D.H. and Stanley, C.J. (1985) Regulation of the loss of frost hardiness in *Pinus radiata* by photoperiod and temperatures. *Plant Cell and Environment* 8, 111-116.

Greer, D.H. and Warrington, I.J. (1982) Effect of photoperiod, night temperature, and frost incidence on development of frost hardiness in *Pinus radiata*. *Australian Journal of Plant Physiology* 9, 333-342.

Greer, D.H., Stanley, C.J. and Warrington, I.J. (1989) Photoperiod control of the initial phase of frost hardiness development in *Pinus radiata*. *Plant Cell and Environment* 12, 661-668.

Hawkins, B.J. (1993) Photoperiod and night frost influence the loss of frost hardiness of *Chamaecyparis nootkatensis* clones. *Canadian Journal of Forest Research* 23, 1408-1414.

Hawkins, B.J. and McDonald, S.E. (1993) Photoperiod influences dehardening of *Chamaecyparis nootkatensis* seedlings. *Canadian Journal of Forest Research* 23, 2452-2454.

Heide, O.M. (1986) Effects of ABA application on cessation of shoot elongation in long-day grown Norway spruce seedlings. *Tree Physiology* 1, 79-83.

Hermann, R.K. (1967) Seasonal variation in sensitivity of Douglas-fir seedlings to exposure of roots. *Forest Science* 13, 140-149.

Hultén, H. and Lindell, M. (1980) *TS-halt ett mått på invintring.* Advances för skogsförnyelse, Sveriges Lantbruksuniversitet, Garpenberg, Sweden.

Johnson, J.R. and Havis, J.R. (1977) Photoperiod and temperature effects on root cold acclimation. *Journal of the American Society for Horticultural Science* 102, 306-308.

Johnson-Flanagan, A.M. and Owens, J.N. (1985) Root growth and root growth capacity of white spruce (*Picea glauca* [Moench] Voss) seedlings. *Canadian Journal of Forest Research* 15, 625-630.

Jonsson, A., Eriksson, G., Dormling, I. and Ifver, J. (1981) Studies on frost hardiness on *Pinus contorta* Douglas seedlings grown in climate chambers. *Studia Forestalia Suecica* 157, 1-47.

Kobayashi, K.D., Fuchigami, L.H. and Weiser, C.J. (1983) Modeling cold hardiness of red-osier dogwood. *Journal of the American Society for Horticultural Science* 108, 376-381.

Koppenaal, R.S. and Colombo, S.J. (1988) Heat tolerance of actively growing, bud-initiated, and dormant black spruce seedlings. *Canadian Journal of Forest Research* 18, 1103-1105.

Kramer, P.J. and Kozlowski, T.T. (1979) *Physiology of Woody Plants.* Academic Press, New York.

Lavender, D.P. (1980) Effects of the environment upon the shoot growth of woody plants. In: Little, C.H.A. (comp. and ed.) *Control of Shoot Growth in Trees.* Proceedings of the Joint Workshop of IUFRO Working Parties on Xylem Physiology and Shoot Growth Physiology (July 20-24, 1980). Fredericton, New Brunswick, Canada, pp. 76-106.

Lavender, D.P. (1981) *Environment and Shoot Growth of Woody Plants.* Forest Research Laboratory, School of Forestry, Oregon State University, Corvallis, Oregon, Research Paper No. 45.

Lavender, D.P. (1985) Bud dormancy. In: Duryea, M.L. (ed.) *Evaluating Seedling Quality: Principles, Procedures, and Predictive Abilities of Major Tests, Proceedings.* (October 16-18, 1984). Forest Research Laboratory, School of Forestry, Oregon State University, Corvallis, Oregon, pp. 7-15.

Lavender, D.P. (1989) Characterization and manipulation of the physiological quality of planting stock. In: Worrall, J., Loo-Dinkins, J. and Lester, D.T. (eds) *Proceedings of the 10th North American Forest Biology Workshop* (July 20-22, 1989). University of British Columbia, Vancouver, British Columbia, Canada, pp. 32-57.

Lavender, D.P. (1991) Measuring phenology and dormancy. In: Lassoie, J.P. and Hinckley, T.M. (eds) *Techniques and Approaches in Forest Tree Ecophysiology.* CRC Press, Boca Raton, Florida, pp. 403-422.

Lavender, D.P. and Hermann, R.K. (1970) Regulation of the growth potential of Douglas-fir seedlings during dormancy. *New Phytologist* 69, 675-694.

Lavender, D.P. and Wareing, P.F. (1972) The effects of daylength and chilling on the responses of Douglas-fir (*Pseudotsuga menziesii* [Mirb.] Franco) seedlings to root damage and storage. *New Phytologist* 71, 1055-1067.

Lavender, D.P., Hermann, R.K. and Zaerr, J.B. (1970) Growth potential of Douglas-fir seedlings during dormancy. In: Luckwill, L.C. and Cutting, C.V. (eds) *Physiology of Tree Crops*. Academic Press, London, pp. 209-222.

Lavender, D.P., Sweet, G.B., Zaerr, J.B. and Hermann, R.K. (1973) Spring shoot growth in Douglas-fir may be initiated by gibberellins exported from the roots. *Science* 182, 838-839.

Levitt, J. (1980) *Responses of Plants to Environmental Stresses*, Vol. I, *Chilling, Freezing, and High Temperature Stresses*. Academic Press, New York.

Little, C.H.A. and Eidt, D.C. (1968) Effect of abscisic acid on bud break and transpiration in woody species. *Nature* 220, 498-499.

Lyr, H. and Hoffmann, G. (1967) Growth rates and growth periodicity of tree roots. *International Review of Forestry Research* 2, 181-206.

MacDonald, J.E. and Owens, J.N. (1993) Bud development in coastal Douglas-fir seedlings in response to different dormancy-induction treatments. *Canadian Journal of Botany* 71, 1280-1290.

Macey, D.E. and Arnott, J.T. (1986) The effect of moderate moisture and nutrient stress on bud formation and growth of container-grown white spruce seedlings. *Canadian Journal of Forest Research* 16, 949-954.

Major, J.E., Grossnickle, S.C. and Arnott, J.T. (1994) Influence of dormancy induction treatments on the photosynthetic response of field planted western hemlock seedlings. *Forest Ecology and Management* 63, 235-246.

Mityga, H.G. and Lanphear, F.O. (1971) Factors influencing the cold hardiness of *Taxus cuspidata* roots. *Journal of the American Society for Horticultural Science* 96, 83-86.

Nelson, E.A. and Lavender, D.P. (1979) The chilling requirement of western hemlock seedlings. *Forest Science* 25, 485-490.

Nienstaedt, H. (1966) Dormancy and dormancy release in white spruce. *Forest Science* 12, 374-384.

Nienstaedt, H. (1967) Chilling requirements in seven Picea species. *Silvae Genetica* 16, 65-68.

Odlum, K.D. and Colombo, S.J. (1988) Short day exposure to induce budset prolongs shoot growth in the following year. In: Landis, T.D. (tech. coord.) *Proceedings of the Combined Meeting of the Western Forest Nursery Associations* (August 8-11, 1988). Vernon, British Columbia, Canada, United States Forest Service, Rocky Mountain Forest and Range Experimental Station, General Technical Report RM-167, pp. 57-59.

Owens, J.N. and Molder, M. (1973) A study of DNA and mitotic activity in the vegetative apex of Douglas-fir during the annual growth cycle. *Canadian Journal of Botany* 51, 1395-1409.

Owens, J.N. and Molder, M. (1977) Bud development in *Picea glauca*. I. Annual growth cycle of vegetative buds and shoot elongation as they relate to date and temperature sums. *Canadian Journal of Botany* 55, 2728-2745.

Parker, J. (1971) Heat resistance and respiratory response in twigs of some common tree species. *Botanical Gazette* 132, 268-273.

Pellett, H. (1971) Comparison of cold hardiness levels of root and stem tissue. *Canadian Journal of Plant Science* 51, 193-195.
Pisek, A., Larcher, W., Pack, I. and Unterholzner, R. (1968) Kardinale Temperaturbereiche der Photosynthese und Grenztemperaturen des Lebens der Blätter verschiedener Spermatophyten. II Temperaturmaximum der Netto-Photosynthese und Hitzeresistenz der Blätter. *Flora (Jena)* 158, 110-128.
Riedmüller-Scholm, H.E. (1974) The temperature resistance of Alaskan plants from the continental boreal zone. *Flora (Jena)* 163, 230-250.
Ritchie, G.A. (1984) Assessing seedling quality. In: Duryea, M.L. and Landis, T.D. (eds) *Forest Nursery Manual: Production of Bareroot Seedlings*. Martinus and Junk, The Hague/Boston/Lancaster, for Forest Research Laboratory, Oregon State University, Corvallis, Oregon, pp. 243-259.
Ritchie, G.A. (1986) Relationships among bud dormancy status, cold hardiness, and stress resistance in 2+0 Douglas-fir. *New Forests* 1, 29-42.
Ritchie, G.A. (1991) Measuring cold hardiness. In: Lassoie, J.P. and Hinckley, T.M. (eds) *Techniques and Approaches in Forest Tree Ecophysiology*. CRC Press, Boca Raton, Florida, pp. 557-582.
Ritchie, G.A. and Dunlap, J.R. (1980) Root growth potential: Its development and expression in forest tree seedlings. *New Zealand Journal of Forestry Science* 10, 218-248.
Ritchie, G.A. and Tanaka, Y. (1990) Root growth potential and the target seedling: Target Seedling Symposium. *Proceedings of the Combined Meeting of the Western Forest Nursery Associations* (August 13-17, 1990). Roseburg, Oregon, Fort Collins, Colorado, United States Department of Agriculture, Forest Service, Rocky Mountain Forest and Range Experimental Station, General Technical Report RM-200, pp. 37-51.
Romberger, J.A. (1963) *Meristems, Growth and Development in Woody Plants*. United States Government Printing Office, Washington, DC, United States Department of Agriculture, Forest Service Technical Bulletin 1293.
Sakai, A. and Larcher, W. (1987) *Frost Survival of Plants*. Springer-Verlag, New York.
Sandvik, M. (1980) Environmental control of winter stress tolerance and growth potential in seedlings of *Picea abies* (L.) Karst. *New Zealand Journal of Forestry Science* 10, 97-104.
Schwarz, W. (1970) Der Einflub des Photoperiode auf das Austreiben, die Frosthärte und die Hitzeresistenz von Zirben und Alpenrosen. *Flora (Jena)* 159, 258-285.
Silim, S.N. and Lavender, D.P. (1991) Relationship between cold hardiness, stress resistance and bud dormancy in white spruce. In: Donnelly, F.P. and Lussenberg, H.W. (eds) *Proceedings of the 1991 Forest Nursery Association of British Columbia Meeting* (September 23-26, 1991). Prince George, British Columbia, Canada, pp. 9-14.
Simpson, D.G. (1993) Root cold hardiness of western Canadian conifers. In: Kooistra, C.M. (tech. coord.) *Proceedings of the 1992 Forest Nursery Association of British Columbia Meeting* (September 28-October 1, 1992). Penticton, British Columbia, Canada, pp. 97-105.
Smit-Spinks, B., Swanson, B.T. and Markhart, A.H., III. (1985) The effect of photoperiod and thermoperiod on cold acclimation and growth of *Pinus sylvestris*. *Canadian Journal of Forest Research* 15, 453-460.

Studer, E.J., Steponkus, P.L., Good, G.L. and Wiest, S.C. (1978) Root hardiness of container-grown ornamentals. *HortScience* 13, 172-174.

Sweet, G.B. (1965) Provenance differences in Pacific Coast Douglas fir. *Silvae Genetica* 14, 46-56.

Timmis, R. (1976) Methods of screening tree seedlings for frost hardiness. In: Cannell, M.G.R. and Last, F.T. (eds) *Tree Physiology and Yield Improvement*. Academic Press, New York, pp. 421-435.

Tung, C.H. and DeYoe, D.R. (1991) Dormancy induction in container-grown *Abies* seedlings: effects of environmental cues and seedling age. *New Forests* 5, 13-22.

van den Driessche, R. (1969) Influence of moisture supply, temperature, and light on frost-hardiness changes in Douglas-fir seedlings. *Canadian Journal of Botany* 47, 1765-1772.

van den Driessche, R. (1970) Influence of light intensity and photoperiod on frost-hardiness development in Douglas-fir seedlings. *Canadian Journal of Botany* 48, 2129-2134.

Warrington, I.J. and Rook, D.A. (1980) Evaluation of techniques used in determining frost tolerance of forest planting stock: a review. *New Zealand Journal of Forestry Science* 10, 116-132.

Weiser, C.J. (1970) Cold resistance and injury in woody plants. *Science* 169, 1269-1278.

Wells, S.P. (1979) *Chilling Requirements for Optimal Growth of Rocky Mountain Douglas-fir Seedlings*. United States Department of Agriculture, Forest Service, Intermountain Forest and Range Experimental Station, Ogden, Utah, Research Note INT-254.

Worrall, J. (1971) Absence of 'rest' in the cambium of Douglas-fir. *Canadian Journal of Forest Research* 1, 84-89.

Young, E. and Hanover, J.W. (1978) Effects of temperature, nutrient, and moisture stresses on dormancy of blue spruce seedlings under continuous light. *Forest Science* 24, 458-467.

Zaerr, J.B. and Lavender, D.P. (1974) The effects of certain cultural and environmental treatments upon the growth of roots of Douglas-fir (*Pseudotsuga menziesii* (Mirb.) Franco) seedlings. *Proceedings of the International Symposium on Ecology and Physiology of Root Growth* (1974). Potsdam, Germany, p. 27.

Zehnder, L.R. and Lanphear, F.O. (1967) The influence of temperature and light on the cold hardiness of *Taxus cuspidata*. *Journal of the American Society for Horticultural Science* 89, 709-713.

12 Early Development of Bud Dormancy in Conifer Seedlings

JOANNE E. MACDONALD
Natural Resources Canada, Canadian Forest Service, PO Box 6028, St John's, Newfoundland A1C 5X8, Canada

Introduction

The development of bud dormancy in conifer seedlings has physiological implications for the successful establishment of forest plantations. It has been demonstrated that seedlings that develop bud dormancy prior to lifting, handling, cold storage, transport and planting are more likely to survive the shock of planting and grow well in subsequent years (Cleary et al., 1978). Consequently, dormancy induction and dormancy development are critical stages of nursery culture.

For woody plants, dormancy is defined as any case in which a tissue predisposed to elongate fails to do so (Doorenbos, 1953). Thus, in forest nursery practice, the cessation of seedling height growth and the visibility of rudimentary terminal buds is an indicator of successful dormancy induction (Cleary et al., 1978). By extension, the presence of well-formed terminal buds (judged only on external appearance) indicates the development of bud dormancy (Cleary et al., 1978). Further, bud dormancy is tested by placing seedlings in conditions conducive to bud break; if buds fail to break, bud dormancy has developed (Lavender, 1985).

Bud dormancy is also delimited by an absence of mitotic activity in the shoot apical meristem (Owens and Molder, 1973). During bud development, bud scale and leaf primordia arise from localized mitotic activity of the shoot apical meristem. Thus, mitotic activity ceases and dormancy is attained only after the shoot apical meristem has completed leaf initiation and the telescoped shoot for next year is preformed (Hallé et al., 1978).

For the last decade, forest harvesting has proceeded from low elevation sites to high elevation sites in coastal British Columbia, Canada. Low elevation sites were traditionally spring planted, but high elevation sites have been difficult to reach in the spring. Consequently, high elevation sites are now autumn-planted. Early development of dormancy is essential in seedlings destined for autumn planting. Based on the external appearance of terminal buds, it has been suggested that dormancy is developed earlier using a short-day treatment as compared with a moderate moisture-stress treatment.

The objectives of this study, conducted on 1-year-old coastal Douglas fir (*Pseudotsuga menziesii* (Mirb.) Franco var. *menziesii*) seedlings, were twofold. The first objective was to determine whether there was a distinct difference in dormancy development in response to a short-day treatment as compared with a moderate moisture-stress treatment. The second objective was to determine the duration in short days required to attain the earliest development of dormancy.

Materials and Methods

The seedling material was grown under cultural practices developed at a commercial nursery on Vancouver Island, British Columbia (49°4'N, 123°55'W). Seeds, from a Vancouver Island seed source (48°49'N, 123°56'W, elevation 610 m), were stratified in early March and sown in greenhouses in early April. Dormancy induction treatments began in early July. For the short-day treatments, an 8-hour day was used. Seedlings were removed after 3, 4, 5 and 6 weeks in short day and returned to natural photoperiods. For the moderate moisture-stress treatment, water was withheld from seedlings until moisture stress (ranging from -1.0 to -1.5 MPa) developed. Moisture stress was induced in seedlings, as often as possible, during an initial 2-week period.

Using sectioned material of shoot tips, changes in organogenesis at the shoot apical meristem were recorded to monitor bud development. To determine mitotic index, cells in a microprojection of the shoot apical meristem were counted. Then, cells in mitosis in the shoot apical meristem were counted under a compound microscope. Mitotic index was calculated as the percentage of mitotically active cells in the shoot apical meristem.

In mid-March, after natural chilling, seedlings were placed in a controlled-environment chamber under forcing conditions (20°C day and night, 16 h day, potting medium moisture maintained near saturation). Every 3 days, terminal buds of seedlings were examined for bud break.

Results and Discussion

Short-day treatments

Prior to the start of the short-day treatments in early July, shoot apical meristems were initiating neoformed leaves (Hallé *et al.*, 1978) (Fig. 12.1).

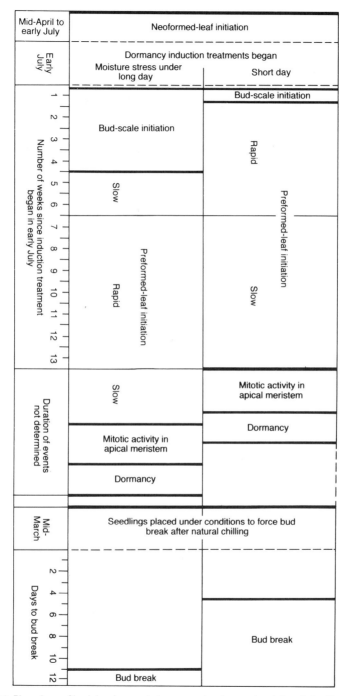

Fig. 12.1. Phenology of bud development, dormancy development and days to bud break for the short-day treatments and the moderate moisture-stress treatment.

During the first week of the short-day treatments, shoot apical meristems ended neoformed leaf initiation and began bud scale initiation (Fig. 12.1). This change in organogenesis indicated the achievement of dormancy induction. Shoot apical meristems also ended bud scale initiation and began leaf initiation during the first week of the short-day treatments. Rudimentary terminal buds were visible 3 weeks after the treatments began. Leaf initiation was completed by late September (10 weeks after the treatment began) for the 6-week duration and by early October (13 weeks after the treatments began) for the 3-, 4- and 5-week durations (Fig. 12.1).

In early October, after the completion of bud development, more shoot apical meristems from the 4- and 5-week durations were mitotically inactive, i.e. had developed dormancy, than had shoot apical meristems from the 3- and 6-week durations (Fig. 12.2). As well, the mitotic index of shoot apical meristems from the 4- and 5-week durations was lower than the mitotic index of shoot apical meristems from the 3- and 6-week durations (Fig. 12.3). These differences suggest that dormancy was attained earlier in shoot apical meristems from the 4- and 5-week durations than in shoot apical meristems from

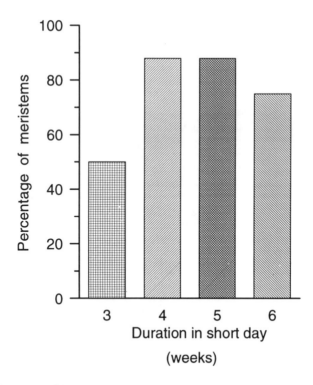

Fig. 12.2. Percentage of dormant (mitotically inactive) meristems in early October after completion of bud development for the short-day treatments.

the 3- and 6-week durations. This inference was supported by results of a 'days to bud break' test.

In mid-March, after natural chilling, days to bud break for the short day treatments ranged from 4.8 to 6.9 (Fig. 12.4). Duration in short day had a highly significant effect on days to bud break ($p = 0.0007$, SS = 46.80, df = 3). Specifically, the order of bud break by duration in short day was 4 weeks, 5 weeks, 3 weeks and 6 weeks (Fig. 12.4). The duration in short day effect on days to bud break fitted a quadratic response model ($p = 0.0046$, SS = 28.80, df = 1) (Fig. 12.4).

Thus, in shoot apical meristems from the 4- and 5-week durations, dormancy developed earlier, chilling was satisfied earlier and thus dormancy was broken earlier than in shoot apical meristems from the 3- and 6-week durations. After breaking dormancy, these shoot apical meristems responded earlier to the environmental factors that were involved in bud break under forcing conditions.

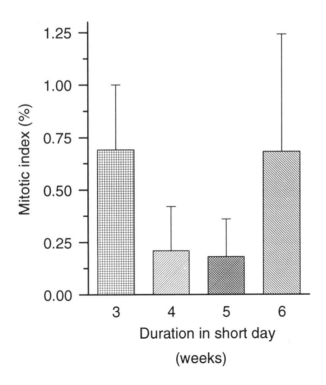

Fig. 12.3. Mitotic index of shoot apical meristems in early October after completion of bud development for the short-day treatments.

Moderate moisture-stress treatment

Prior to the start of the moderate moisture stress treatment in early July, shoot apical meristems were initiating neoformed leaves (Fig. 12.1). During the first week of moisture stress, shoot apical meristems ended neoformed leaf initiation and began bud scale initiation (Fig. 12.1). Four weeks after the treatment began, shoot apical meristems ended bud scale initiation and began leaf initiation (Fig. 12.1). Rudimentary terminal buds were visible 5 weeks after the start of treatment. Leaf initiation continued beyond early October (Fig. 12.1). Thus, compared with the short-day treatments, bud development was completed later (Fig. 12.1).

Shoot apical meristems from the moderate moisture stress treatment developed dormancy later, began accumulating chilling later, broke dormancy later and began responding to the cues for bud break later than the shoot apical meristems from the short-day treatments. Consequently, in mid-March, bud break was later than bud break for the short-day treatments (Fig. 12.1). Buds broke after 11.0 days for the moderate moisture-stress treatment as compared with 4.8 to 6.9 days for the short-day treatments (Fig. 12.1, Fig. 12.4).

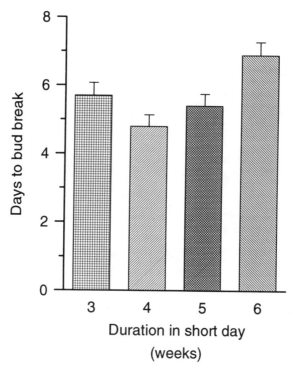

Fig. 12.4 Days to bud break after natural chilling under forcing conditions in mid-March for the short-day treatments.

Conclusions

Successful dormancy induction – as indicated by the transition from neo-formed leaf initiation to bud scale initiation – was achieved during the first week of both the short-day and the moderate moisture-stress treatments. However, the phenology of the completion of leaf initiation and thus the phenology of dormancy development were different. Dormancy developed earlier in response to the short day treatments as compared with the moderate moisture-stress treatment. Of the short-day treatments, dormancy development was earliest for the 4-week duration.

Monitoring the completion of leaf initiation and the cessation of mitotic activity in shoot apical meristems of seedlings would provide forest tree nurseries with a clearer demarcation of the dormancy status of seedlings than current methods, which rely only on the external appearance of terminal buds. The use of sectioned material is not suggested, because it is a lengthy procedure. Instead, squashed preparations are recommended as they provide a quicker determination of mitotic index. Grob and Owens (1994) provide a rapid and standardized technique for determination of mitotic index in squash preparations of shoot apical meristems.

References

Cleary, B.D., Greaves, R.D. and Owston. P.W. (1978) Seedlings. In: Cleary, B.D., Greaves, R.D. and Hermann, R.K. (eds) *Regenerating Oregon's Forests*. Oregon State University Extension Service, Corvallis, pp. 63-97.

Doorenbos, J. (1953) Review of the literature on dormancy in buds of woody plants. *Medelinger van de Landbouwhogeschool te Wageningin, Nederland* 53, 1-24.

Grob, J.A. and Owens, J.N. (1994) Techniques to study the cell-cycle in conifer shoot apical meristems. *Canadian Journal of Forest Research* 23, 472-482.

Hallé, F., Oldeman, R.A.A. and Tomlinson, P.B. (1978) *Tropical Trees and Forests: an Architectural Analysis*. Springer-Verlag, Berlin.

Lavender, D.P. (1985) Bud dormancy. In: Duryea, M.L. (ed.) *Evaluating Seedling Quality: Principles, Procedures, and Predictive Abilities of Major Tests*. Forest Research Laboratory, Oregon State University, Corvallis, pp. 7-15.

Owens, J.N. and Molder, M. (1973) A study of DNA and mitotic activity in the vegetative apex of Douglas fir during the annual growth cycle. *Canadian Journal of Botany* 51, 1395-1409.

13 Near-Lethal Stress and Bud Dormancy in Woody Plants

Michael Wisniewski[1], Leslie H. Fuchigami[2], Jörg J. Sauter[3], Abbas Shirazi[2] and Liping Zhen[2]
[1]USDA-ARS, 45 Wiltshire Road, Kearneysville, WV 25430, USA; [2]Department of Horticulture, Oregon State University, Corvallis, OR 97331, USA; [3]Botanical Institut, Christian Albrecht Universität, Kiel, Germany

Introduction

The interaction of dormancy with other physiological processes, such as cold hardiness, heat tolerance and plant recovery from near-lethal stresses, can have an impact on temperate plant production and may explain the phenomenon of plant dieback in temperate regions of the world. A variety of near-lethal stresses, such as sub- or near-lethal heat, freezing and chemical stress, release woody plant buds from both endo- and ecodormancy (Doorenbos, 1953; Erez and Lavee, 1974; Fuchigami and Nee, 1987; Wang and Faust, 1994; Shirazi and Fuchigami, 1995) and decrease or eliminate deep supercooling of flower buds (Chalker-Scott, 1988; Wisniewski et al., 1993). The physiological and biochemical events that occur in response to near-lethal stress, however, are not well characterized.

Fuchigami and Nee (1987) speculated that the reduced form of glutathione and other thiol compounds are involved in overcoming endodormancy. Siller-Cepeda et al. (1992) speculated that the breaking of endodormancy by natural chill accumulation was due to chilling stress resulting in the production of oxidizing agents followed by reducing agents. Recently, Wang and Faust (1994) reported that the rapid breaking of endodormancy and paradormancy increased production of ascorbic acid, the reduced form of glutathione, total non-protein thiol and non-glutathione thiol, whereas dehydroascorbic acid and oxidized glutathione decreased. The activities of enzymes involved in production of the above compounds were also affected.

They speculated that bud break is associated with the removal of free radicals through activated peroxide-scavenging systems. Zhen (1994) also reported that hydrogen cyanamide, a dormancy breaking agent, induced similar changes in the free radical-scavenging system of plant cells. In response to near-lethal heat stress, production of heat shock proteins has also been observed (Wisniewski et al., 1994a).

In this chapter the degree growth stage (°GS) model (Fuchigami et al., 1982) will be used to characterize the developmental status of plant tissues at the time of near-lethal or dormancy-altering treatment applications. The °GS model describes numerically the various developmental stages of dormancy in woody plants. The model is divided into 360°GS, illustrated as either a sine curve or a circle, that serve as a time line for the cyclical passage of woody plants through five distinct growth stages: (i) spring bud break, 0°GS; (ii) maturity induction, 90°GS; (iii) vegetative maturity or the onset of endodormancy, 180°GS; (iv) maximum endodormancy, 270°GS; and (v) end of endodormancy, 315°GS. The sine curve quantifies the relative degree of development during each growth phase, such as 0 to 90°GS, rapid growth phase; 90 to 180°GS, paradormancy phase; 180 to 270°GS, deepening endodormancy phase; 180 to 315°GS, decreasing endodormancy phase; and 315 to 360°GS, ecodormancy phase.

Near-Lethal Stress: Endodormancy and Ecodormancy

In our laboratories, we have used high temperature (e.g. 47°C for 1 h) as a simple and rapid method to expose buds to near-lethal stress in a controlled manner to overcome dormancy (Shirazi and Fuchigami, 1993; Wisniewski et al., 1993). Wisniewski et al. (1993) found that exposure to moist heat (i.e. shoots in porous bags immersed in water at 40–45°C) overcame endodormancy more effectively than dry heat (i.e. shoots immersed at the same temperatures using impervious plastic bags), whereas immersion in water at 20°C had little effect. Exposure to near-lethal stress increases ethylene release by tissues and leakage of electrolytes through cell membranes (Nee, 1986). Electrolyte leakage measurements (i.e. electrical conductivity) have been used to quantitate and standardize near-lethal stress treatments (Shirazi and Fuchigami, 1993). The heat-stress treatment that overcame endodormancy in red-osier dogwood (*Cornus sericea* L.) at 200°GS increased electrical conductivity by 12 to 15% above the controls. Near-lethal stress-induced electrolyte leakage differs among plant species and stages of development within the same species. The degree of near-lethal stress required to overcome endodormancy, therefore, varies with species and growth stage (Shirazi, 1992). It is possible that quantitative tests, such as electrical conductivity, may have commercial use for determination of chemical dosages required to overcome endodormancy in deciduous fruit crops.

The effects of near-lethal stress on bud break and shoot elongation are °GS-dependent. Fuchigami and Nee (1987) divided endodormancy into two distinct stages: (i) between 200 and 270°GS, near-lethal stress released bud dormancy but did not promote shoot elongation (shoot growth was rosette-like); and (ii) between 270°GS and 315°GS, near-lethal stress released bud dormancy and resulted in normal shoot extension.

Near-lethal heat stress releases endodormancy and apparently decreases the thermal units needed for bud break during ecodormancy in poplar (*Populus nigra* and *P.* × *canadensis* Moench 'Robusta'), peach (*Prunus persica*) (Wisniewski et al., 1994) and apple (*Malus domestica*) (Wang and Faust, 1994). In general, heat stress is most effective during the early and later stages of endodormancy (180 to 200°GS and 300 to 315°GS, respectively). Heat and chemical treatments are least effective during the deeper stages of endodormancy, from 200 to 300°GS (Nee, 1986; Shirazi, 1992; Siller-Cepeda et al., 1992; Wisniewski et al., 1994). By the late stages of ecodormancy, 330 to 360°GS, near-lethal stress either has no effect or delays bud break (Wisniewski et al., 1994a, b; Shirazi and Fuchigami, 1995). Delayed bud break after chemical application may be due to chemotoxicity, whereas the response to near-lethal heat stress may be affected by intrinsic levels of thermal sensitivity.

In poplar, buds exhibited a definitive shift in thermal sensitivity during the dormant period (Wisniewski et al., 1994). In the early stages of endodormancy, 180 to 200°GS, 40°C (wet) for 2 h was necessary to induce bud break. During deep endodormancy, 45°C was required. During ecodormancy, percentage bud break differed little between the 40 and 45°C heat treatment. By late ecodormancy, near 360°GS, 45°C was either inhibitory to bud break or lethal. In peach, the response of ecodormant buds to near-lethal heat stress was more complex. Increasing the time of 40°C exposure (optimum 4 h) was more effective at releasing buds from dormancy than increasing the absolute temperature for a set time (Wisniewski et al., 1993). By late January (~ 320°GS), however, percentage bud break differed little between the 2 and 4 h treatments at 40°C. Releasing flower buds from endodormancy was more difficult, and flower buds were more sensitive to injury. This may be related to the stage of flower bud development and/or the thermal sensitivity of flower bud tissues.

The response of woody plant dormant buds to near-lethal heat and chemical stress is dynamic and dependent on the stage of development, the intrinsic sensitivity to the stress treatment and whether the buds are floral or vegetative. Burr et al. (1993) found that the relationship between thermotolerance and dormancy in conifers was species-specific and that the greatest thermotolerance was observed in mature needles from actively growing plants rather than in needles from dormant, cold-hardy plants. In our studies, extensive damage generally occurs ~ 5°C above the optimum temperature for releasing dormant buds. Electrolyte leakage from the cells of stem tissues has been useful for quantitating the level of near-lethal stress required to overcome endodormancy.

One of our objectives has been to determine the effect of near-lethal stress on time of bud break and cold hardiness under natural conditions in temperate regions (Shirazi and Fuchigami, 1993). Endodormant red-osier dogwood plants exposed to near-lethal heat stress at three stages of development (200, 270 and 300°GS) and subsequently placed in natural autumn to spring conditions at Corvallis, Oregon, resulted in earlier (by 2 months) bud break only at a late stage of endodormancy (300°GS) (Shirazi and Fuchigami, 1993). This response may explain some instances of early temperate plant bud break that cannot be explained by accumulation of chill units and/or heat units alone. The periodic failure of chill-unit and other dormancy models to accurately predict time of bud break in certain climates suggests that near-lethal stresses may be potential factors that should be incorporated into these models.

Recovery from Near-Lethal Stress

Near-lethal stress generally is not injurious to plant tissues on a long term basis. The only documented changes are increases in ethylene and electrolyte leakage immediately following the stress. Phytotoxicity occurs when electrolyte leakage exceeds a 12 to 15% increase above untreated tissues. However, because sensitivity of plant tissues to near-lethal stress changes with developmental stage, and possibly with environment, development of a standard dosage to overcome dormancy under commercial conditions has not been possible.

Furthermore, the post-stress environment can dramatically affect plant response to near-lethal stress and potential injury. We documented an interaction between near-lethal stress treatment and post-stress temperature (PST) on the recovery (i.e. survival) of plants (Shirazi and Fuchigami, 1993). Exposure of endodormant red-osier dogwood stem and bud tissues to near-lethal stress treatments (i.e. heat, cold or chemical) followed by storage at $\geq 10°C$ PST resulted in no physical signs of injury. In contrast, at storage $< 10°C$ PST, stem and bud tissues were killed. Treated plants placed under natural autumn and winter conditions at Corvallis recovered with no evidence of injury. The interaction between PST and near-lethal stress was greatest during the early stages of endodormancy (i.e. 200°GS). In excised stem tissues, the injurious effect of low PST was observed within 48 h and in intact plants the effect was observed within 2 weeks. In contrast, recovery (as measured by a decrease in electrolyte leakage) at warm PST in intact plants occurred within 1 week.

The reason for the interaction of near-lethal stress treatments and low PST on plant dieback is not understood. Gates (1980) suggested that the ability of an organism to survive extreme temperature is dependent on its repair mechanisms and that the extent of temperature-induced cellular damage determines the efficiency of repair. We found that the level of glutathione decreases immediately following exposure to near-lethal stresses and then, after a few hours, increases to a level higher than the original (Siller-Cepeda et al., 1992; Shirazi and Fuchigami, 1993). The increase in glutathione level is temperature dependent and occurs only at warm temperatures (e.g. 23°C), suggesting a possible role of glutathione production, and perhaps other peroxide-scavenging systems, in the recovery of plants exposed to near-lethal stress. This may account for the observed changes in free radical-scavenging systems coupled to the pentose phosphate cycle during dormancy-breaking treatments and cold acclimatization (Kuroda et al., 1990a, b; Wang and Faust, 1994; Zhen, 1994).

The dieback of dormant plants exposed to near-lethal stresses and a low PST may help explain temperate-zone plant diebacks in nature and commerce. Various near-lethal stresses (e.g. pollution, desiccation, freeze, high temperature, etc.) may result in substantial plant dieback following exposure to low PST for even a short duration. Depending on the environment, however, delayed expression of the damage can make detection of the cause of dieback difficult.

Breaking Endodormancy and Cold Hardiness

Although endodormancy and cold hardiness appear to be regulated somewhat independently (Fuchigami et al.; 1982), the two processes tend to overlap in woody plants. Likewise, near-lethal stress both overcomes endodormancy and reduces flower bud cold hardiness (Chalker-Scott, 1988; Wisniewski et al., 1991). The near-lethal stress treatment required to release buds from dormancy and remove the ice nucleation barrier of supercooled buds, causing immediate loss of bud hardiness, is similar. In peach, use of monoclonal antibodies that recognize the homogalacturonic sequences of pectin, in conjunction with immunogold electron microscopy, indicated that highly esterified epitopes of pectin are distributed throughout the pit membrane and primary cell walls of xylem and floral bud tissues; non-esterified epitopes are localized in middle lamellae, along the margin of the cell wall, lining empty intercellular spaces, and within filled intercellular spaces (Wisniewski and Davis, 1995). These data suggest that pectins may influence both water movement and intrusive growth of ice crystals at freezing temperatures. In *Rhododendron*, the area between the bud scales and floral buds contains phenolic compounds, which may serve as a barrier to water movement and ice penetration (Chalker-Scott, 1989). Heat and other near-lethal stresses may remove these barriers, at any stage of development, by oxidizing the compounds that make up the barrier, resulting in an immediate loss of hardiness. The relationship of these ice nucleation barriers and endodormancy is unknown. The fact that endodormancy is regulated in buds and that similar levels of stress are required to overcome both endodormancy and the ability of buds to deep supercool suggests a possible relationship. Perhaps the restriction of water to the dormant bud by the formation of a barrier is involved in endodormancy.

Bud endodormancy status does not appear to be directly involved in stem tissue hardiness; however, the effect of near-lethal stress on stem tissue hardiness is °GS-dependent (Shirazi and Fuchigami, 1995). Generally, plants begin to acclimatize at 180°GS and hardiness increases in response to decreasing temperatures (Kobayashi *et al.*, 1983). During the endodormancy period from 180 to 300°GS, cold hardiness is maintained even at warm temperatures. Maximum hardiness is achieved after the chilling requirement for overcoming endodormancy is satisfied. Thereafter, the rate of deacclimatization is a function of temperature and growth stage. The temperature range for deacclimatization increases with later growth stages (Fuchigami *et al.*, 1982). During the development of endodormancy (180 to 270°GS), overcoming endodormancy with heat stress did not affect hardiness development even at 23°C PST and natural conditions (Shirazi and Fuchigami, 1995). At maximum endodormancy, 270°GS, near-lethal stress caused loss of hardiness at 23°C PST, but not at 5°C PST or in natural conditions. At later growth stages (i.e. 300°GS), near-lethal stress caused earlier, rapid deacclimatization both at 23°C PST and in natural conditions.

We have found that various near-lethal stresses have the same effect. Therefore, natural stresses (e.g. freezing temperatures, desiccation, etc.) are likely to cause an immediate loss of endodormancy and possible loss of plant hardiness. The significance of these data for stress resistance and survival of plants under natural conditions in temperate regions is still not known.

Protein Changes and Dormancy

Biochemical and molecular studies have shown that cold acclimatization in higher plants induces the synthesis and/or accumulation of specific proteins as a result of altered gene expression (Guy, 1990). The functional role of these proteins in imparting or maintaining cold hardiness, however, has not been demonstrated clearly. Indeed, dormancy and cold hardiness are often superimposed in woody plants, making it unclear whether specific changes in proteins are associated with changes in dormancy or cold hardiness. In recent years, however, there have been several attempts to correlate changes in polypeptide patterns with dormancy and/or cold hardiness status. Thus far, specific proteins that have been associated with the onset and decline of either dormancy (endodormancy and ecodormancy combined) or cold hardiness fall into two groups. The first are referred to as vegetative or bark storage proteins (Sauter *et al.*, 1989; Wetzel *et al.*, 1989; Coleman *et al.*, 1991; Arora and Wisniewski, 1992; Stepien and Martin, 1992), which have been reviewed recently (Stanswick, 1994; Stepien *et al.*, 1994). Although it is believed that these proteins function principally as a means of nitrogen storage, other functional roles in cryoprotection, plant defence and general stress tolerance have not been ruled out (Stepien *et al.*, 1994). How these proteins affect or reflect dormancy status is unknown. At present, the relationship of these proteins to dormancy is only one of association. They may, however, serve as a useful biochemical marker for dormancy.

The other group of proteins that have been associated with cold hardiness or dormancy in woody plants is the group 2 late embryogenesis abundant (LEA) proteins (Arora and Wisniewski, 1994; Muthalif and Rowland, 1994), also known as dehydrins. Dehydrins are stress-induced proteins that are characterized by the consensus 15 amino acid sequence EKKGIMDKI-KEKLPG near the carboxy terminus (Close *et al.*, 1993a). They are glycine-rich, hydrophilic and heat-stable (i.e. they remain water-soluble after boiling). A putative function of dehydrins is that they alter the thermodynamic interactions between macromolecules and water via solute exclusion or direct binding (Close *et al.*, 1993b). Thus, they may provide stability to macromolecules, such as proteins and nucleic acids, during desiccation by preventing denaturation or inhibiting ice crystal formation at freezing temperatures. Muthalif and Rowland (1994) reported the association of 14, 60 and 65 kDa proteins in blueberry buds (*Vaccinium corymbosum* and *V. ashei*) that were related immunologically to dehydrin and whose increase was correlated with chill unit accumulation. Arora and Wisniewski (1994) reported the presence of a 60 kDa protein in peach (*Prunus persica*) that was both immunologically and sequence-related to dehydrin as well as sequence-related to other stress-induced proteins reported in the literature. Again, how these proteins are causally related to dormancy, chill-unit accumulation or cold hardiness is an open question.

Regarding the ability of near-lethal stress to overcome dormancy, and the proteins and/or genes that are induced as a result of such stress, little definitive knowledge is presently available. Preliminary evidence has indicated that Hsp 70 and other low-molecular-weight heat-shock proteins are induced in dormant poplar buds in response to a near-lethal heat treatment that releases buds from dormancy (Wisniewski *et al.*, 1994). The use of near-lethal stresses (especially heat, which can be controlled tightly) should offer an excellent system to study proteins and changes in gene regulation associated with the onset and release of dormancy, and it is expected that a great deal of information will be forthcoming in this area over the next few years. Additionally, Arora and Wisniewski (1992, 1994) have used sibling genotypes of deciduous and evergreen peach as a model system for studying dormancy. In this system, the evergreen genotype is characterized by shoots that do not set a terminal bud and whose lateral buds exhibit only a very short period of endodormancy. The evergreen genotype will, however, cold-acclimatize (Arora and Wisniewski, 1992). Use of this model system should enable progress to be made in separating changes in proteins that are related to dormancy from those that are associated with cold acclimatization.

Conclusions

A variety of near-lethal stresses can release buds of woody plants from endodormancy (Doorenbos, 1953; Erez and Lavee, 1974; Fuchigami and Nee, 1987). How this is accomplished and the biochemical and/or molecular events associated with the onset and release from dormancy are unknown. The abrupt release of dormancy with near-lethal stress produces oxidizing compounds and free radicals, followed by antioxidant production. Oxidizing compounds (e.g. hydrogen cyanamide) also induce dormancy release and antioxidant production. Likewise, Siller-Cepeda et al. (1992) reported that release of endodormancy by chilling treatment induced oxidizing agents followed by glutathione production.

Exposure of woody plants to near-lethal stresses during endodormancy can reduce the level of cold hardiness and result in plant dieback and/or earlier bud break. The impact of near-lethal stresses on temperate woody plants under natural conditions is understood poorly. Our research indicates that near-lethal stress may help to explain, in some instances, occurrences of plant dieback, the failure of predictive models for bud break and the unexpected early and/or delayed spring bud break of plants in temperate zones.

Although numerous stress-induced proteins, as well as their corresponding genes, have been reported in the literature for herbaceous plants, little is known about their existence in woody plants. The use of near-lethal stress offers an excellent opportunity to make progress in understanding the biochemical and molecular basis for dormancy in woody plants. Although several proteins have been associated with changes in dormancy status, the direct relation of these proteins to either endodormancy or near-lethal stress physiology remains to be elucidated.

References

Arora, R. and Wisniewski, M.E. (1992) Characterization of proteins in sibling deciduous and evergreen peach using free-solution isoelectric focusing and SDS-PAGE. *Plant Physiology* 99 (S), 126.

Arora, R. and Wisniewski, M.E. (1994) Cold acclimation in genetically related (sibling) deciduous and evergreen peach (*Prunus persica* L. Batsch). *Plant Physiology* 105, 95–101.

Burr, K.E., Wallner, S.J. and Tinus, R.W. (1993) Heat tolerance, cold hardiness, and bud dormancy relationships in seedlings of selected conifers. *Journal of the American Society for Horticultural Science* 118, 840-844.

Chalker-Scott, L. (1988) Relationships between endogenous phenolic compounds and *Rhododendron* tissues and organs and cold hardiness. PhD dissertation, Oregon State University, Corvallis.

Close, T.J., Fenton, R.D. and Moonan, F. (1993a) A view of plant dehydrins using antibodies specific to the carboxy terminal peptide. *Plant Molecular Biology* 23, 279-286.

Close, T.J., Fenton, R.D., Yang, A., Asghar, R., DeMason, D.A., Crone, D.E., Meyer, N.C. and Moonan, F. (1993b) Dehydrin: the protein. In: Close, T.J. and Bray, E.A. (eds) *Plant Responses to Cellular Dehydration During Environmental Stress*. American Society of Plant Physiologists, Rockville, pp. 104-114.

Coleman, G.D., Chen, T.H.H., Ernst, S.G. and Fuchigami, L.H. (1991) Photoperiod control for poplar bark storage protein accumulation. *Plant Physiology* 96, 686-692.

Doorenbos, J. (1953) Review of the literature of dormancy in buds of woody plants. *Mededelingen Landbouwhogeschool Wageningen, Netherlands* 53, 1-24.

Erez, A. and Lavee, S. (1974) Recent advances in breaking the dormancy of deciduous fruit trees. *19th International Horticultural Congress, Warszawa* 11-18, 69-78.

Fuchigami, L.H. and Nee, C.C. (1987) Degree growth stage model and rest-breaking mechanisms in temperate woody perennials. *HortScience* 22, 836-845.

Fuchigami, L.H., Weiser, C.J., Kobayashi, K., Timmis, R. and Gusta, L.V. (1982) A degree growth stage (°GS) model and cold acclimation in temperate woody plants. In: Li, P.H. and Sakai, A. (eds) *Plant Cold Hardiness and Freezing Stress*, Vol. 2. Academic Press, New York, pp. 93-116.

Gates, D.M. (1980) *Biophysical Plant Ecology*. Springer-Verlag, New York.

Guy, C.L. (1990) Cold acclimation and freezing tolerance: role of protein metabolism. *Annual Review in Plant Physiology and Plant Molecular Biology* 41, 187-223.

Kobayashi, K.D., Fuchigami, L.H. and Weiser, C.J. (1983) Modeling cold hardiness of red-osier dogwood. *Journal of the American Society for Horticultural Science* 108, 376-381.

Kuroda, H., Sagisaka, S. and Chiba, K. (1990a) Seasonal changes in peroxide-scavenging systems of apple trees in relation of cold hardiness. *Journal of the Japanese Society for Horticultural Science* 59, 399-408.

Kuroda, H., Sagisaka, S. and Chiba, K. (1990b) Frost induces cold acclimation and peroxide-scavenging systems coupled with the pentose phosphate cycle in apple twigs under natural conditions. *Journal of the Japanese Society for Horticultural Science* 59, 409-416.

Muthalif, M.M. and Rowland, L.J. (1994) Identification of dehydrin-like proteins responsive to chilling in floral buds of blueberry. *Plant Physiology* 104, 1439-1447.

Nee, C.C. (1986) Overcoming bud dormancy with hydrogen cyanamide. Timing and mechanisms. PhD dissertation, Oregon State University, Corvallis.

Sauter, J.J., Van Cleve, B. and Wellencamp, S. (1989) Ultrastructural and biochemical results on the localization and distribution of storage proteins in a poplar tree and in twigs of other plant species. *Holzforschung* 43, 1-6.

Shirazi, A.M. (1992) Relationship of 'near-lethal' stress on dormancy, cold hardiness and recovery of red-osier dogwood. PhD dissertation, Oregon State University. Corvallis.

Shirazi, A.M. and Fuchigami, L.H. (1993) Recovery of plants from 'near-lethal' stress. *Oecologia* 93, 429-434.

Shirazi, A.M. and Fuchigami, L.H. (1995) The relationship of 'near-lethal' stress on dormancy and stem cold hardiness in Red-osier dogwood. *Tree Physiology* 15, 275-279.

Siller-Cepeda, J.H., Fuchigami, L.H. and Chen, T.H.H. (1992) Glutathione content in peach buds in relation to development and release of rest. *Plant and Cell Physiology* 33, 867-872.

Stanswick, P.E. (1994) Storage proteins of vegetative plant tissues. *Annual Review of Plant Physiology and Plant Molecular Biology* 45, 303-322.

Stepien, V. and Martin, F. (1992) Purification, characterization and localization of the bark proteins of poplar. *Plant Physiology and Biochemistry* 30, 399-407.

Stepien, V., Sauter, J.J. and Martin, F. (1994) Vegetative storage proteins in woody plants. *Plant Physiology and Biochemistry* 32, 185-192.

Wang, S.Y. and Faust, M. (1994) Changes in the antioxidant system associated with bud break in 'Anna' apple (*Malus domestica* Borkh.) buds. *Journal of the American Society for Horticultural Science* 119, 735-741.

Wetzel, S., Demmers, C. and Greenwood, J.S. (1989) Seasonally fluctuating bark proteins are a potential form of nitrogen storage. *Planta* 178, 275-281.

Wisniewski, M. and Davis, G. (1995) Immunogold localization of pectins and glycoproteins in tissues of peach with reference to deep supercooling. *Trees* 9, 253-260.

Wisniewski, M., Davis. G. and Arora, R. (1991) Effect of macerase, oxalic acid and EGTA on deep supercooling and pit membrane structure of xylem parenchyma of peach. *Plant Physiology* 96, 1354-1359.

Wisniewski, M., Davis, G., Arora, R. and Fuchigami, L.H. (1993) Tissue injury and supercooling response of peach buds to near-lethal heat stress. *HortScience* 28, 584 (abstract).

Wisniewski, M., Santer, J.J., Stepien, V. and Fuchigami, L.H. (1994) Effects of near-lethal heat stress on endodormancy and ecodormancy of peach and hybrid poplar. *HortScience* 29, 511.

Zhen, L. (1994) The free radical scavenging system in overcoming phytotoxicity and breaking dormancy. PhD dissertation, Oregon State University, Corvallis.

IV Biochemistry

14 Structural Requirements of the ABA Molecule for Maintenance of Dormancy in Excised Wheat Embryos

Suzanne R. Abrams[1], Patricia A. Rose[1] and M.K. Walker-Simmons[2]
[1]*Plant Biotechnology Institute, National Research Council of Canada, Saskatoon, Saskatchewan, S7N 0W9, Canada;*
[2]*United States Department of Agriculture, Agricultural Research Service, Washington State University, Pullman, WA 99164-6420, USA*

Introduction

Abscisic acid (ABA) is a potent regulator of seed development and dormancy (Black, 1991). For example, applied ABA blocks germination and maintains dormancy of excised embryos from mature wheat seeds (Fig. 14.1; Walker-Simmons *et al.*, 1992). However, how the hormone is perceived in the wheat embryo is not yet known. In spite of extensive research on the ability of ABA to regulate many processes in plant growth and development, many questions remain about the existence, location and multiplicity of receptors for the hormone.

This chapter examines the structure of the active site of the putative ABA receptor involved in dormancy maintenance in wheat embryos. To probe the ABA response system, ABA analogues have been designed and tested. These analogues are useful tools for relating maintenance of dormancy to expression of a subset of ABA-inducible genes. It is anticipated that, in addition to defining the steric and electronic requirements of the ABA response system in dormancy maintenance, the structure/activity results in embryo dormancy could be related to the requirements of other ABA-inducible systems.

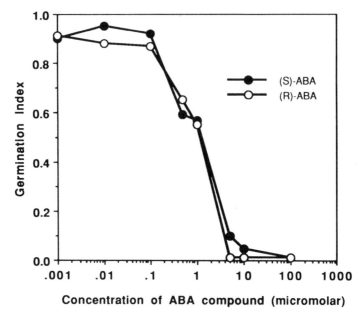

Fig. 14.1. Biological activity of natural (S)-ABA and unnatural (R)-ABA as inhibitors of wheat embryo germination.

Methodology

Embryos from dormant seeds of wheat (*Triticum aestivum* L. cv. Brevor) were used for germination assays and Northern blot analysis of ABA-responsive transcripts, as described previously (Walker-Simmons *et al.*, 1992). Analyses were conducted 24 h after imbibition with ABA or ABA analogues. Racemic ABA was resolved as described by Dunstan *et al.* (1992). Optically pure analogues (pairs of optical isomers of analogues of ABA that had modifications in the side-chain (triple bond replacing the *trans* double bond) or ring (double bond reduced)) were synthesized according to Lamb and Abrams (1990), Rose *et al.* (1992) and Walker-Simmons *et al.* (1992).

Molecular modelling was carried out on a Silicon Graphics Indigo^2 workstation, using Tripos Sybyl 6.03 software. Calculations of low energy conformers of ABA and analogues were performed starting from X-ray crystallographic data for ABA, with the functional groups changed as appropriate, followed by Powell energy minimization calculations.

ABA, Analogues, Receptors and Physiological Activity

The structural formula of the naturally occurring form of ABA is shown in Fig. 14.2. The molecule has one chiral centre, at the junction of the ring and side-

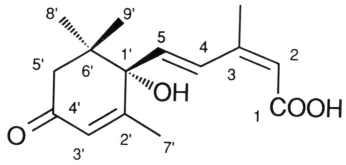

Fig. 14.2. The structural formula of abscisic acid.

chain. The (S)-enantiomer is found in plants; the (R)-form is not. The *cis,trans*-diene side-chain terminates in a carboxylic acid, and the six-membered ring has a double bond conjugated to the ketone. The shape that ABA adopts in the crystal structure is that with the side-chain axial and the hydroxyl group at the chiral centre equatorial and in the plane of the ring (Ueda and Tanaka, 1977; Fig. 14.3).

The three dimensional structure of the ABA molecule in the active site of the receptor is not known. In nuclear magnetic resonance (NMR) studies of ABA in solution, even at low temperature, we see evidence only for the conformer with the side-chain axial. Willows and Milborrow (1993) reported similar results. However, calculations of the energy of an alternate form, with the side-chain equatorial, suggest that both are of similar energy and either conformer could occupy the active site of the ABA receptor (Fig. 14.3).

Work previous to our studies with dormant wheat embryos had shown that seed germination could be inhibited by ABA and chemical variants of the structure, i.e. analogues of ABA. Radical changes to the size and shape of the

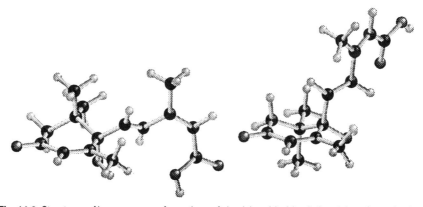

Fig. 14.3. Structures of low-energy conformations of abscisic acid; side-chain axial conformation is shown on the right, side-chain equatorial on the left.

ABA molecule, especially the carbon skeleton, resulted in loss of activity (Walton, 1983). Changes in oxidation level at the carboxylic acid carbon (C-1 in the conventional nomenclature system) gave compounds that were more potent inhibitors than ABA itself, probably because these compounds were converted to ABA in the tissue over time, providing a long-lasting pulse of the hormone and prolonging dormancy. Not all analogues are active because they are converted to ABA. An acetylenic aldehyde analogue is very much superior to ABA in germination inhibition and dormancy maintenance in lettuce as well as cress (Walton, 1983), and it is not metabolized to ABA in plant cells, as far as has been determined (S.R. Abrams, unpublished).

With a few notable exceptions, most investigators studying the effects of exogenous ABA in plants have used racemic forms of the hormone. The optically pure forms of ABA were not readily accessible until the development of HPLC methods for resolving quantities of methyl-ABA (Dunstan *et al.*, 1992). In our studies with dormant wheat embryos, the natural (S)-ABA and unnatural (R)-ABA inhibit germination equally (Fig. 14.1). Sondheimer *et al.* (1971) found that both ABA mirror-image forms inhibited growth of roots and shoots of germinating barley seeds, as did Milborrow (1978) in experiments on coleoptile growth of excised wheat embryos. If the conformation of both ABA isomers in the active site is that with the side-chain axial (Fig. 14.4), then these results lead to the conclusion that the receptor can tolerate alterations to the lower face of the molecule as shown.

The ABA mirror-image forms are not active equally in all ABA-responsive plant processes, suggesting that there may be differences in the active sites of ABA receptors. The classic example is the stomatal guard cell assay, in which

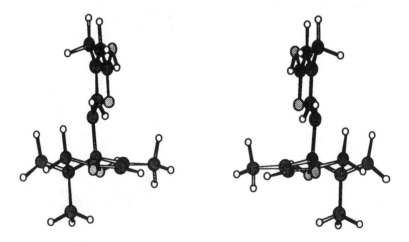

Fig. 14.4. Structures of (S)-ABA (left) and (R)-ABA (right), viewed with the carbonyl group in front and the side-chain behind. Structures are very similar, except for the position of the axial methyl group.

(R)-ABA is inactive (Walton, 1983). Cress seed germination is inhibited by the natural (S)-form of ABA, but is much less affected by the (R)-form (Gusta *et al.*, 1992). Genes coding for storage proteins in microspore-derived embryos of canola are induced more strongly by natural ABA than by unnatural ABA (Wilen *et al.*, 1993).

For our wheat embryo dormancy studies, three parts of the molecule were altered systematically, producing a set of eight analogues with which to probe the active site. The C-1 carbon was maintained as the acid, since analogue metabolism over the course of the experiment would confound the results. The stereochemistry of the chiral centre was altered, as discussed above. The two other modifications were a reduction of the ring double bond and a changing of the *trans* double bond of the side-chain to a triple bond. The molecules with a triple bond in the side-chain and the double bond in the ring were not stable, but the remaining six were compared for activity in dormancy maintenance. The analogues with the same relationship at the chiral centre as natural ABA (designated as N-forms) were all potent germination inhibitors, comparable to ABA (Fig. 14.5). Their counterparts, resembling unnatural (R)-ABA (U-forms), were completely inactive, either because they fitted poorly into the active site or due to some non-receptor-influenced explanation, such as more rapid metabolism.

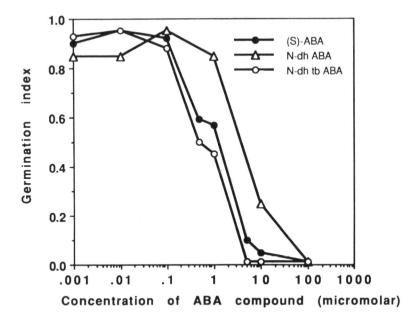

Fig. 14.5. Biological activity of natural (S)-ABA, dihydro-ABA (N-dh ABA), and dihydroacetylenic ABA (N-dh tb ABA) (all having similar stereochemistry) as inhibitors of wheat embryo germination.

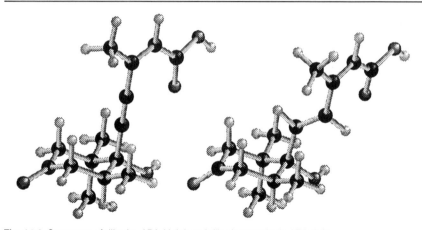

Fig. 14.6. Structures of dihydro-ABA (right) and dihydroacetylenic ABA (left).

The low-energy conformations of the active compounds were calculated and their shapes compared with that of ABA. The carbon skeleton of the analogue with the ABA-like side-chain, dihydro-ABA (Fig. 14.6, structure on right), is nearly superimposable with that of ABA. The dihydro-ABA compound has hydrogen atoms axial to the ring on the C-2' and C-3' positions. The germination inhibition results show that these do not interfere with perception of the molecule. The active analogue with the triple bond in the side-chain, dihydroacetylenic ABA (Fig. 14.6, structure on the left), has the same features in the ring, but the side-chain with the four linear carbon atoms is perpendicular to the ring. The activity of the acetylenic molecule suggests that the receptor has flexibility in this region.

The set of physiologically active compounds described above was used by Walker-Simmons *et al.* (1992) to study the expression of certain ABA-inducible genes in relation to dormancy maintenance. The message for the *Em* gene was increased by natural ABA, but much less so by its mirror image (R)-form. This result suggests that *Em* gene expression is not related to dormancy, as the (R)-ABA was active in maintaining dormancy. The levels of message for the other two ABA-responsive genes, *dhn* (*rab*) and *lea* (group 3), increased on treatment of embryos with the biologically active compounds. This suggests that these two ABA-responsive genes may be involved in dormancy maintenance.

The ring dihydro analogue also provides a probe for studying the activity of ABA and its metabolites in dormancy. The principal oxidative metabolite of ABA is phaseic acid, formed in the plant by hydroxylation of the 8' methyl group and subsequent attack of the new oxygen atom on the double bond of the ABA ring to form a new ether ring. The dihydro compound is missing the ABA ring double bond, and thus cannot be converted into phaseic acid. The activity of the dihydro analogue suggests that ABA or its hydroxylated, but not cyclized, form is the hormonal signal in dormancy maintenance. In a study on freezing tolerance in bromegrass cells, the dihydro-ABA analogue substituted for ABA (Lamb *et al.*, 1993). However, its metabolite with a hydroxyl group at the 8' position did not, suggesting that, in this ABA-induced physiological stress response, ABA itself is the active signal compound, rather than an ABA metabolite.

Conclusions

These studies have shown that certain ABA-induced genes are associated with dormancy maintenance in wheat embryos. The set of ABA analogues discussed above has provided some indications of the dormancy-related ABA receptor's tolerance for structural changes. Future ABA structural research will determine sites on the molecule that could be used for attachment of photoaffinity and other labels to further advance the study of ABA receptors and gene expression associated with dormancy.

References

Black, M. (1991) Involvement of ABA in the physiology and development of mature seeds. In: Davies, W.J. and Jones, H.G. (eds) *Abscisic Acid, Physiology and Biochemistry*. Bios Scientific Publishers, London, pp. 99-124.

Dunstan, D.I., Abrams, G.D., Bock, C.A. and Abrams, S.R. (1992) Metabolism of (+)- and (−)-abscisic acid by somatic embryo cultures of white spruce. *Phytochemistry* 31, 1451-1454.

Gusta, L.V., Ewan, B., Reaney, M.J.T. and Abrams, S.R. (1992) The effects of abscisic acid metabolites on the germination of cress seed. *Canadian Journal of Botany* 70, 1550-1555.

Lamb, N. and Abrams, S.R. (1990) Synthesis of optically active cyclohexanone analogs of the plant hormone abscisic acid. *Canadian Journal of Chemistry* 68, 1151-1162.

Lamb, N., Shaw, A.C., Abrams, S.R., Reaney, M.J.T., Robertson, A.J. and Gusta, L.V. (1993) Oxidation of the 8'-position of a biologically active abscisic acid analog. *Phytochemistry* 34, 905-917.

Milborrow, B.V. (1978) Abscisic acid. In: Letham, D.S., Goodwin, P.B. and Higgins, T.J.V. (eds) *Phytohormones, a Comprehensive Treatise*. Elsevier/North Holland Biochemical Press, Amsterdam, pp. 295-344.

Rose, P.A., Abrams, S.R. and Shaw, A.C. (1992) Synthesis of chiral acetylenic analogs of the plant hormone abscisic acid. *Tetrahedron: Asymmetry* 3, 343-350.

Sondheimer, E., Galson, E.C., Chang, Y.P. and Walton, D.C. (1971) Asymmetry, its importance to the action and metabolism of abscisic acid. *Science* 174, 829-831.

Ueda, H. and Tanaka, J. (1977) The crystal and molecular structure of *d*,l-2-*cis*-4-*trans*-abscisic acid. *Bulletin of the Chemical Society of Japan* 50, 1506-1509.

Walker-Simmons, M.K., Anderburg, R.J., Rose, P.A. and Abrams, S.R. (1992) Optically pure ABA analogs - tools for relating germination inhibition and gene expression in wheat embryos. *Plant Physiology* 99, 501-507.

Walton, D.C. (1983) Structure-activity relationships of abscisic acid analogs and metabolites. In: Addicott, F.T. (ed.) *Abscisic Acid*. Praeger Press, New York, pp. 113-146.

Wilen, R.W., Hays, D.B., Mandel, R.M., Abrams, S.R. and Moloney, M.M. (1993) Competitive inhibition of ABA-regulated gene expression by stereoisomeric acetyleneic analogs of abscisic acid. *Plant Physiology* 101, 469-476.

Willows, R.D. and Milborrow, B.V. (1993) Configurations and conformations of abscisic acid. *Phytochemistry* 34, 233-237.

15 Changes in Hormone Sensitivity in Relation to Onset and Breaking of Sunflower Embryo Dormancy

MARIE-THÉRÈSE LE PAGE-DEGIVRY, JACQUELINE BIANCO, PHILIPPE BARTHE AND GINETTE GARELLO
Laboratoire de Physiologie Végétale, Université de Nice – Sophia Antipolis, 06108 Nice Cedex 2, France

Introduction

Hormone sensitivity is defined as a change in the observed response to an exogenously supplied hormone. Since in many cases no relationship has been found between levels of hormones in seeds and the level of dormancy, discussions initiated by Trewavas (1982) have resulted in a greater awareness of the importance of physiological sensitivity to hormones. Moreover, control exerted through sensitivity and via hormone levels need not be mutually exclusive. An important consideration is that the level of endogenous hormone in the tissue always interferes with the sensitivity to the applied hormone (Roberts and Hooley, 1988). A picture of embryo abscisic acid (ABA) status, taking into account not only ABA levels but also ABA synthesis capacity, allows this type of interaction to be discussed.

Sensitivity to hormones can be modified genetically, e.g. developing wheat embryos from two cultivars, sprouting-resistant and sprouting-susceptible, displayed a differential sensitivity to ABA (Walker-Simmons, 1987). Many hormone-sensitive mutants (e.g. peas, barley, *Arabidopsis*) have been screened in relation to their responsiveness to gibberellins (GAs) or ABA (reviewed in Reid, 1990). Sensitivity can also be modified by ontogenic changes. In rape (Finkelstein *et al.*, 1987), soybean (Eisenberg and Mascarenhas, 1985) or wheat (Walker-Simmons, 1987), a decline in embryo sensitivity to exogenous ABA was observed late in development. Moreover, environmental stimuli,

© CAB INTERNATIONAL 1996. *Plant Dormancy* (ed. G.A. Lang)

especially water availability, can affect sensitivity. A reduced sensitivity to ABA in castor bean embryos, as a consequence of desiccation, may be an important factor in eliciting the switch to germination (Kermode *et al.*, 1989). An increase in sensitivity to GA in wheat aleurone tissue could be induced by subjecting the grains to a period of dehydration (Armstrong *et al.*, 1982). Dry storage, a dormancy-releasing treatment for *Arabidopsis* seeds, increased their sensitivity to GAs (Karssen and Laçka, 1986).

Changes in sensitivity to either ABA or GAs have been described during either seed maturation or dehydration. To obtain a more complete picture of the involvement of these changes in the dormancy process, changes in sensitivity to both ABA and GAs were examined during the onset and breaking of dormancy in sunflower, *Helianthus annuus*. In the following studies, the observed response for ABA sensitivity was the induction of dormancy and for GA sensitivity the manifestation of germination.

Sensitivity to Hormones During Embryo Development

Sensitivity to ABA

When applied to young, non-dormant sunflower embryos, ABA inhibited germination. However, this inhibition required the continued presence of the hormone since, whatever the dose (5 μM or 50 μM) used or application duration (5 to 10 days), germination occurred when the embryos were transferred to the control medium. This temporary inhibition was distinct from dormancy, which, during development *in situ*, was induced by ABA in less than 1 week. Thus, these young embryos were considered to be insensitive (relative to dormancy induction) to ABA.

In embryos at developmental ages of 13 and 17 days after pollination (DAP), i.e. immediately prior to the natural induction of dormancy (Fig. 15.1), 5 days of exogenous ABA (50 μM) effectively induced dormancy to develop. This suggests that ABA induces dormancy only if it is supplied during a critical time period in which tissues are ABA-sensitive.

Sensitivity to GAs

The effect of GAs on germination during sunflower embryo development is shown in Fig. 15.2. Young embryos (10 to 14 DAP) germinated readily in the absence of exogenous GAs, but at an embryo developmental age of 17 DAP, GA_3 applied at a low concentration (5 μM) caused both an increase in the germination percentage and an improvement in the mode of development towards a more normal type. An important step in embryo development was the period when embryos appeared insensitive to low, and even to higher, GA_3 concentrations according to the year of experiment, e.g. in 1989, a GA_3 concentration as high as 500 μM did not increase the germination percentage

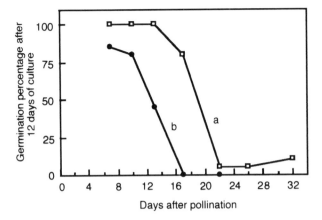

Fig. 15.1. Changes in sunflower embryo sensitivity to ABA (50 μM) throughout embryo development. Embryos were cultured on control medium either (a) directly or (b) after a 5-day pretreatment on exogenous 50 μM ABA.

for 21 DAP embryos. Later, the sensitivity to GAs increased as maturation progressed.

Changes in uptake and metabolism of tritiated hormones

When tritiated hormones were fed to sunflower embryos through their cotyledons, the uptake of both hormones reached maximal values during the same period, between 10 and 20 DAP (Fig. 15.3). Metabolism of the labelled

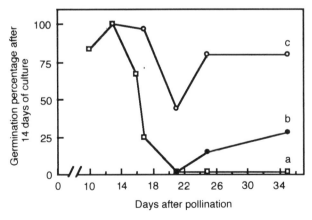

Fig. 15.2. Changes in sunflower embryo sensitivity to GAs throughout embryo development. Embryos were cultured on (a) control medium, (b) medium supplemented with 5 μM GA_3, or (c) medium supplemented with 50 μM GA_3.

exogenous hormones was also studied. Whatever the age of the embryo, the main metabolic products were the glucoside of dihydrophaseic acid for (+)-ABA (Barthe *et al.*, 1993) and GA conjugates for GA_4.

Sensitivity to Hormones During Dormancy Release

Changes in sensitivity to both hormones were also studied during the release of dormancy by different treatments. Slow drying of isolated sunflower embryos carried out for 3 days over saturated salt solutions resulted in a slight increase in the germination percentage; however, dry storage for several weeks was necessary for the complete release of dormancy (Bianco *et al.*, 1994). On the other hand, incubation in agitated water for 3 days allowed dormancy to be released (Le Page-Degivry *et al.*, 1990). If the duration of incubation was reduced to 1 day, the release of dormancy was only partial. The changes in sensitivity to ABA and to GAs reported below were studied in these conditions.

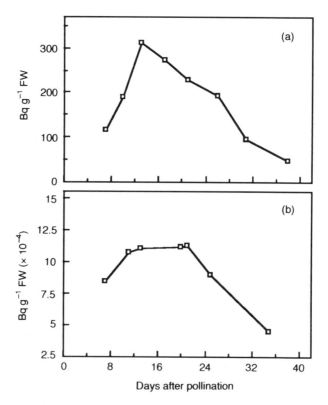

Fig. 15.3. Changes in uptake of (a) [^{3}H]ABA and (b) [^{3}H]GA_4 throughout sunflower embryo development.

Sensitivity to ABA

When dormant sunflower embryos were cultured for 5 or 10 days on a medium containing 5 or 50 μM ABA, no germination occurred after transfer on to a control medium (Fig. 15.4a). However, after a long dry storage leading to the complete release of dormancy, germination was slowed only slightly in the presence of 5 μM ABA (Fig. 15.4b). Even in the presence of 50 μM ABA, inhibition was only partial; as soon as non-germinated embryos were transferred to a control medium, germination occurred. After this dormancy releasing treatment, embryos became completely insensitive to ABA.

Sensitivity to GAs

When GA_3 was applied at a low concentration (5 μM) to the culture medium of freshly harvested dormant (26 DAP) sunflower embryos, germination was not increased compared with the controls. When the same GA_3 concentration

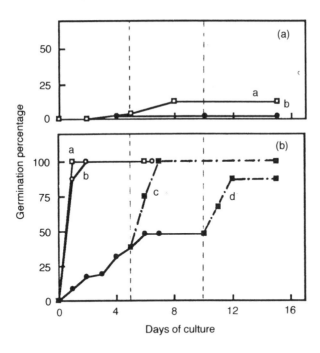

Fig. 15.4. Comparison of sensitivity to ABA between (a) dormant sunflower embryos and (b) embryos in which dormancy was completely released by a long dry-storage period. Embryos were cultured on (a) a control medium, (b) a medium supplemented with 5 μM ABA, or a medium supplemented with 50 μM ABA for either (c) 5 or (d) 10 days before transfer to the control medium.

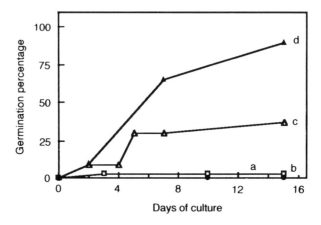

Fig. 15.5. Comparison of sensitivity to GA$_3$ before (a and b) and after (c and d) a 3-day drying treatment that partially released sunflower dormancy. Embryos were cultured on (a) and (c) a control medium or (b) and (d) a medium supplemented with 5 μM GA$_3$.

was applied just at the end of the 3-day drying treatment (that, by itself, led to an only partial release from dormancy), high germination percentages were obtained (Fig. 15.5). Thus, the drying treatment elicited a response by immature embryos to GA. The same low concentration (5 μM) of exogenous GAs became similarly effective after a treatment (24 h incubation in agitated

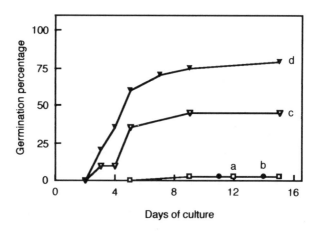

Fig. 15.6. Comparison of sensitivity to GA$_3$ before (a and b) and after (c and d) a 1-day leaching treatment that partially released sunflower dormancy. Embryos were cultured on (a) and (c) a control medium or (b) and (d) a medium supplemented with 5 μM GA$_3$.

water) which, without GA, only partially released embryo dormancy (Fig. 15.6).

ABA Status During Embryo Dormancy Onset and Release

ABA levels

No correlation could be established between the level of ABA in sunflower embryos and their physiological behaviour. During seed development, embryos germinated well at the time when the endogenous ABA level was at its highest (Le Page-Degivry *et al.*, 1990); thereafter, ABA decreased to a low value when embryo dormancy became established. During a 3-day drying treatment that promoted only a partial dormancy release, levels of embryo ABA decreased (Bianco *et al.*, 1994). Dry storage, which led to complete dormancy release, did not induce any additional decline in ABA.

ABA synthesis capacity in the axis

When sunflower embryos were cultured on control medium, their ABA levels decreased dramatically due to catabolism of pre-existent ABA, mainly in cotyledons. When axes were analysed separately after 1 day of culture, their level of ABA had increased. After an application of fluridone, ABA level decreased, showing that in the axis itself, catabolism of pre-existent ABA occurred during culture. Therefore, the comparison between axis ABA content, estimated either in the presence or absence of fluridone after 24 h of culture, allowed the amount of ABA newly synthesized by an axis to be estimated.

During embryo development and the onset of dormancy, maximal ABA synthesis capacity was found in 26 DAP dormant embryos (Table 15.1). This capacity decreased thereafter during late maturation. After a dormancy-breaking treatment (dry storage), no ABA synthesis was detectable (Table 15.2). Thus, there appears to be a very high potential correlation between

Table 15.1. Changes in embryonic axis ABA synthesis capacity during embryo development and the onset of dormancy in sunflower, *Helianthus annuus*.

	ABA (pg axis^{-1})	
	25-day-old embryo	Mature embyro
ABA synthesis during 24 h culture	355	106
After 24 h culture		
Without fluridone	423 ± 42	162 ± 20
With fluridone	68 ± 7	56 ± 5

Table 15.2. Changes in embryonic axis ABA synthesis capacity during dormancy release (by dry storage) in sunflower, *Helianthus annuus*.

	ABA (pg axis^{-1})	
	At time of isolation	After 6 weeks' dry storage
ABA synthesis during 24 h culture	355	18
After 24 h culture		
Without fluridone	423 ± 42	113 ± 12
With fluridone	68 ± 7	96 ± 9

ABA synthesis capacity in the axis and the physiological behaviour of embryos. This suggests that *in situ* ABA synthesis is necessary to impose and maintain embryo dormancy.

Discussion

Changes in sensitivity to hormones appeared strongly correlated with the physiological state of sunflower embryos. Responsiveness to GA$_3$ decreased during onset of dormancy and increased during release of dormancy. Similar behaviour was reported by Derkx and Karssen (1993) in *Arabidopsis*. Moreover, responsiveness to ABA increased during dormancy onset, whereas the opposite occurs during dormancy release. Thus, it appeared that at each step, embryo sensitivity to GAs and to ABA evolved in an opposite way. During seed development on the mother plant, sensitivity to ABA was at its highest and sensitivity to GAs at its lowest, just at the time when dormancy was established. During dormancy release by various treatments, embryos lost their sensitivity to ABA while their sensitivity to GAs increased.

The term 'sensitivity' has been used here in its most general sense, i.e. a change in the observed response to a hormone. Such changes can include the level of perception (receptor levels or affinity), the direct sequence of events in the transduction chain leading to the physiological response and/or any indirect processes that modify the ability of the plant to respond. Changes in the level of a receptor molecule or in its affinity have been a most attractive and sought after hypothesis. Among hormone-insensitive mutants screened for a change in the observed response to a hormone, some (such as mutant D8 in maize or the slender mutants in peas and barley) are considered to be putative GA receptor mutants; however, no definitive proof has been put forward. The enhanced GA sensitivity caused by dry storage in both wild-type and GA-deficient mutant seeds of *Arabidopsis* was proposed by Hilhorst and Karssen (1992) to be correlated with activation of a GA receptor. Indeed, drying might alter membrane composition and structure, which could result

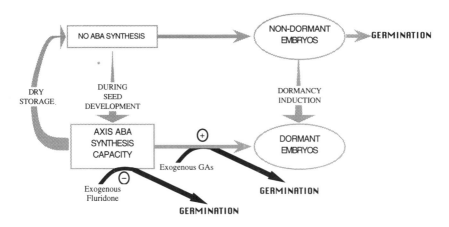

Fig. 15.7. The key role of *in situ* ABA synthesis in the induction and maintenance of sunflower embryo dormancy. (+) At 23°C, in control conditions, ABA synthesis takes place. (−) Application of fluridone inhibits this ABA synthesis.

in structural changes of a membrane-bound or associated protein. Such a hypothesis is difficult to conceptualize in *Helianthus* since the sensitivity to GAs and ABA changed in an opposite way following the same treatment.

However, changes in sensitivity appeared to relate strongly to the physiological state of the embryo. In agreement with the hypothesis proposed by Karssen *et al.* (1983), ABA was demonstrated to play a key role (e.g. see Fig. 15.7) in the induction of dormancy (Le Page-Degivry *et al.*, 1990). Moreover, *in situ* ABA synthesis appeared to be necessary not only to induce, but also to maintain, dormancy (Le Page-Degivry and Garello, 1992). In natural conditions, this synthesis appeared during seed development, along with the onset of dormancy; it was suppressed during dry storage, which led to the release of dormancy. Experimentally, the suppression of this synthesis by fluridone also led to dormancy release. It must be noted that exogenous ABA cannot serve as a substitute for this synthesis. For young, non-dormant embryos in which synthesis is not already induced, exogenous ABA was unable to induce dormancy. Since embryos became sensitive to ABA thereafter, it could be suggested that the sensitivity to ABA was acquired at a given time during development (Hole *et al.*, 1989). However, this is not the case since, when these sensitive embryos were fluridone-treated, the sensitivity to exogenous ABA disappeared, along with ABA synthesis suppression (Le Page-Degivry and Garello, 1992). It therefore appears that ABA responsiveness is correlated directly to the capacity for ABA synthesis.

Research with hormone-deficient mutants of *Arabidopsis* led Karssen and Laçka (1986) to propose that ABA and GA never acted simultaneously during any stage of seed biology, i.e. while ABA would induce dormancy during seed development, GAs would stimulate germination at later stages. The case of sunflower embryo dormancy reported above supports this hypothesis, as far as endogenous hormones are concerned. At any step during dormancy (onset, maintenance, release), ABA is the primary hormone involved; GAs are present at sufficient levels to promote germination as soon as ABA synthesis is suppressed. However, for embryos able to synthesize ABA, a supply of exogenous GAs can induce germination. This GA requirement (i.e. the sensitivity to GAs) depends on the degree of dormancy, which is determined by the level of ABA synthesis capacity. It thus appears that sensitivity to GA is correlated indirectly to ABA synthesis capacity.

In the perspective of a requirement for continued ABA synthesis to maintain embryo dormancy, Morris *et al.* (1991) proposed the hypothesis that dormancy is under positive control, resulting in continued expression of specific ABA-responsive genes. Experiments are now in progress to determine, in *Helianthus*, whether such genes are involved in the processes of dormancy and whether their expression is repressed by GAs.

Acknowledgements

We thank Hélène Le Bris for excellent technical assistance.

References

Armstrong, C., Black, M., Chapman, J.M., Norman, H.A. and Angold, R. (1982) The induction of sensitivity to gibberellin in aleurone tissue of developing wheat grains. I. The effect of dehydration. *Planta* 154, 573–577.

Barthe, P., Hogge, L.R., Abrams, S.R. and Le Page-Degivry, M.T. (1993) Metabolism of (+) abscisic acid to dihydrophaseic acid-4'-β-D-glucopyranoside by sunflower embryos. *Phytochemistry* 34, 645–648.

Bianco, J., Garello, G. and Le Page-Degivry, M.T. (1994) Release of dormancy in sunflower embryos by a dry storage: involvement of gibberellins and abscisic acid. *Seed Science Research* 4, 57–62.

Derkx, M.P.M. and Karssen, C.M. (1993) Effects of light and temperature on seed dormancy and gibberellin-stimulated germination in *Arabidopsis thaliana*: studies with gibberellin-deficient and insensitive mutants. *Physiologia Plantarum* 89, 360–368.

Eisenberg, A.J. and Mascarenhas, J.P. (1985) Abscisic acid and the regulation of synthesis of specific seed proteins and their messenger RNAs during culture of soybean embryos. *Planta* 166, 505–514.

Finkelstein, R.R., De Lisle, A.J., Simon, A.E. and Crouch, M.L. (1987) Role of abscisic acid and restricted water uptake during embryogeny in *Brassica*. In: Fox, J.E. and Jacobs, M. (eds) *Molecular Biology of Plant Growth*. Alan R. Liss, New York, pp. 73–84.

Hilhorst, H.W.M. and Karssen, C.M. (1992) Seed dormancy and germination: the role of abscisic acid and gibberellins and the importance of hormone mutants. *Plant Growth Regulation* 11, 225-238.

Hole, D.J., Smith, J.D. and Cobb, B.G. (1989) Regulation of embryo dormancy by manipulation of abscisic acid in kernels and associated cob tissue of *Zea mays* L. cultured *in vitro*. *Plant Physiology* 91, 101-105.

Karssen, C.M. and Laçka, E. (1986) A revision of the hormone balance theory of seed dormancy: studies on gibberellin and/or abscisic acid-deficient mutants of *Arabidopsis thaliana*. In: Bopp, M. (ed.) *Plant Growth Substances*, Springer-Verlag, Berlin, Heidelberg, pp. 315-323.

Karssen, C.M., Brinkhorst-Van der Swan, D.L.C., Breekland, A.E. and Koornneef, M. (1983) Induction of dormancy during seed development by endogenous abscisic acid: studies on abscisic acid deficient genotypes of *Arabidopsis thaliana* (L.) Heynh. *Planta* 157, 158-165.

Kermode, A.R., Dumbroff, E.B. and Bewley, J.D. (1989) The role of maturation drying in the transition from seed development to germination. VII. Effects of partial and complete desiccation on abscisic acid levels and sensitivity in *Ricinus communis* L. seeds. *Journal of Experimental Botany* 40, 303-313.

Le Page-Degivry, M.T. and Garello, G. (1992) *In situ* abscisic acid synthesis: a requirement for induction of embryo dormancy in *Helianthus annuus*. *Plant Physiology* 98, 1386-1390.

Le Page-Degivry, M.T., Barthe, P. and Garello, G. (1990) Involvement of endogenous abscisic acid in onset and release of *Helianthus annuus* embryo dormancy. *Plant Physiology* 92, 1164-1168.

Morris, C.F., Anderberg, R.J., Goldmark, P.J. and Walker-Simmons, M.K. (1991) Molecular cloning and expression of abscisic acid-responsive genes in embryos of dormant wheat seeds. *Plant Physiology* 95, 814-821.

Reid, J.B. (1990) Phytohormone mutants in plant research. *Journal of Plant Growth Regulators* 9, 97-111.

Roberts, J.A. and Hooley, R. (1988) *Plant Growth Regulators*. Blackie and Son, Glasgow, pp. 49-67.

Trewavas, A.J. (1982) Growth substance sensitivity: the limiting factor in plant development. *Physiologia Plantarum* 55, 60-72.

Walker-Simmons, M.K. (1987) ABA levels and sensitivity in developing wheat embryos of sprouting resistant and susceptible cultivars. *Plant Physiology* 84, 61-66.

16 Processes at the Plasma Membrane and Plasmalemma ATPase during Dormancy

GILLES PÉTEL AND MICHEL GENDRAUD
Unité Associée Bioclimatologie – PIAF (INRA – Université Blaise Pascal), 4 rue Ledru, 63038 Clermont-Ferrand Cedex 01, France

Introduction

Our investigations focus on paradormancy, as defined by Lang *et al.* (1987), or more precisely, the short distance relationships between the bud and its underlying tissue in relation to expression of the bud's morphogenetic potential. It has been shown, using Jerusalem artichoke (*Helianthus tuberosus* L.), that behaviour of the bud depends, at least partially, on the properties of its underlying parenchyma. For instance, parenchyma from dormant tissues exhibits higher sucrose absorption (Gendraud and Lafleuriel, 1983) and cytoplasmic pH (Gendraud, 1981) than that from non-dormant tissue. These observations led to the trophic hypothesis (Gendraud and Pétel, 1990), which proposes that dormant parenchyma acts as a 'nutrient sink', thus inhibiting bud growth. Modification of parenchyma properties implies that parenchyma cell metabolism is modified in relation to dormancy, leading to the regulation of nutrient uptake. Several phenomena involved in these modifications have been identified with different models. This chapter will describe these modifications and establish a primary scheme of succession for cellular events.

Plant Models

Original investigations were made with mature, dormant Jerusalem artichoke tubers that were harvested in October. Tubers stored in the dark at 15°C remained dormant, while those stored at 4°C for 16 weeks became non-dormant (Courduroux, 1967). This model allowed comparisons of dormant and non-dormant situations and investigation of cellular properties during the breaking of dormancy. Other studies were made with filiate tubers grown *in vitro* in a neutral medium, which were then submitted to ethanol vapours, for 24 h in the dark at 24°C (Courduroux, 1967). In these conditions, the apical bud elongated within 24 h after ethanol treatment. Moreover, the growing period never exceeded 5 days after the end of ethanol exposure. This means that ethanol mimics a dormancy-breaking and induction cycle in a very short time, which constitutes a good model for experimentation (Pétel *et al.*, 1993). Other results have been obtained using a peach tree model (*Prunus persica* L.) described extensively by Rageau *et al.* (1994).

Cellular Modifications in Underlying Tissues

In experimental conditions, the first event noted during treatments that lead to the breaking of dormancy is a modification of plasmalemma fluidity. During cold treatment of tubers, a decrease of fluidity was noted by the third week of cold storage (Fig. 16.1) (Pétel *et al.*, 1992b). This change could be linked to a decrease of the phosphatidylcholine/phosphatidylethanolamine (PC/PE) ratio at the plasmalemma (Fig. 16.1). A similar early modification of membrane viscosity was observed in dormant filiate tubers 6 h after the beginning of ethanol exposure (Pétel *et al.*, 1993), before bud elongation was observable.

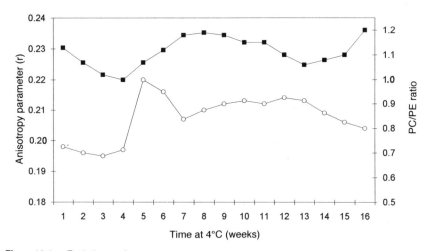

Fig. 16.1. Evolution of plasma membrane fluidity (o) and phosphatidylcholine/phosphatidylethanolamine (PC/PE) ratio (■) of Jerusalem artichoke tuber parenchyma cells during dormancy release induced by cold treatment. (From Pétel *et al.*, 1992a.)

Modification of membrane fluidity is known to influence the activity of membrane proteins. This has been demonstrated mainly for plasmalemma adenosine triphosphatase (ATPase) in various plant species, such as *Citrus* (Douglas and Walker, 1984), mung bean (Kasamo and Noutchi, 1987; Kasamo and Yamanishi, 1991), oat (Palmgren *et al.*, 1988) and maize (Braueur *et al.*, 1989). In mature Jerusalem artichoke tubers, activities of two bound proteins were measured during cold treatment of tubers: plasmalemma ATPase (Pétel *et al.*, 1992a) and nicotinamide adenine dinucleotide (NADH) dehydrogenase (Pétel *et al.*, 1992b). Both activities decreased significantly after 6 or 7 weeks of cold exposure (Fig. 16.2), i.e. after the noted modification of membrane fluidity. In the filiate tuber model, no enzymatic activities could be measured because of insufficient sample size. Indirect evidence of plasmalemma ATPase activity was achieved by measuring the cytosolic adenosine triphosphate (ATP) content of parenchyma cells (Pétel *et al.*, 1993). Indeed, it was demonstrated previously that cytoplasmic ATP content was linked to plasmalemma ATPase activity (Gendraud and Lafleuriel, 1984); in the presence of fusicoccin, a stimulator of plasma membrane ATPase, ATP efflux out of parenchyma cells was decreased after poly-L-lysine treatment. Evolution of cytosolic ATP content of filiate tuber parenchyma cells is shown in Fig. 16.3. ATP content increases 12 h after ethanol treatment ends, i.e. after the noted modification in membrane fluidity.

The above results indicate that dormant parenchyma cells exhibit a high plasmalemma ATPase activity, which leads to the generation of an important proton-motive force, involving proton extrusion by this enzyme (Pétel and

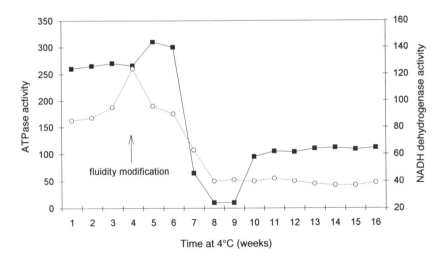

Fig. 16.2. Evolution of plasmalemma ATPase (■) and NADH dehydrogenase (○) activities of Jerusalem artichoke tuber parenchyma cells during dormancy release induced by cold treatment. (from Pétel *et al.*, 1992a, b.)

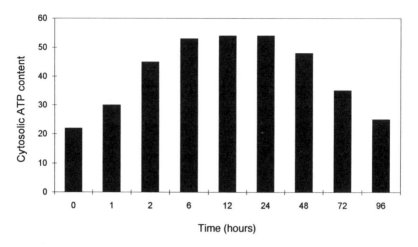

Fig. 16.3. Evolution of cytosolic ATP content in parenchyma cells from dormant filiate tubers after ethanol exposure. Time 0 represents the ATP content of dormant material. (From Pétel et al., 1993.)

Gendraud, 1986, 1993). Similar observations have been made concerning NADH dehydrogenase activity (Pétel and Gendraud, 1987). As nutrient, especially sugar, uptake into plant cells depends on proton extrusion, results indicate that nutrient uptake potentials are higher in parenchyma cells from dormant, compared with non-dormant, tubers (Bush, 1989; Lemoine and Delrot, 1989). In other words, dormant parenchyma would represent a 'nutrient sink', preventing nutrient fluxes from reaching the bud and thus inhibiting its growth.

Several phenomena can be involved in regulation of plasmalemma AT Pase activity. The *in situ* localization of ATPase in dormant and non-dormant tuber parenchyma has been investigated (Chaubron *et al.*, 1994). Immunolabelling of plasma membrane ATPase was performed on transverse sections of parenchyma (Fig. 16.4A, B), using plasmalemma ATPase antibodies from *Arabidopsis thaliana* (gift from R. Serrano, University of Valencia, Spain). A greater immunological reaction was observed on dormant parenchyma. Similar results were also obtained in dot–blot experiments using the same plasma membrane ATPase antibodies (Fig. 16.4C). According to these results, it appears that the breaking of dormancy in Jerusalem artichoke is linked to a modification of the amount of ATPase in parenchyma cells. Similar observations were made by Hase (1993), who investigated callus induction from Jerusalem artichoke tuber tissue. Hase's result was interpreted as a stimulation of nutrient transport processes needed for callus induction, i.e. cells became a 'nutrient sink', exhibiting higher plasmalemma ATPase amount and activity. This peculiar situation can be compared to that

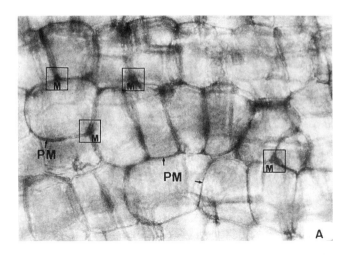
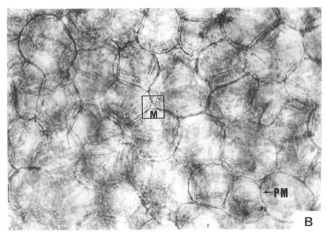
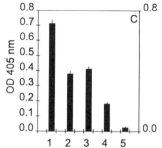

Fig. 16.4. Cytolocalization of plasma membrane ATPase in parenchyma cells from dormant (A) and non-dormant (B) Jerusalem artichoke tubers (M: meatus; PM: plasma membrane). Dot blot assays (C) were performed using crude plasmalemma fractions from dormant (lane 1, 40 μg protein) and non-dormant (lane 2, 40 μg protein) tubers and the corresponding solubilized fractions (lane 3, dormant material, 10 μg protein; lane 4, non-dormant material, 10 μg protein) (lane 5: control). OD = optical density. (From Chaubron et al., 1994.)

Table 16.1. Michaelis constant of plasmalemma ATPase from dormant and non-dormant tubers parenchyma cells. (From Pétel and Gendraud, 1988.)

	Membranes		Solubilized enzyme	
	Dormant	Non-dormant	Dormant	Non-dormant
V_{max}	444.0	185.0	1250.0	500.0
K_m	1.25	0.65	0.36	0.19

of dormant tubers, wherein parenchyma cells represent a 'nutrient sink' that inhibits bud growth.

Other significant results were obtained in studies of the biochemical properties of plasma membrane ATPase from dormant and non-dormant tuber parenchyma (Pétel and Gendraud, 1988). Higher maximum ATP hydrolysis rate (V_{max}) and Michaelis constant (K_m) values were found in preparations from dormant tubers (Table 16.1), suggesting that the enzymes differ between dormant and non-dormant materials. This is likely to be correlated with the appearance of new messenger ribonucleic acid (mRNA) during dormancy-breaking cold treatment of tubers (Mussigman, 1988).

Consequences for Nutrient Flux Orientation

As described previously, ATP hydrolysis by plasmalemma ATPase leads to proton extrusion out of the cell. Consequences of this activity are: (i) an alkalization of these cells, linked to the enzyme activity (Pétel *et al.*, 1992a); and (ii) the formation of a transmembrane pH gradient, enabling H^+-sucrose cotransporters to work. Indirect evidence was obtained on Jerusalem artichoke and peach tree models. Concerning Jerusalem artichoke, it was previously observed that intracytoplasmic penetration of sucrose and absorption of the lipophilic cation tetraphenylphosphonium (TPP^+) were higher in dormant parenchyma, compared with non-dormant parenchyma (Gendraud and Lafleuriel, 1983). The latter parameter (TPP^+ absorption) was increased by fusicoccin, a plasmalemma ATPase stimulator.

In the peach tree model, estimation of active sucrose absorption was performed on the vegetative bud and its underlying tissue (defined here as the 'cushion'), after H^+-solute cotransporter characterization (Marquat and Pétel, 1994). Dormant samples exhibited a higher active sucrose absorption in the cushion, compared with the bud (Fig. 16.5), whereas the contrary was observed on non-dormant samples (higher absorption in the bud than in the cushion). These results are correlated to a cytoplasmic acidification of cushion cells during dormancy-breaking (Rageau *et al.*, 1994). In both Jerusalem artichoke and peach, it appears that bud growth can only be expressed when nutrient absorption capabilities of underlying tissues are

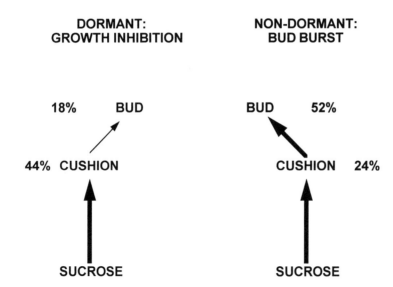

Fig. 16.5. Estimation of active sucrose absorption in bud and subtending tissue (cushion) from peach tree single node cuttings, expressed in percentage. (From Pétel et al., 1994.)

lowered. In dormant samples, underlying tissues can be considered as a 'nutrient sink', preventing bud growth. Dormancy-breaking is so characterized, at least partially, by the reacquisition of 'nutrient source' possibilities in the tissues underlying buds.

Conclusions

Figure 16.6 summarizes the known events that occur during the breaking of dormancy by cold treatment. The first event, in our experimental conditions, is a modification of plasmalemma fluidity. This could be due to physical alteration of the membrane by specific environmental parameters, such as low temperature or ethanol vapours, or to modification of phospholipid synthesis. Next, plasmalemma ATPase activity decreases, which could be caused by the changes in membrane fluidity evolution or new synthesis of this protein. Consequences of these changes are an enhancement of cytosolic ATP content, cytoplasmic acidification induced by the lowering of proton extrusion and thus a decrease of the transmembrane pH gradient. The final step is a decrease of H^+-solute cotransporter activities in the tissue underlying buds, which promotes nutrient flux to the bud tissues and thereby enables sprout generation. This scheme is developed from the results discussed above, using Jerusalem artichoke tubers, as well as those from the breaking of dormancy in peach vegetative buds (Rageau et al., 1994).

Unfortunately, these models are less suitable for investigation of dormancy induction, which is best addressed by the exposure of *in vitro* filiate tubers to ethanol vapours. Indeed, in these experimental conditions, bud growth is only temporary and tuberization of the generated sprout is observed 96 h after the end of ethanol treatment, mimicking dormancy induction. Moreover, the events associated with the transient breaking of dormancy in this model fit well with those observed on mature tubers: membrane fluidity modification, increase of cytosolic ATP content and cytoplasmic acidification. Conversely, during the reacquisition of dormant characteristics, the following events (Fig. 16.7) were noted: (i) membrane viscosity is enhanced; (ii) cytosolic ATP content decreases, which could be correlated to an enhancement of plasmalemma ATPase activity, as described previously; and (iii) parenchyma cell alkalization occurs. This latter observation could be interpreted as an incremental change in the transmembrane pH gradient, which leads to 'sink' characteristics in filiate tuber parenchyma cells.

The sum of the observations related here indicate that the specific short distance relationships involved, at least partially, in dormancy induction and release are linked to the 'sink strength' of tissues which subtend the buds.

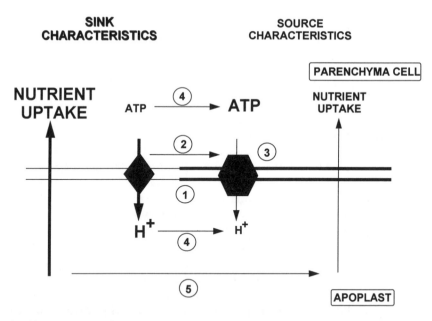

Fig. 16.6. Summary of the cellular events that occur in parenchyma during dormancy release by low temperature. Numbers represent event chronology: (1) change in membrane fluidity; (2) modification of plasmalemma ATPase characteristics and amount; (3) decrease of ATPase activity; (4) increase of cytosolic ATP content and cytoplasmic acidification; (5) decrease of nutrient uptake potential.

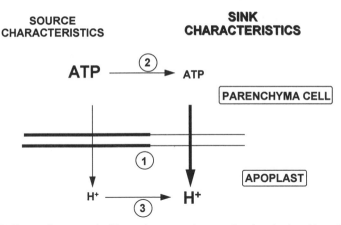

Fig. 16.7. Events that occur in filiate tuber parenchyma cells after the breaking of transient dormancy induced by ethanol exposure. Numbers represent event chronology: (1) change in membrane fluidity; (2) decrease of cytosolic ATP content; (3) cytoplasmic alkalization.

'Sink' or 'source' characteristics are caused by several cellular modifications, which are induced by certain environmental conditions. Their very first target is still unknown, but modification of gene expression during the period of bud dormancy is suspected.

References

Braueur, D., Tu, S.I., Hsu, A.F. and Thomas, C.E. (1989) Kinetic analysis of proton transport by the vanadate-sensitive ATPase from maize root microsomes. *Plant Physiology* 89, 464-471.

Bush, D.R. (1989) Proton-coupled sucrose transport in plasmalemma vesicles isolated from sugar beet (*Beta vulgaris* L. cv. Great Western) leaves. *Plant Physiology* 89, 1318-1323.

Chaubron, F., Robert, F., Gendraud, M. and Pétel, G. (1994) Partial purification and immunocharacterization of the plasma membrane ATPase of Jerusalem artichoke (*Helianthus tuberosus* L.) tubers in relation to dormancy. *Plant Cell Physiology* 35, 1179-1184.

Courduroux, J.C. (1967) Etude du mécanisme physiologique de la tubérisation chez le topinambour (*Helianthus tuberosus* L.). *Annales des Sciences Naturelles, Botanique* 8, 215-356.

Douglas, J.T. and Walker, R.R. (1984) Phospholipids, free sterols and adenosine triphosphatase of plasma membrane-enriched preparations from roots of *Citrus* genotypes differing in chloride exclusion ability. *Physiologia Plantarum* 62, 51-58.

Gendraud, M. (1981) Etude de quelques propriétés des parenchymes de topinambour cultivés *in vitro* en relation avec leurs propriétés morphogénétiques. *Physiologie Végétale* 19, 473-481.

Gendraud, M. and Lafleuriel, J. (1983) Caractéristiques de l'absorption du saccharose et du tétraphénylphosphonium par les parenchymes de topinambour, dormants et non-dormants, cultivés *in vitro*. *Physiologie végétale* 21, 1125-1133.

Gendraud, M. and Lafleuriel, J. (1984) Intracellular compartmentation of ATP in dormant and non-dormant tubers of Jerusalem artichoke (*Helianthus tuberosus* L.) grown *in vitro*. *Journal of Plant Physiology* 118, 251-258.

Gendraud, M. and Pétel, G. (1990) Modifications in intercellular communications, cellular characteristics and change in morphogenetic potentialities of Jerusalem artichoke (*Helianthus tuberosus* L.). In: Millet, B. and Greppin, H. (eds) *Intra- and Intercellular Communications in Higher Plants: Reception, Transmission, Storage and Expression of Messages*. INRA, Paris, pp. 170-175.

Hase, A. (1993) Changes in protein composition and the H^+-ATPase activity of the plasma membrane during induction of callus from tuber tissues of Jerusalem artichoke. *Plant Cell Physiology* 34, 67-74.

Kasamo, K. and Noutchi, I. (1987) The role of phospholipids in plasma membrane ATPase activity in *Vigna radiata* (mung bean) roots and hypocotyl. *Plant Physiology* 83, 323-328.

Kasamo, K. and Yamanishi, H. (1991) Functional reconstitution of plasma membrane H^+-ATPase from mung bean (*Vigna radiata* L.) hypocotyls in liposomes prepared with various molecular species of phospholipids. *Plant Cell Physiology* 32, 1219-1225.

Lang, G.A., Early, J.D., Martin, G.C. and Darnell, R.L. (1987) Endo-, para- and eco-dormancy: physiological terminology and classification for dormancy research. *HortScience* 22, 371-377.

Lemoine, R. and Delrot, S. (1989) Proton-motive force-driven sucrose uptake in sugar beet plasma membrane vesicles. *FEBS Letters* 249, 129-133.

Marquat, C. and Pétel, G. (1994) Involvement of H^+-solute co-transport in peach tree vegetative buds dormancy. *First International Symposium on Plant Dormancy*, Corvallis (Oregon), USA, August 1994, p. 164.

Mussigman, C. (1988) Aspects de l'expression génomique en relation avec les potentialités morphogénétiques de tubercules de topinambour (*Helianthus tuberosus* L.). PhD thesis, University of Clermont-Ferrand.

Palmgren, M.G., Sommarin, M., Ulvskov, P. and Jorgensen, P.L. (1988) Modulation of plasma membrane H^+-ATPase from oat roots by lysophosphatidylcholine, free fatty acids and phospholipase A2. *Physiologia Plantarum* 74, 11-19.

Pétel, G. and Gendraud, M. (1986) Contribution to the study of ATPase activity in plasmalemma enriched fractions from Jerusalem artichoke tubers (*Helianthus tuberosus* L.) in relation to their morphogenetic properties. *Journal of Plant Physiology* 123, 373-380.

Pétel, G. and Gendraud, M. (1987) Activité d'un système oxydoréducteur au niveau du plasmalemme des cellules de parenchyme de topinambour (*Helianthus tuberosus* L.) en relation avec les propriétés morphogénétiques des tubercules. *Comptes Rendus de l'Académie des Sciences de Paris* 305, 51-54.

Pétel, G. and Gendraud, M. (1988) Biochemical properties of the plasmalemma ATPase of Jerusalem artichoke tubers (*Helianthus tuberosus* L.) in relation to dormancy. *Plant Cell Physiology* 29, 739-741.

Pétel, G. and Gendraud, M. (1993) ATP- and NADH-dependent membrane potential generation in plasmalemma enriched vesicles from parenchyma of dormant and non-dormant Jerusalem artichoke tubers. *Biologia Plantarum* 35, 161-167.

Pétel, G., Lafleuriel, J., Dauphin, G. and Gendraud, M. (1992a) Cytoplasmic pH and plasmalemma ATPase activity of parenchyma cells during the release of dormancy of Jerusalem artichoke tubers. *Plant Physiology and Biochemistry* 30, 379-382.

Pétel, G., Sueldo, R., Coudret, A. and Gendraud, M. (1992b) Plasmalemma fluidity in parenchyma cells from Jerusalem artichoke (*Helianthus tuberosus* L.) tubers during the break of dormancy. *Biologia Plantarum* 34, 373-380.

Pétel, G., Candelier, P. and Gendraud, M. (1993) Effect of ethanol on filiate tubers of Jerusalem artichoke: a new tool to study dormancy. *Plant Physiology and Biochemistry* 31, 67-71.

Pétel, G., Marquat, C. and Gendraud, M. (1994) Some aspects of bud growth inhibition related to dormancy. *Life Science Advances - Plant Physiology* 13, 279-285.

Rageau, R., Laroche, A., Pétel, G., Gendraud, M., Bonhomme, M. and Balandier, P. (1994) Simultaneous use of biological and biochemical methods for characterization of bud dormancy: application to cold deprived peach trees. *Abstracts, First International Symposium on Plant Dormancy*, Corvallis (Oregon), USA, August 1994.

17 Carbohydrate Metabolism as a Physiological Regulator of Seed Dormancy

Michael E. Foley
Purdue University, Department of Botany and Plant Pathology, West Lafayette, IN 47907-1155, USA

Introduction

Dormancy in wild oat caryopses (*Avena fatua* L.) has been proposed to involve two restrictions in sugar metabolism (Naylor and Simpson, 1961). This idea arose following the observation that exogenous gibberellic acid (GA) promoted the germination of dormant (D) seeds (Lona, 1956), including those of wild oat (Black and Naylor, 1959). The first restriction was thought to be a block in sugar production in the endosperm; the second was a block in utilization of sugar by the embryo. Naylor and Simpson (1961) demonstrated that exogenous GA overcame both blocks in highly D wild oat caryopses. They also noted that the restriction of sugar production had a 1000-fold higher requirement for GA than the requirement to overcome the restriction in sugar utilization. It was demonstrated early on that the GA-induced production of low-molecular-weight sugars from wild oat endosperm reserves was a postgermination event not involved with the dormancy-breaking activity of GA (Drennan and Berrie, 1961). Other indirect, sometimes conflicting, evidence for the restriction of sugar metabolism has been reported (Simpson and Naylor, 1962; Simpson 1965, 1966; Chen and Varner, 1969). However, further investigation of sugar metabolism restrictions waned as testing of a new hypothesis, that dormancy breakage may be connected to increased activity of the pentose phosphate pathway, began (Simmons and Simpson, 1971).

Recently, it has been shown that highly D excised embryos (from the M73 genetic line of wild oat) display true embryo dormancy and that their germination can be induced with several soluble sugars, e.g. fructose (Fru) (Foley, 1992). Excised embryos from after-ripened (AR) (under warm, dry conditions) M73 caryopses have a rapid onset and rate of germination in the absence of exogenous soluble sugars (Foley et al., 1992a). Based on these observations, we hypothesize that there is a restriction in sugar production, but not utilization, in D embryos and after-ripening overcomes this restriction. Although Naylor and Simpson (1961) proposed that breaking wild oat seed dormancy by after-ripening was controlled through changes in the content of an endogenous inhibitor that antagonized GA, little evidence exists to support this particular mechanism.

Steady-state levels of non-structural carbohydrates in imbibed D and AR excised wild oat embryos, as well as embryos from intact caryopses, were measured to ascertain the profile of carbohydrates in D embryos and determine how after-ripening changes those profiles. Similar experiments were conducted using various combinations of Fru and/or GA to determine their effect on carbohydrate profiles in D and AR embryos (Foley et al., 1992a, 1993; Nichols et al., 1993). The results of these experiments suggest that after-ripening overcomes a restriction in raffinose (Raf) family oligosaccharide metabolism, which is a prelude to normal germination. In contrast, Fru and other soluble sugars do not break dormancy *per se*; rather, they circumvent the block to germination in D embryos. Although it is an apparent paradox, soluble sugar-induced germination (in D embryos) or GA-induced germination (in D caryopses) is, in effect, germination of a D embryo. After-ripening is the only treatment that can break dormancy, i.e. remove the restriction in carbohydrate metabolism.

Changes in carbohydrate metabolism have been correlated with the termination of developmental arrest in other plant systems. Induction of loblolly pine seed germination by cold stratification may result from a change in the ability of the embryo to absorb and metabolize sucrose (Suc) (Carpita et al., 1983). In stratified sugar pine embryos, an increase in Suc levels prior to germination coincides with decreased levels of stachyose (Stach) (Murphy and Hammer, 1988). Levels of Raf and Stach, which are correlated with bud hardiness as it relates to freezing-induced drought stress, decrease as buds deacclimatize prior to growth (Lasheen and Chaplin, 1977; Flinn and Ashworth, 1995). Temperatures at which dormancy ends and bud development proceeds in *Betula pendula* lead to a decline of Raf and Stach levels to less than 5 and 18% of their original levels, respectively (Sauter and Ambrosius, 1986). Germination of non-dormant *Spirodela polyrrhiza* turions is the same with or without exogenous soluble sugars. In contrast, germination of D turions can be induced with exogenous glucose (Glc), Suc and maltose (Mal) (Xylander et al., 1992). A preliminary report by Lehle et al. (1978) indicates that normal germination of D excised *Setaria lutescens* embryos on tissue culture medium requires Suc, Glc, Mal or Fru.

Preliminary pulse-chase experiments to investigate after-ripening-induced changes in Raf family oligosaccharide metabolism indicated that ^{14}C-Fru was incorporated into Raf in D and AR embryos and Stach in D embryos (Nichols *et al.*, 1993). The label moved out of Raf in AR, but not D, embryos between 12 and 18 h after imbibition. The levels of galactose (Gal) in the embryo of AR caryopses decreased significantly with the onset of germination (between 18 and 24 h), whereas the level of Gal remained relatively high in the embryo of D caryopses. Germination of AR, but not D, excised embryos on media containing Raf or Gal suggests that after-ripening affects the embryo's capacity to metabolize these sugars. It is apparent that Gal does not induce germination of D embryos like Fru, Glc, Suc and Mal. The inability of D embryos to utilize Raf family oligosaccharides in the same manner as AR embryos could be due to a block in *in vivo* α-galactosidase activity, or a restriction at a metabolic step at some point further along the pathway, i.e. Gal metabolism.

To resolve further whether the restriction to germination in D embryos is associated with carbohydrate metabolism, we investigated the effect of 2-deoxy-D-glucose (d-Glc), a non-metabolic analogue of Glc, on D and AR wild oat embryo germination, and measured the levels of some intermediates in the synthesis and breakdown of Raf in D and AR wild oat embryos.

Materials and Methods

Seed from the genetically pure inbred line M73 wild oat (*Avena fatua* L.) were collected, maintained and surface-sterilized, as described previously (Foley *et al.*, 1993). Excised embryos for germination studies and carbohydrate determinations were prepared by surface-sterilizing D or AR caryopses and imbibing them in water at 16°C for 3 to 4 h prior to embryo excision. Embryos were excised from caryopses under sterile conditions and transferred to Petri dishes which contained sterile N6 tissue culture medium solidified with 0.25% (w/v) gellum (Foley, 1992). There were three or four replications per treatment with 10 and 20 embryos per replication for inhibitor studies and carbohydrate determinations, respectively.

Treatments to determine the effect of d-Glc on germination were: D or AR embryos incubated for 6 days on medium containing no soluble sugars; D or AR embryos incubated for 6 days on medium containing 88 mM d-Glc; D embryos incubated for 6 days on medium with 88 mM d-Glc plus 10 μM GA; D embryos incubated for 3 days on medium containing no soluble sugars or with 88 mM d-Glc, followed by 3 days on medium containing 88 mM Fru plus 10 μM GA; and D embryos incubated for 3 days on medium containing 88 mM d-Glc plus 10 μM GA, followed by 3 days on medium containing 88 mM d-Glc plus 10 μM GA and 88 mM Fru. Petri dishes were maintained at 16°C in the dark in boxes lined with wet paper towel to maintain high relative humidity. The embryos were evaluated daily for germination, i.e. protrusion of the coleorhiza through the testa.

Carbohydrates were determined in excised embryos after incubation on solidified N6 medium amended with or without 88 mM Fru. The three treatments for excised embryos were: D embryos cultured for 36 h with 88 mM Fru (~ 20% germination); D embryos cultured for 36 h without Fru (0% germination); and AR embryos cultured for 18 h without Fru (~ 20% germination). The three treatments for intact caryopses (imbibed on germination buffer) were: AR for 18 h (12% germination); AR for 36 h (92% germination); and D for 36 h (0% germination). Carbohydrates were extracted using a modification of the procedures of Kanabus *et al.* (1986). Embryos were frozen at $-80°C$, homogenized in a microcentrifuge tube containing 600 μl of cold 10% trichloroacetic acid (TCA) (w/v) with a hand-held tissue homogenizer for 2 min at full speed. Following 10 min of incubation at $4°C$ with intermittent agitation, homogenates were microcentrifuged at $4°C$ for 5 min at 13,000 rpm. The supernatants were neutralized by shaking for 1 min with 800 μl of a mixture of 1,1,2-trichlorotrifluoroethane (Freon) and *n*-trioctylamine (3:1,v/v). Suspensions were microcentrifuged as above, and the upper aqueous phases were collected for analysis. In some experiments, standards were added to the TCA solutions used for extraction to assess recovery. Soluble sugars and sugar alcohols in embryos from intact caryopses were extracted with hot 80% ethanol as described previously (Nichols *et al.*, 1993).

HPLC analytical methods for sugar alcohols are similar to those described by Foley (1992) for soluble sugars. Briefly, analyses were performed using a Dionex BioLC Gradient Pump Module HPLC with pulsed amperometric detector (PAD). Modifications were as follows. Sample extracts were frozen at $-80°C$, lyophilized and resuspended in 1 ml of HPLC-grade water. Analysis was carried out utilizing an isocratic elution with 5 mM NaOH. PAD gold working electrode settings were $E1 = 0.05$ ($t1 = 480$ ms); $E2 = 0.70$ ($t2 = 300$ ms); $E3 = -0.70$ ($t3 = 240$ ms). Retention times and response factors were established from commercially available standards and the linearity of the response was checked at 250 pM to 20 nM concentration range for a detector sensitivity of 1 μA. HPLC analytical methods for sugar phosphates were modified from those described above in that analysis was carried out with a 100 mM NaOH isocratic elution against a linear gradient of Na acetate from 100 to 500 mM over 50 min of elution.

Results

The Glc analogue d-Glc is taken up by wild oat embryos but does not induce germination of D excised embryos, with or without GA (Table 17.1). It inhibits both germination of AR embryos and the induction of D embryo germination by Fru plus GA. The inhibitory effect of d-Glc on induction of D embryo germination is reversible. Inhibition of germination by d-Glc supports

Table 17.1. Germination of dormant and after-ripened M73 excised wild oat embryos cultured at 16°C on N6 medium with (+) and without (−) various combinations of 88 mM fructose (Fru), 88 mM 2-deoxy-D-glucose (d-Glc) and/or 10 μM gibberellic acid (GA). Amendments in the medium are in parentheses. The arrow (→) indicates that, after 3 days of incubation, embryos were moved to media with different amendments.

	Days of germination					
	1	2	3	4	5	6
Treatment	Germination (%)					
After-ripened (−Fru, −GA)	90	100	100	100	100	100
After-ripened (+d-Glc)	0	0	0	0	0	0
Dormant (−Fru, −GA)	0	0	0	0	0	0
Dormant (+d-Glc)	0	0	0	0	0	0
Dormant (+d-Glc, +GA)	0	0	0	0	0	0
Dormant (+d-Glc)→(+Fru, +GA)	0	0	0	40	87	97
Dormant (−Fru, −GA)→(+Fru, +GA)	0	0	0	77	100	100
Dormant (+d-Glc, +GA)→ (+d-Glc, +GA, +Fru)	0	0	0	0	7	7

our hypothesis that carbohydrate metabolism plays a key role in induction of germination in both AR and D embryos.

The steady-state levels of soluble sugars, sugar phosphates and sugar alcohols in the embryo of imbibed D (36 h) and AR (18 and 36 h) caryopses, and in D (36 h), D Fru-induced (36 h) and AR (18 h) excised embryos were determined. For excised embryos, carbohydrate levels were determined shortly after the onset of after-ripening and Fru-induced germination (Figs. 17.1 and 17.2). Dormant excised embryos cultured without Fru for 36 h served as a control. Although D Fru-induced and AR embryos are in the same stage of germination (∼ 20%), the steady-state levels of soluble sugars and intermediates, with the exception of glucose-1-phosphate (Glc-1-P) and Raf, are all significantly different. A comparison of the steady-state levels between D and D Fru-induced excised embryos at 36 h indicates that galactose-1-phosphate (Gal-1-P), galactinol (Gal-ol), Glc, Fru and Suc levels are significantly different, whereas the levels of Raf and Stach are similar. This supports a previous report that induction of D embryos with Fru has a minor effect on the levels of Raf family oligosaccharides in D embryos (Foley *et al.*, 1992a, 1993). A comparison of D (36 h) and AR (18 h) excised embryos indicates that steady-state levels of Glc-1-P, Gal-ol and Stach are significantly different.

Carbohydrate levels in the embryo of AR intact caryopses were determined at the onset (18 h) and near the completion (36 h) of germination (Figs

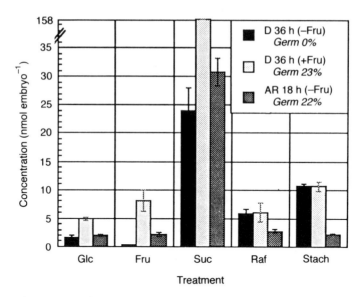

Fig. 17.1. Concentration of soluble sugars in dormant (D) and after-ripened (AR) excised M73 wild oat embryos cultured on N6 medium amended with (+) and without (−) 88 mM Fru. The duration of incubation and germination are indicated in the key. Vertical bars represent the SE (0.05%).

17.3 and 17.4). Embryos from D caryopses at 36 h serve as a control. Germination of AR caryopses led to significant differences in the steady-state levels of glucose-6-phosphate (Glc-6-P), Gal-ol, Glc, Raf and Stach. A comparison of D and AR embryos at 36 h indicates significant changes in the steady state levels of Glc-6-P, fructose-6-phosphate (Fru-6-P), Gal-ol, Gal, Glc, Fru, Raf and Stach.

Discussion

The above changes in steady-state levels of soluble sugars support previous observations (Foley *et al.*, 1992a, 1993; Nichols *et al.*, 1993). Differences in the absolute levels of some soluble sugars were probably due to the use of different extraction procedures. While changes in carbohydrate metabolism could not be evaluated directly from steady-state measurements, several observations indicated metabolic differences. Steady-state levels of soluble sugars and intermediates were very different in D Fru-induced and AR excised embryos. There appear to be significant differences in carbohydrate metabolism depending on the manner of germination induction, i.e. after-ripening vs. Fru. These observations support our proposal that after-ripening removes the restriction to germination and Fru circumvents it.

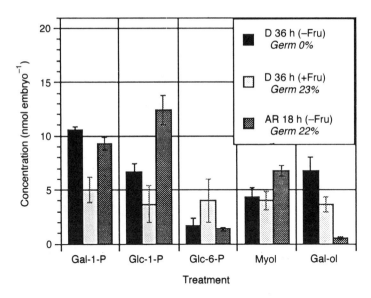

Fig. 17.2. Concentration of sugar phosphates and sugar alcohols in dormant (D) and after-ripened (AR) excised M73 wild oat embryos cultured on N6 medium amended with (+) and without (−) 88 mM Fru. The duration of incubation and germination are indicated in the key. Vertical bars represent the SE (0.05%).

The steady-state levels of Gal-ol, a key intermediate in Raf family oligosaccharide synthesis, follow the same trend as do Raf family oligosaccharides in D, D Fru-induced and AR embryos. Since it has been proposed that the restriction to germination involves Raf family oligosaccharide metabolism, it is not surprising that various germination induction treatments have different effects on the level of Gal-ol. As with Raf and Stach (Foley et al., 1992a, 1993; Nichols et al., 1993), changes in steady-state levels of Gal-ol occurred prior to and coincident with germination (Figs. 17.2 and 17.4). These changes were not associated with germination *per se* because Fru-induced germination of D embryos does not lower Gal-ol levels in the same way as AR-induced germination. Changes in the levels of Gal-1-P and myoinositol (MyoI), which are also involved directly with metabolism of Raf family oligosaccharides, showed no obvious trends in relation to induction of germination. Unlike Gal-ol, which is involved primarily in synthesis of Raf and Stach, these compounds are involved in both synthesis and breakdown of Raf family oligosaccharides. This complicates potential inferences about their metabolism based on steady-state measurements.

Previous evidence in support of the hypothesis that a restriction in Raf family oligosaccharide metabolism, perhaps related to Gal metabolism, inhibits germination of D wild oat embryos is as follows: the steady-state levels

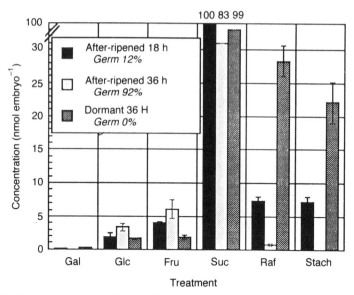

Fig. 17.3. Concentration of soluble sugar in the embryo of dormant and after-ripened intact M73 wild oat caryopses. The duration of incubation and germination are indicated in the key. Levels of Suc are indicated above the graph. Vertical bars represent the SE (0.05%).

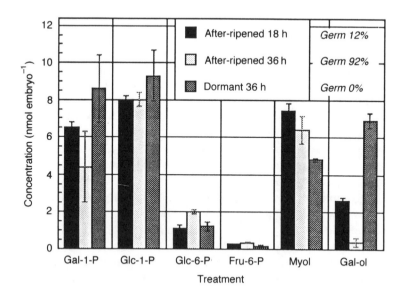

Fig. 17.4. Concentration of sugar phosphates and sugar alcohols in the embryo of dormant and after-ripened intact M73 wild oat caryopses. The duration of incubation and germination are indicated in the key. Vertical bars represent the SE (0.05%).

of Raf family oligosaccharides and Gal in the embryo of AR caryopses decline to nearly undetectable levels prior to and coincident with the onset of germination (Nichols *et al.*, 1993). In contrast, the level of Raf family oligosaccharides remains constant for at least 7 days in the embryo of D caryopses (Nichols *et al.*, 1993). Raf family oligosaccharide levels in the embryo also remain constant when 7-day-imbibed D caryopses are subjected to a dehydration/rehydration treatment (M.E. Foley, unpublished data). Raf family oligosaccharide and Gal levels in the embryo of D GA-induced caryopses remain constant prior to and coincident with the induction of germination, and they remain at moderate levels even after D GA-induced caryopses are 100% germinated (Foley *et al.*, 1993). These data demonstrate that Raf family oligosaccharide metabolism is not a result of germination *per se*. Pulse-chase experiments demonstrate that D and AR embryos metabolize Raf family oligosaccharides in a different manner well before the onset of visual germination (Nichols *et al.*, 1993). After-ripened, but not D, embryos germinate on N6 medium amended with Gal (Nichols *et al.*, 1993).

The distribution of α-galactosyl-containing oligosaccharides in the plant kingdom is extensive (Dey, 1985). Raf family oligosaccharides, which are particularly abundant in the embryos of seeds, have been ascribed various roles, including soluble storage carbohydrates and regulators of the free levels of Gal or its metabolites, which are known to be toxic to plant cells. It has also been proposed that, in conjunction with Suc, Raf family oligosaccharides may impart desiccation tolerance to seeds by protecting cellular membranes during dehydration stress (Leopold and Vertucci, 1986; Koster and Leopold, 1988; Blackman *et al.*, 1992). While the primary role of Raf family oligosaccharides in seeds has not been resolved, their accumulation in maturing seeds and decreased levels during germination are a general occurrence (Dey, 1985). In this regard, the non-dormant line SH430 and the AR line M73 wild oat embryos are similar (Foley *et al.*, 1992b).

In contrast, imbibed D embryos maintain relatively high levels of Raf family oligosaccharides. Why? If it is not their lack of germination (as the data on Fru and GA-induced germination suggest), are there alternative explanations? Clearly all orthodox seed, whether D or not, undergo dehydration stress during maturation drying. Dormant seed with long after-ripening requirements experience additional stress in that they undergo repeated cycles of hydration/dehydration in the soil. We propose that a restriction in the metabolism of Raf family oligosaccharides serves to maintain their levels to protect the embryo during the repeated cycles of desiccation during after-ripening. Then, upon after-ripening (which gradually decreases the restriction to germination), monosaccharide components of Raf and Stach are the foundation of energy and cell wall production for germination. The level of the restriction to germination will dictate the vigour, i.e. onset and rate, of germination. While a restriction in the metabolism of Raf family oligosaccharides maintains seeds in a D state, it does not prevent germination, as demonstrated by the capacity of soluble sugars and GA to induce germination of D embryos and caryopses, respectively. This hypothesis, which relates dormancy and longevity, is logical since dormancy would have no adaptive advantage unless the seed had the capacity to remain viable in the soil when germination was restricted (Naylor, 1983).

The basis for seed/embryo dormancy in wild oat may be multifaceted, given the following properties governing it.

1. The transition from the D to the non-dormant state during warm, dry AR is not an 'all-or-nothing' occurrence. During AR, individual caryopses become progressively less dormant. This progressive elimination of the restriction to germination increases the onset and rate of germination, and the seeds become less sensitive to temperatures at which germination can occur.

2. The restriction to germination can be reinduced in imbibed caryopses (secondary D) if they displayed primary D and are not yet fully AR (Symons *et al.*, 1986). Caryopses from non-dormant lines cannot be induced into secondary D (e.g. by anoxia and high-temperature stress). Since secondary D caryopses respond to AR in the same way as primary D caryopses, the restriction to germination is probably the same in both instances (Tilsner and Upadhyaya, 1985).

3. The environment during development affects the depth of dormancy (Sexsmith, 1969). For example, relatively cool temperatures and/or moist soil conditions increase the depth of dormancy, whereas the opposite conditions decrease it.

4. The genetic model for dormancy in wild oat indicates that dormancy is a recessive trait with at least three genes controlling the rate of AR (Jana *et al.*, 1979).

Any mechanism to explain dormancy (Trewavas, 1988), including our physiological mechanism, should be regarded as hypothetical and just one 'piece of the puzzle' regarding seed dormancy in wild oat. While the mechanism we propose is not inconsistent with the aforementioned properties, a great deal of additional physiological, biochemical, molecular and genetic research is needed to develop an integrated model of true wild oat embryo dormancy/after-ripening.

Acknowledgements

The assistance of Maxine Nichols and Wanda Foley is gratefully acknowledged. This study was supported by the Purdue University Agricultural Research Programs. This is journal series paper number 14336.

References

Black, M. and Naylor, J.M. (1959) Prevention of the onset of seed dormancy by gibberellic acid. *Nature* 184, 468-469.

Blackman, S.A., Obendorf, R.L. and Leopold, A.C. (1992) Maturation proteins and sugars in desiccation tolerance of developing soybean seeds. *Physiologia Plantarum* 100, 225-230.

Carpita, N.C., Skaria, A., Barnett, J.P. and Dunlap, J.R. (1983) Cold stratification and growth of radicles of loblolly pine (*Pinus taeda*) embryos. *Physiologia Plantarum* 59, 601-606.

Chen, S.S.C. and Varner, J.E. (1969) Metabolism of ^{14}C-maltose in *Avena fatua* seeds during germination. *Plant Physiology* 44, 770-774.

Dey, P.M. (1985) D-Galactose-containing oligosacchairdes. In: Dey, P.M. and Dixon, R.A. (eds) *Biochemistry of Storage Carbohydrates in Green Plants*. Academic Press, London, pp. 53-129.

Drennan, D.S.H. and Berrie, A.M.M. (1961) Physiological studies on germination in the genus *Avena*. I. The development of amylase activity. *New Phytologist* 61, 1-9.

Flinn, C.L. and Ashworth, E.N. (1995) The relationship between carbohydrates and flower bud hardiness among *Forsythia* taxa. *Journal of the American Society for Horticultural Science* 120, 607-613.

Foley, M.E. (1992) Effect of soluble sugars and gibberellic acid in breaking of dormancy of excised wild oat (*Avena fatua*) embryos. *Weed Science* 40, 208-214.

Foley, M.E., Bancal, M. and Nichols, M.B. (1992a) Carbohydrate status in dormant and afterripened excised wild oat embryos. *Physiologia Plantarum* 85, 461-466.

Foley, M.E., Nichols, M.B., Bancal, M., Myers, S.P. and Volenec, J.J. (1992b) Raffinose family oligosaccharides in dormant and afterripened wild oat (*Avena fatua* L.) caryopses. In: Come, D. and Corbineau, F. (eds) *Fourth International Workshop on Seeds - Basic and Applied Aspects of Seed Biology*. ASFIS, Paris, France, pp. 591-598.

Foley, M.E., Nichols, M.B. and Myers, S.P. (1993) Carbohydrate concentrations and interactions in afterripening-responsive dormant *Avena fatua* caryopses induced to germinate by gibberellic acid. *Seed Science Research* 3, 271-278.

Jana, S., Acharya, S.N. and Naylor, J.M. (1979) Dormancy studies in seed of *Avena fatua*. 10. On the inheritance of germination behavior. *Canadian Journal of Botany* 57, 1663-1667.

Kanabus, J., Bressan, R.A. and Carpita, N.C. (1986) Carbon assimilation in carrot cells in liquid culture. *Plant Physiology* 82, 363-368.

Koster, K.L. and Leopold, A.C. (1988) Sugars and desiccation tolerance in seeds. *Plant Physiology* 88, 829-832.

Lasheen, A.M. and Chaplin, C.E. (1977) Seasonal sugar concentration in two peach cultivars differing in cold hardiness. *Journal of the American Society for Horticultural Science* 102, 171-174.

Lehle, F.R., Staniforth, D.W. and Stewart, C.R. (1978) Caryopsis dormancy in *Setaria lutescens*: endosperm starch susceptibility to amylase digestion. *Plant Physiology* 61 (Suppl.), 16.

Leopold, A.C. and Vertucci, C.W. (1986) Physical attributes of desiccated seeds. In: Leopold, A.C. (ed.) *Membranes, Metabolism and Dry Organisms.* Cornell University Press, Ithaca, pp. 22-34.

Lona, F. (1956) Acido gibberellico determina la germinazione dei semi di *Latuca scariola* in fase scoto-inhibizione. *Atheneo Parmense* 27, 641-644.

Murphy, J.B. and Hammer, M.F. (1988) Respiration and soluble sugar metabolism in sugar pine embryos. *Physiologia Plantarum* 74, 95-100.

Naylor, J.M. (1983) Studies on the genetic control of some physiological processes in seeds. *Canadian Journal of Botany* 61, 3561-3567.

Naylor, J.M. and Simpson, G.M. (1961) Dormancy studies in seed of *Avena fatua*. 2. A gibberellin-sensitive inhibitory mechanism in the embryo. *Canadian Journal of Botany* 39, 281-295.

Nichols, M.B., Bancal, M., Foley, M.E. and Volenec, J.J. (1993) Non-structural carbohydrates in dormant and afterripened wild oat caryopses. *Physiologia Plantarum* 88, 221-228.

Sauter, J.J. and Ambrosius, T. (1986) Changes in the partitioning of carbohydrates in the wood during bud break in *Betula pendula* Roth. *Journal of Plant Physiology* 124, 31-43.

Sexsmith, J.J. (1969) Dormancy of wild oat seed production under various temperature and moisture conditions. *Weed Science* 17, 405-407.

Simmons, J.A. and Simpson, G.M. (1971) Increased participation of pentose phosphate pathway in response to afterripening and gibberellic acid treatments in caryopses of *Avena fatua*. *Canadian Journal of Botany* 49, 1833-1840.

Simpson, G.M. (1965) Dormancy studies in seed of *Avena fatua*. 4. The role of gibberellin in embryo dormancy. *Canadian Journal of Botany* 43, 793-816.

Simpson, G.M. (1966) The suppression by (2-chloroethyl)trimethylammoniumchloride of synthesis of a gibberellin-like substance by embryos of *Avena fatua*. *Canadian Journal of Botany* 44, 115-117.

Simpson, G.M. and Naylor, J.M. (1962) Dormancy studies in seed of *Avena fatua*. 3. A relationship between maltase, amylases, and gibberellins. *Canadian Journal of Botany* 40, 1659-1673.

Symons, S.J., Naylor, J.M., Simpson, G.M. and Adkins, S.W. (1986) Secondary dormancy in *Avena fatua*: induction and characteristics in genetically pure lines. *Physiologia Plantarum* 68, 27-33.

Tilsner, H.R. and Upadhyaya, M.K. (1985) Induction and release of secondary seed dormancy in genetically pure lines of *Avena fatua*. *Physiologia Plantarum* 64, 377-382.

Trewavas, A.J. (1988) Timing and memory processes in seed embryo dormancy - a conceptual paradigm for plant development questions. *Bioessays* 6, 87-93.

Xylander, M., Augsten, H. and Appenroth, K.J. (1992) Photophysiology of turion germination in *Spirodela polyrhiza* (L.) Schleiden. XII. Role of carbohydrate supply. *Plant Physiology (Life Science Advances)* 11, 241-245.

18 Chemical Mechanisms of Breaking Seed Dormancy

MARC ALAN COHN
*Department of Plant Pathology and Crop Physiology,
Louisiana State University Agricultural Center, Baton Rouge,
LA 70803, USA*

Introduction

Seed dormancy-breaking chemicals are useful for attainment of at least two major objectives: (i) they serve as molecular probes of the mechanisms involved in the transition from developmental arrest to growth and (ii) they can increase the efficacy of weed control and crop establishment. Despite the complementary utility of these research areas, successful field applications have been limited due to insufficient potency of available chemicals, and the mechanisms of chemical action are poorly understood (Cohn, 1987).

Using red rice (*Oryza sativa* L.) - an economically important weed of southern USA, Central and South American rice production regions - as a model system, we have attempted to define the chemical features required for the activity of non-hormonal dormancy-breaking compounds (Cohn, 1989). This work has also advanced some elucidation of the involved mechanisms of action. On both counts, significant success has been achieved, albeit with rather slow progress towards the ultimate objectives.

In this chapter, results from structure–activity studies are summarized and integrated, where possible, with studies addressing physiological mechanisms of seed dormancy. The data demonstrate clearly that uptake of dormancy-breaking chemicals by seeds is extremely rapid, and the physiological response by the seeds can be equally fast. However, structurally related dormancy-breaking chemicals can elicit dramatically different levels of the same physiological marker. In addition, the structural integrity of a dormancy-breaking chemical inside the seed is not static, i.e. applied dormancy-breaking chemicals are quickly metabolized.

Chemical Structure and Physiological Activity

The relative response of seeds to dormancy-breaking substances is generally correlated with the lipophilicity of the applied chemical (expressed as the $\log_{10}$ octanol/water partition coefficient) (Cohn et al., 1989). The higher the lipid solubility of a compound, the lower the concentration required to obtain a standard germination response. This relationship has been extended to more than 40 chemicals including acids, aldehydes, alcohols and esters. For weak acids the degree of dissociation in the contact solution influences the dormancy-breaking activity. Weak acids are required in the undissociated form for the best response (Cohn et al., 1983, 1987; Cohn and Hughes, 1986). Similar results have been obtained using other species (e.g. Van Mulders et al., 1986; Brooks and Mitchell, 1988; Tilsner and Upadhyaya, 1989). For low molecular weight weak acids such as nitrite, azide and cyanide, activity can be related better to molecular size than to lipophilicity. While these data have generated some controversy in the realm of dormancy research, the trends summarized above are not novel in the historical context of structure–activity studies. Relationships of chemical properties to uptake and physiological activity generally have been recognized for decades (e.g. Clark, 1898; Overton, 1901; Harvey, 1911; Loeb, 1913).

Activity of dormancy-breaking chemicals is also modulated by the nature and location of functional groups. Monocarboxylic acids act at lower concentrations than the corresponding aldehyde, followed by the alcohol, in a manner not fully accounted for by lipophilicity alone. Furthermore, alcohol activity is dramatically altered by hydroxyl group placement, e.g. while 1-pentanol yields 95% germination, 2-pentanol and 3-pentanol have no effect (Cohn et al., 1991; M.A. Cohn, unpublished). Parallel relationships have also been observed for linear and cyclic 6-carbon alcohols. Modulation of alcohol activity by hydroxyl group position is unexpected in the context of the anaesthetic action hypothesis (Taylorson and Hendricks, 1979) because the partition coefficients of primary vs. secondary alcohols are similar, and one would, therefore, expect any short-chain alcohol to break dormancy. In addition, pressure reversibility of alcohol action – one of the cornerstones of the anaesthetic action hypothesis – has not been the consistent criterion it appeared to be initially (Taylorson, 1991).

How Do All Those Different Chemicals Break Dormancy?

Changes in cell pH

In both the animal and plant kingdoms, the termination of developmental arrest is associated with changes in cell pH, and exogenously applied weak acids activate the resumption of development in some cases (reviewed in

Footitt and Cohn, 1992a). Since seed dormancy can be broken with similar chemicals, a reduction in cytoplasmic pH has been proposed as an important trigger and/or marker in the dormancy-breaking process. Data consistent with this idea have been obtained using nitrite, propionic acid, methyl propionate, propionaldehyde, and propanol as dormancy-breaking treatments (Footitt and Cohn, 1992a). As part of a rigorous historical review conducted after completion of these experiments, data demonstrating decreased embryo pH during cold stratification of initially dormant seeds were uncovered as well (Eckerson, 1913; Rose, 1919; Jones, 1920; Pack, 1921). Recently, Petel *et al.* (1993) have shown a potentially analogous decrease in tissue pH after the application of ethanol as a dormancy-breaking stimulus in Jerusalem artichoke tubers.

While alteration of tissue or cellular pH is associated with a change in developmental pattern in many diverse biological systems, there is no clear consensus as to the function of the pH change. Is it cause, effect, or correlation only? Even in a well studied system such as sea urchin eggs, the answer is not clear (Epel, 1990). Any number of possibilities come to mind. Among these, one could easily propose a role for increased proton concentration: (i) acting as a secondary messenger integrated with control of intracellular calcium levels; (ii) eliciting a mass action shift in the equilibrium of an inhibitor (ABA?)-receptor complex; (iii) inducing cell activity via stimulation of plasmalemma proton pumps; (iv) altering an equilibrium between constitutive bound inactive vs. soluble active enzymes; or (v) eliciting responses similar to the acid-growth effect upon cell elongation. More information must be developed before any of these hypotheses can be considered seriously.

Timing is everything

Physiological evidence suggests that dormancy-breaking is not an instantaneous phenomenon. A finite time interval elapses prior to germination-related events, e.g. (i) a photoreversible period precedes the irreversible commitment (escape time) to germination in light-requiring species; (ii) growth kinetics of initially dormant seeds are slower than those of non-dormant seeds after the introduction of a dormancy-breaking chemical (nitrogen dioxide) in the absence of a liquid solvent (Cohn and Castle, 1984); and (iii) growth kinetics of germinable cereal embryos isolated from dormant seeds are slower than those of embryos isolated from non-dormant seeds (e.g. Van Beckum *et al.*, 1993). Therefore, if one desires to increase the chances of successfully identifying events associated with the dormancy-breaking process, it is essential to conduct time course studies (e.g. Cohn and Footitt, 1993) using chemical pulse treatments applied to a uniform and vigorous seed sample. Markers for the progression from dormancy to germination processes, i.e. those occurring prior to visible embryo growth, are almost nonexistent.

Uptake of dormancy-breaking chemicals is rapid

Direct and indirect evidence indicates that uptake of dormancy-breaking chemicals proceeds quickly even when seeds are first imbibed in water (e.g. see protocols and results for Fig. 18.1). When nitrite or propionate was applied under conditions that should produce an acid load on the seed, the tissue pH was readily lowered after 1 h of chemical contact, indicating uptake and dissociation of the applied weak acid (Footitt and Cohn, 1992a). Direct measurements of ^{13}C-labelled propionate and propanol via carbon-13 nuclear magnetic resonance spectroscopy show uptake within 2 h of chemical application (Footitt *et al.*, 1995). Rapid uptake of nitrite, propionaldehyde and propanol within 2 h also can be discerned by the metabolic increase in fructose 2,6-bisphosphate (F2,6-BP) (Footitt and Cohn, 1992b, 1995).

Dormancy-breaking chemicals can be metabolized rapidly

After 4 h of ^{13}C-propanol or ^{13}C-propionate uptake, measurable levels of 3-hydroxypropionate were detected (Fig. 18.2). No other metabolites were evident during the contact period. Metabolism occurred prior to completion of the pulse-dose interval required to initiate the breaking of dormancy in the population (between 8 to 12 h for each chemical) (Footitt, 1993; Footitt *et al.*, 1995). Metabolic conversion of ethanol into acetaldehyde and vice versa has also been documented in unimbibed, after-ripening cocklebur seeds (Esashi *et al.*, 1994). However, the time course of these conversions was not obtained, and the physiological role attributed to these conversions in relation to dormancy was interpreted very differently.

Decreased pH is directly related to acid uptake or metabolism to an acid

The decrease in tissue pH in response to dormancy-breaking chemicals (Footitt and Cohn, 1992a) is correlated highly with the uptake of, and metabolism to, weak acids ($y = -0.61x + 4.61$, $r = -0.98$, $P < 0.001$). The greater the sum of the ^{13}C acid signals observed by NMR, the lower the tissue pH. It is interesting to note that collection of data for each half of this correlation was separated by at least two years.

Similar chemicals can elicit different biochemical responses

Just because a group of structurally related chemicals readily break dormancy does not mean they will elicit the same degree of a single physiological response. For example, propionaldehyde and propanol stimulate an increase

Chemical Mechanisms of Breaking Seed Dormancy

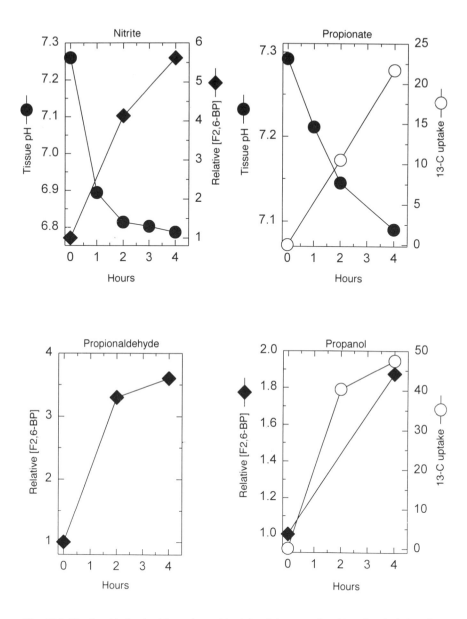

Fig. 18.1. Direct and indirect evidence for rapid uptake of dormancy-breaking chemicals in red rice. Dehulled, dormant seeds were imbibed in water for 24 h and then transferred to nitrite (ca. 3 mM free acid), propionate (10 mM free acid), propionaldehyde (40 mM), or propanol (75 mM) for up to 4 h at 30°C. When applied for 24 h, these treatments result in > 90% germination after 7 additional days in water at 30°C. Seeds were harvested, embryos excised and extracts were analysed for pH (Footitt and Cohn, 1992a), fructose-2,6-bisphosphate (after Van Schaftingen, 1984), or scanned by ^{13}C-NMR (Footitt et al., 1995). Each data point represents the mean of at least three independent experiments. Each homogenate was prepared from 100 embryos.

in F2,6-BP by 4 h of chemical contact. In contrast, propionic acid and methyl propionate break dormancy but do not increase F2,6-BP above control levels. A similar divergence can be observed comparing two weak acids, propionic acid vs. nitrite, which induces F2,6-BP levels comparable to the aldehyde (Footitt, 1993; Footitt and Cohn, 1992b, 1995). It is likely that low F2,6-BP levels are due, in part, to the organic acid anions' inhibitory properties on glycolysis (e.g. Francois *et al.*, 1988; Warth, 1991). It is important, therefore, to employ a group of dormancy-breaking chemicals, rather than a single substance, as physiological probes because there is always a chance that key events could be missed due to unforseen pharmacological complications.

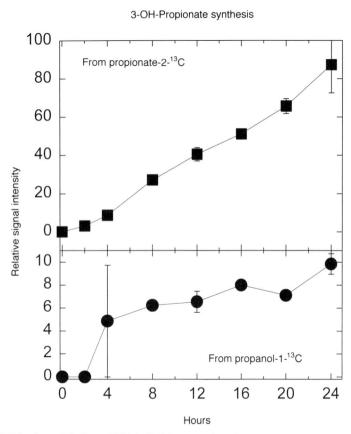

Fig. 18.2. *In vivo* metabolism of ^{13}C-labelled dormancy-breaking chemicals in red rice. Dehulled, dormant seeds were imbibed in water for 24 h and then transferred to carrier-free propionate-2-^{13}C (10 mM free acid) or propanol-1-^{13}C (75 mM) for up to 24 h at 30°C. Briefly, 100 embryos were excised, powdered in liquid N_2, and extracted with 2 M perchloric acid. After clarification and neutralization, spectra were recorded on a Bruker AM 400 spectrometer operating at 100.62 MHz with a 5 mm broad band probe and ^{2}H lock signal. Chemical shifts: 3-hydroxypropionate-2-^{13}C, 41 ppm; 3-hydroxypropionate-1-^{13}C, 181.6 ppm.

Conclusions

Most of the cellular responses reported here for the red rice system occurred within 4 h after application of a dormancy-breaking chemical. These events are taking place at least 16 to 24 h before the first indications of the visible germination event. The rapid tissue responses to dormancy-breaking chemicals warrant orientation of future experiments to the earliest events during the dormancy-to-germination timecourse. Any success of future molecular studies will depend upon avoiding confusion between normal germination events and those attributed uniquely to the dormancy-breaking process. Therefore, the frequency and timing of sampling during kinetic experiments must be considered carefully to avoid compromising otherwise excellent physiological, biochemical or molecular studies in plant dormancy.

References

Brooks, C.A. and Mitchell, C.A. (1988) Effect of salicylhydroxamic acid on endosperm strength and embryo growth of *Lactuca sativa* L. cv. Waldmann's Green seeds. *Plant Physiology* 86, 826-829.

Clark, J.F. (1898) Electrolytic dissociation and toxic effect. *Journal of Physical Chemistry* 3, 263-316.

Cohn, M.A. (1987) Mechanisms of physiological dormancy. In: Frasier, G.W. and Evans, R.A. (eds) *Seed and Seedbed Ecology of Rangeland Plants*. USDA-ARS, Washington, pp. 14-20.

Cohn, M.A. (1989) Factors influencing the efficacy of dormancy-breaking chemicals. In: Taylorson, R.B. (ed.) *Recent Advances in Development and Germination of Seeds*. Plenum Press, New York, pp. 261-267.

Cohn, M.A. and Castle, L. (1984) Dormancy in red rice. IV. Response of unimbibed and imbibing seeds to nitrogen dioxide. *Physiologia Plantarum* 60, 552-556.

Cohn, M.A. and Footitt, S. (1993) Initial signal transduction steps during the dormancy-breaking process. In: Come, D. and Corbineau, F. (eds) *Proceedings of the Fourth International Workshop on Seeds: Basic and Applied Aspects of Seed Biology, Volume 2*. Association pour la Formation Professionnelle de l'Interprofession Semences, Paris, pp. 599-605.

Cohn, M.A. and Hughes, J.A. (1986) Seed dormancy in red rice. V. Response to azide, hydroxylamine, and cyanide. *Plant Physiology* 80, 531-533.

Cohn, M.A., Butera, D.L. and Hughes, J.A. (1983) Seed dormancy in red rice. III. Response to nitrite, nitrate, and ammonium ions. *Plant Physiology* 73, 381-384.

Cohn, M.A., Chiles, L.A., Hughes, J.A. and Boullion, K.J. (1987) Seed dormancy in red rice. VI. Monocarboxylic acids: a new class of pH-dependent germination stimulants. *Plant Physiology* 84, 716-719.

Cohn, M.A., Jones, K.L., Chiles, L.A. and Church, D.F. (1989) Seed dormancy in red rice. VII. Structure-activity studies of germination stimulants. *Plant Physiology* 89, 879-882.

Cohn, M.A., Church, D.F., Ranken, J. and Sanchez, V. (1991) Hydroxyl group position governs activity of dormancy-breaking chemicals (abstract No. 411). *Plant Physiology* 96, S-63.

Eckerson, S. (1913) A physiological and chemical study of afterripening. *Botanical Gazette* 55, 286-299.

Epel, D. (1990) The initiation of development at fertilization. *Cell Differentiation and Development* 29, 1-12.

Esashi, Y., Zhang, M., Segawa, K., Furihata, T., Nakaya, M. and Maeda, Y. (1994) Possible involvement of volatile compounds in the after-ripening of cocklebur seeds. *Physiologia Plantarum* 90, 577-583.

Footitt, S. (1993) Seed dormancy in red rice (*Oryza sativa*): changes in embryo pH and metabolism during the dormancy-breaking process. PhD thesis, Louisiana State University, Baton Rouge.

Footitt, S. and Cohn, M.A. (1992a) Seed dormancy in red rice. VIII. Embryo acidification during dormancy-breaking and subsequent germination. *Plant Physiology* 100, 1196-1202.

Footitt, S. and Cohn, M.A. (1992b) Levels of fructose 2,6-bisphosphate in red rice embryos during dormancy-breaking and germination (abstract No. 64). *Sixth International Symposium on Pre-Harvest Sprouting in Cereals*, mimeograph.

Footitt, S. and Cohn, M.A. (1995) Seed dormancy in red rice. IX. Embryo fructose-2,6-bisphosphate during dormancy breaking and subsequent germination. *Plant Physiology* 107, 1365-1370.

Footitt, S., Vargas, D. and Cohn, M.A. (1995) Seed dormancy in red rice. X.A. ^{13}C-NMR study of the metabolism of dormancy-breaking chemicals. *Physiologia Plantarum* 94, 667-671.

Francois, J., Van Schaftingen, E. and Hers, H.G. (1988) Characterization of phosphofructokinase 2 and of enzymes involved in the degradation of fructose 2,6-bisphosphate in yeast. *European Journal of Biochemistry* 171, 599-608.

Harvey, E.N. (1911) Studies on the permeability of cells. *Journal of Experimental Zoology* 10, 507-556.

Jones, H.A. (1920) Physiological study of maple seeds. *Botanical Gazette* 69, 127-152.

Loeb, J. (1913) *Artificial Parthenogenesis and Fertilization*. University of Chicago Press, Chicago.

Overton, C.E. (1901) *Studien Uber Die Narkose*. G. Fischer, Jena. Translation: Lipnick, R.L. *Studies of Narcosis* (1991), Chapman and Hall, New York.

Pack, D.A. (1921) After-ripening and germination of *Juniperus* seeds. *Botanical Gazette* 71, 32-60.

Petel, G., Candelier, P. and Gendraud, M. (1993) Effect of ethanol on filiate tubers of Jerusalem artichoke: a new tool to study tuber dormancy. *Plant Physiology and Biochemistry* 31, 67-71.

Rose, R.C. (1919) After-ripening and germination of seeds of *Tilia, Sambucus,* and *Rubus. Botanical Gazette* 67, 281-308.

Taylorson, R.B. (1991) Interactions of alcohols and increased air pressure on *Echinochloa crus-galli* (L.) Beauv. seed germination. *Annals of Botany* 68, 337-340.

Taylorson, R.B. and Hendricks, S.B. (1979) Overcoming dormancy in seeds with ethanol and other anesthetics. *Planta* 145, 507-510.

Tilsner, H.R. and Upadhyaya, M.K. (1989) The effect of pH on the action of respiratory inhibitors in *Avena fatua* seeds. *Annals of Botany* 64, 707-711.

Van Beckum, J.M.M., Libbenga, K.R. and Wang, M. (1993) Abscisic acid and gibberellic acid-regulated responses of embryos and aleurone layers isolated from dormant and nondormant barley grains. *Physiologia Plantarum* 89, 483-489

Van Mulders, R.M., Van Laere, A.J. and Verbeke, M.N. (1986) Effects of pH and cations on the germination induction of *Phycomyces* spores with carboxylic acids. *Biochemie und Physiologie der Pflanzen* 181, 103-115.

Van Schaftingen, E. (1984) D-fructose 2,6-bisphosphate. In: Bergmeyer, H.U., Bergmeyer, J. and Grossl, M. (eds) *Methods in Enzymatic Analysis, Vol 16, Metabolism 1: Carbohydrates*. Verlag Chemie, Weinheim, pp. 335-341.

Warth, A.D. (1991) Effect of benzoic acid on glycolytic metabolite levels and intracellular pH in *Saccharomyces cerevisiae*. *Applied and Environmental Microbiology* 57, 3415-3417.

V Molecular Biology

19 Molecular Analysis of Turion Formation in *Spirodela polyrrhiza*: a Model System for Dormant Bud Induction

Cheryl C. Smart
Institute of Plant Sciences, Swiss Federal Institute of Technology (ETH) Zürich, Universitätsstrasse 2, CH-8092 Zürich, Switzerland

Introduction

During normal vegetative growth, plants continually adapt their development in order to cope with any stress-inducing or adverse changes in the environment. Under extreme conditions, in which the survival of the plant itself may be at risk, immediate growth and development can be suspended and a state of dormancy assumed. The arrest of growth may be imposed by environmental factors, but in tissues showing true dormancy growth is inhibited by factors residing within the dormant tissues themselves.

Temperate woody perennials have evolved a dormancy mechanism which helps them to survive winter frosts. The shoot apices cease active growth and become enclosed in protective bud scales to form dormant buds. Dormant buds are frost-resistant, unlike their actively growing counterparts, and the bud scales help to reduce water loss. Although growth inhibition plays an obvious and essential role, the formation of dormant buds is a distinct developmental phase which involves the telescoping of bud scales and leaf primordia around the apical region due to an arrest of internode extension. The bud scales can be formed from stipules or from modified leaves which exhibit suppressed lamina development.

When first formed, terminal buds are not innately dormant (endodormant), as their growth is apparently inhibited by the mature leaves on the shoot (i.e. paradormancy). However, later the buds enter true dormancy (endodormancy) and cannot resume growth under favourable environmental conditions, even if the mature leaves are removed. Chilling temperatures are required for the elimination of this dormancy, with the required amount of chilling varying according to the species and even variety. Often the chilling requirement is satisfied quite early in winter and dormancy is then only imposed by the cold climate (i.e. ecodormancy). Bud burst and shoot elongation begin with the start of warmer spring temperatures.

Although the formation of dormant buds is of scientific, agricultural and horticultural interest, their investigation poses a number of experimental problems. Chief among these must be the relatively slow growth and long life cycle of woody perennials, the difficulty of maintaining constant growth conditions, the space required for growth and the problems of biochemical analysis of plant tissues that are often woody, phenolic and tannin-rich. One approach to these problems is to use a model system for dormant bud induction which circumvents some of the difficulties listed above. *Spirodela polyrrhiza* may provide such a system.

Overwintering Structures (Turions) in *Spirodela polyrrhiza*

Spirodela polyrrhiza is a member of the Lemnaceae, a monocotyledonous family of floating aquatic angiosperms characterized by a peculiar mode of vegetative reproduction (Hillman, 1961; Landolt, 1986; Bell, 1991). The *Lemnaceae* grow as plantlets of clonally-related fronds on the surface of freshwater lakes and ponds where, under appropriate conditions, they may come to cover large areas of this special habitat. The structure of the *S. polyrrhiza* plant body or frond is shown in Fig. 19.1(a). A mother frond produces daughter fronds from two meristematic pockets at its proximal end. The daughter frond itself contains two meristematic pockets from which new frond primordia develop. This repetition of frond primordia development within fronds leads to the groups of clonally-related plants observed in nature (Fig. 19.1(b)).

Vegetative fronds of *S. polyrrhiza* cannot tolerate temperatures of less than about 7°C, dying during late autumn in the natural environment. As an overwintering device, the plants produce dormant buds – turions – during mid- to late-summer via a specialization of the normal developmental process (Fig. 19.1(c)). A turion is an anatomically distinct, dormant overwintering structure from which a vegetative frond later arises. Ontogenically, a turion is a modified frond. A turion is distinguished from a vegetative frond by its smaller size (although it is thicker than the vegetative frond), reniform appearance, the lack of aerenchyma, the presence of thicker cell walls, and the accumulation of both starch and anthocyanin (Jacobs, 1947; Smart and Trewavas, 1983a,b). Each turion also contains two meristematic pockets

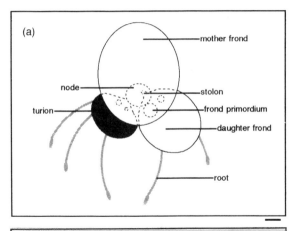

Fig. 19.1. Structure and development of the plant body of *S. polyrrhiza*; (a) Schematic diagram of a dorsal view of a mother frond with a vegetative frond in its right-hand pocket and a turion in the left (bar = 1 mm); (b) Photograph of an untreated culture of *S. polyrrhiza* grown at 25°C. Control cultures only produce vegetative fronds and never turions (bar = 5 mm); (c) Photograph of a culture 7 days after treatment with 100 nM ABA. The fronds are producing turions and several turions have abscised (bar = 5 mm).

from which new vegetative frond primordia can develop following 'germination'. Unlike the vegetative fronds, which tend to remain linked together for some time via the stolon, the turion abscises from the mother frond as soon as it is mature and sinks to the bottom of the pond or lake and lies dormant before rising to the surface and producing a new vegetative frond in spring (Jacobs, 1947).

Spirodela polyrrhiza offers many potential advantages as a model system for investigating dormant bud induction and development. Large numbers of plants can be maintained in a small space *in vitro*, where they can be grown under totally controlled medium and environmental conditions. Certain environmental and chemical triggers can be used to induce turion formation, the functional analogue to a dormant bud (see below). Biochemical analysis of *S. polyrrhiza* (and related *Lemnaceae*) is well established (see Hillman, 1961), including molecular biological investigations (Silverthorne and Tobin, 1990). Turion formation in *S. polyrrhiza* could thus be manipulated as a model system to gain molecular insights into the induction and development of dormant buds. However, this is based on the assumption that the mechanism of turion induction is analogous to that of dormant bud induction in other plants.

Similarities between Turions and Dormant Buds

The development of turions has many features in common with the development of dormant buds in woody plants. From a morphological view, the 'telescoping' of the bud scales and leaf primordia in the apical region of woody shoots, due to the arrest of normal internode development, could be seen as analogous to the cessation of elongation in the stolon connecting the turion to the mother frond. The suppression of lamina development of the dormant tree bud scales can be compared to the inhibition of cell expansion and lack of normal aerenchyma development in the turion (Smart and Trewavas, 1983b). Turions are also cold- and frost-resistant, and can survive temperatures of 4°C for more than 2 years, -4°C for more than 3 weeks, and -8°C for several days. Turions formed in nature and by certain signals in the laboratory (see below) develop innate dormancy and require a chilling period before 'germination'. Like dormant buds, the actual 'germination' (analogous to bud burst) only begins in spring with the advent of warmer temperatures and in the laboratory above 15°C (Jacobs, 1947; Landolt, 1986). Thus, functionally and anatomically, turions and dormant buds are quite similar.

This similarity in function also extends to the environmental factors involved in the induction of turions and dormant buds. For example, it has been suggested that a lack of nutrients may play an important role in bud dormancy induction (Powell, 1988), and turion formation can also be induced by general nutrient starvation brought about by overcrowding or deficiency of nitrate, phosphate or sulphate (Jacobs, 1947; reviewed by Landolt, 1986). Also, photoperiod has been reported to effect both turion and dormant bud formation in the laboratory (Wareing, 1956; Perry, 1968; Saks *et al.*, 1980), although in both cases the interpretation of the data may not be trivial (Wareing and Phillips, 1983; Powell, 1988). In both plant systems in the natural environment, the induction of the dormant structure generally occurs immediately following periods of vigorous growth, when daylength and temperature would still be expected to be conducive to growth (Jacobs, 1947; Powell, 1988). This implies that both turion and dormant bud formation are programmed active events in plant development, rather than simple responses to any one environmental stimulus. There appears to be an interplay between endogenous and exogenous signals, the efficacy of any one signal being dependent on the balance of the developmental state of the tissue and the environment.

Does Abscisic Acid Play a Role in the Induction of Dormancy?

Debate still continues as to the role of abscisic acid (ABA) in dormant bud induction. The demonstration of photoperiodic induction of dormant buds in laboratory seedlings led to the isolation of ABA from sycamore (Cornforth *et al.*, 1965). A growth inhibitory activity was found in extracts of sycamore seedlings which was higher (by bioassay) under short days than long days (Robinson and Wareing, 1964) and Eagles and Wareing (1964) were able to induce the formation of dormant buds on birch seedlings grown under marginal daylength by treatment of the leaves with an extract from leaves of plants grown under short days. When the inhibitor activity was shown to be ABA, El-Antably *et al.* (1967) repeated the experiment of Eagles and Wareing (1964) using synthetic ABA and found that ABA supplied to the leaves and shoot apices induced the formation of dormant buds in birch, blackcurrant and sycamore.

However, repeated attempts to confirm these findings were unsuccessful (Perry and Hellmers, 1973; Saunders *et al.*, 1974; Hocking and Hillman, 1975). Moreover, with the advent of physical methods of ABA analysis, no differences were found between the ABA content of shoot tips of various species grown under short or long days (Lenton *et al.*, 1972; Powell, 1976; Alvim *et al.*, 1978, 1979; Phillips *et al.*, 1980; Barros and Neill, 1986; Johansen *et al.*, 1986). More recently, however, increases in the ABA content of pine needles (Qamaruddin *et al.*, 1993) and apical and lateral buds of birch (Rinne *et al.*, 1994) under short days have been reported. A convincing role for ABA in the induction of dormant buds thus has yet to be demonstrated. If ABA does have a role in the induction of dormant buds, it is more likely to manifest itself either in changes in sensitivity to the hormone (Barros and Neill, 1986), or in the fine-tuning of the balance of activities between ABA and growth-promoting regulators such as gibberellins and cytokinins.

There is no doubt, however, that exogenous ABA induces turion formation in *S. polyrrhiza* (Perry and Byrne, 1969; Stewart, 1969; Smart and Trewavas, 1983a). In my laboratory, this induction occurs after 7 days at low, physiological concentrations of ABA from about 100 nM to 1 μM, although concentrations as low as 50 pM and as high as 15 μM have been reported by other researchers using different strains, different growth conditions and different criteria for turion identification (Stewart, 1969; Perry and Byrne, 1969). Growth inhibition is an integral and essential part of the turion-forming response to ABA (Stewart, 1969; Smart and Trewavas, 1983a).

What is not quite so clear is the role of endogenous ABA in turion formation. Involvement of endogenous ABA in the regulation of turion induction was first suggested by Stewart (1969) on the basis of his observation that both ABA-induced and spontaneous turion formation could be suppressed by kinetin. These findings have been confirmed recently (Chaloupková and Smart, 1994; C.C. Smart, unpublished data). Early attempts to detect a turion-inducing substance in the media of cultures undergoing short day–cold night induced turion formation were unsuccessful (Perry, 1968). Nevertheless, using a combination of short photoperiod and sucrose supplementation to induce turion formation in overcrowded conditions, Saks *et al.* (1980) reported that cultures producing turions secreted ABA into the medium and that this secreted ABA (about 60 nM) could induce turion formation in cultures grown under non-inductive conditions. The abscisic acid was identified by TLC and GC-ECD, and it was found in the medium about 6-8 days before turion abscission, thus correlating with the onset of turion induction. In order to detect these low amounts in the medium, however, the experiments were conducted with large numbers of fronds in very small volumes, which resulted in very low rates of growth and probably serious nutrient deficiency. The authors proposed that these conditions might mimic the situation in the natural environment. In subsequent experiments (Chaloupková and Smart, unpublished results), using both HPLC-GC-ECD techniques and an ELISA assay, no increase in ABA in the medium could be found during any stage of turion formation induced by low temperatures (11-15°C) during the 14-18 day induction period.

It is thus still unclear whether ABA plays a physiological role in the induction of either turions or dormant buds. At least in *Spirodela*, ABA can be used as a simple reliable trigger to induce turion formation for examination of the molecular events following this induction. Given the premise that these molecular events may be analogous to the events involved in dormant bud formation, the next section analyses turion induction to reveal possible mediators of dormant bud formation.

Turion Induction and Novel Gene Product Expression

The developmental process leading to the formation of the ABA-induced turion is accompanied by a repression of nucleic acid and protein synthesis in the young primordia destined to become turions. DNA synthesis in the developing turion is inhibited within three hours of ABA addition, followed by a repression of protein synthesis after 24 h. The inhibitory effect of ABA on protein synthesis is general, but the translatable mRNA pattern changes during ABA-induced turion formation; several turion-specific novel proteins are produced in the first 24 h of ABA addition (Smart and Trewavas, 1984). The rapid general inhibition of protein synthesis at early stages of turion formation are not accountable by the change in levels of translatable RNA, indicating an effect of ABA at the translational level. Only after 3 days in ABA, when the developmental primordium is committed to the turion developmental pathway, is there an apparent total inhibition of the production of extractable translatable mRNA leading to the onset of the dormant state. These observations could suggest that, in the later stages of turion development, an inhibitor of *in vitro* translation is present in the extracted RNA. Whether this inhibition is an artifact of extraction or actually plays a role *in vivo* is not known. In my laboratory, the yield of total RNA decreases consistently and dramatically during turion development (this could be an effect on extractability), although the RNAs extracted at these later stages of turion development show comparable levels of mRNA transcripts, as revealed by northern blot hybridization.

With the aim of understanding the molecular basis of early turion development, a cDNA library has been constructed from mRNA extracted from ABA-treated *S. polyrrhiza* plants in my laboratory. Using a differential screening strategy, a number of cDNAs have been identified whose respective transcripts are up-regulated early during ABA-induced turion formation. The clones analysed thus far by Northern blotting fall into four groups (Fig. 19.2). The first of these *tur* (turion-*u*p-*r*egulated) genes, *tur1*, represents a medium-sized transcript of 1.9 kb which is rapidly induced by ABA to reach a maximum at 2 h. It remains high throughout turion development, declining only when the first turions have abscised from the mother fronds, and it is not induced significantly during cold-temperature induction of turion formation (Smart and Fleming, 1993). Interestingly, the rapid up-regulation of *tur1* was shown (by *in situ* hybridization) to be localized to the stolon, a tissue which is altered in the turion developmental programme. Since a similar induction of *tur1* gene expression is not observed during low-temperature induction of turion formation, we conclude that the rise in *tur1* transcript level is not a general feature of turion formation *per se*; it is either linked specifically to ABA induction of this morphogenic event or is a specific effect of ABA on the stolon tissue unrelated to turion formation.

Tur1 is highly homologous to a yeast gene (*ino1*) that codes for D-*myo*-inositol-3-phosphate synthase (Ins3P synthase), the enzyme catalysing the first committed step in inositol biosynthesis (Loewus, 1990). Inositol is not only a growth controlling factor in plant cell and tissue cultures, but is a structural precursor of the membrane lipid phosphatidylinositol and its phosphorylated derivatives, phosphatidylinositol phosphate and phosphatidylinositol bisphosphate which, in animal cells, have an important role in signal transduction (Berridge and Irvine, 1989). However, inositol can be metabolized in plants to a wide variety of other products (Fig. 19.3). These include phytic acid, which is a phosphate-storing molecule; glucuronic acid, which is a precursor for the pectic components of the cell wall; and pinitol, which is thought to be involved in plant responses to osmotic stress (Loewus *et al.*, 1990). Based on the spatial expression of the *tur1* gene in ABA-treated plants, one could hypothesize that ABA-induced changes in cell wall inositol derivatives might represent an important mechanism by which morphogenesis is affected. Research in my laboratory has been initiated to follow the fate of inositol in *S. polyrrhiza* during ABA-induced turion development and in transgenic *Arabidopsis* plants that over- and under-express the *tur1* gene.

cDNA clone	Insert size (kbp)	Transcript size (kb)	Northern 2h Con	24h ABA	24h ABA	cold
tur1	1.8	1.9				
tur2	4.2	5.6				
tur3	1.2	1.4				
tur4	1.2	1.3				
control	0.7	1.1				

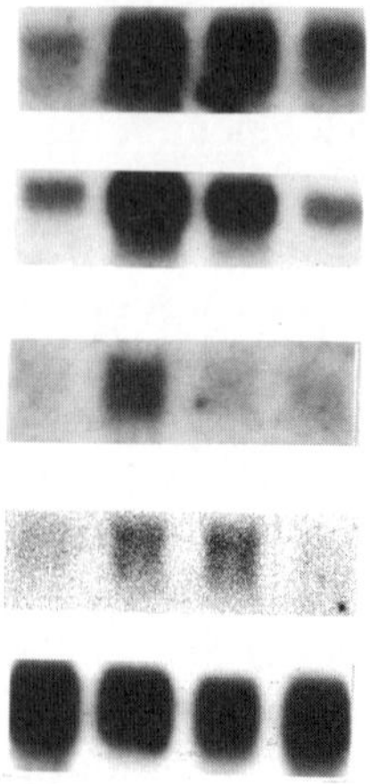

Fig. 19.2. Characteristics of ABA-up-regulated cDNA *tur* clones. Each *tur* clone is listed along with the size of the isolated cDNA insert, the estimated size of the transcript from Northern analysis and an autoradiograph showing the effect of 250 nM ABA and cold (15°C) on the induction of the transcript. The control clone was picked at random from the cDNA library, is not ABA-regulated, and is shown to indicate equivalent loading of total RNA (5 μg) in all lanes.

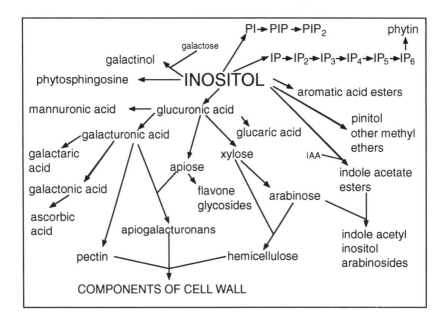

Fig. 19.3. Summary of inositol metabolism in higher plants.

Tur2 is an extremely large cDNA (4.2 kbp) which hybridizes to an even larger transcript of 5.6 kb. *Tur2* is rapidly induced by ABA, but the induction is only transient and by 24 h the level has declined and reaches resting levels within three days (C.C. Smart, unpublished results). During cold-induced turion formation, the transcript level rises steadily after 5 days at 15°C.

Tur3 hybridizes to a smaller transcript of 1.4 kb. Its induction is similar to that of *tur2*, but it seems to be a low abundance mRNA which makes further analysis of the induction difficult. The protein encoded by *tur3* shows no significant homology to any sequence in the GenEMBL data bank and therefore represents a novel ABA-induced gene (C.C. Smart, unpublished data). The first 19 amino acids of the amino terminus fulfil the criteria for a consensus sequence called the TonB box. The TonB box is a sequence found at the amino terminus of bacterial outer membrane receptor proteins involved in high affinity binding and energy-dependent uptake of various substrates into the periplasmic space.

Tur4 represents an ABA-up-regulated transcript which shows the same pattern of induction as *tur2*, but is similar to *tur1* in that it is not up-regulated during cold treatment. *Tur4* encodes a novel basic peroxidase which is likely to be localized to the cell wall (Chaloupková and Smart, 1994). The maintained increase of the level of *tur4* peroxidase mRNA by ABA can be inhibited by the concomitant addition of kinetin to the growth medium. This phenomenon correlates with the observed inhibition of ABA-induced-turion formation by this same growth regulator. The regulation of *tur4* represents the first example of antagonism of an ABA-induced gene by a cytokinin which correlates with an antagonistic effect of these two hormones on plant morphogenesis. These data contribute to a growing body of evidence linking growth regulators with changes in peroxidase gene expression, and to the concept of pairs of hormones playing antagonistic roles during plant development. In view of the proposed functions of many peroxidases in affecting the rate of cell elongation (Fry, 1986; Bradley *et al.*, 1992), it is possible that the ABA-induced *tur4* peroxidase functions at least partly by increasing the extent of cell wall polymer cross-linking and thus decreasing cell wall extensibility, which seems to be an important component of turion development (Smart *et al.*, 1987; Trewavas and Jones, 1991).

Future Perspectives

So far, two of the genes that we have investigated (*tur1* and *tur4*) encode proteins which could function to affect the nature of the cell wall. Although their precise role *in vivo* has yet to be proven, it is intriguing to note that, in both dormant buds and turions, an altered leaf morphogenesis occurs which is characterized by a general decrease in cell elongation. Such changes in cell extension could be either a result of the dormancy inducing process or linked intrinsically with it in a causal fashion. The studies reported above with *S. polyrrhiza* suggest that an investigation of the changes in cell wall characteristics during dormant bud induction could reveal novel insights into the mechanism and regulation of this process.

Acknowledgements

I would like to thank the Swiss Nationalfonds for supporting this research, Professor Nikolaus Amrhein for his encouragement and Dr Andrew Fleming for critical reading of the manuscript.

References

Alvim, R., Thomas, S. and Saunders, P.F. (1978) Seasonal variation in the hormone content of willow. II. Effect of photoperiod on growth and abscisic acid content of trees under field conditions. *Plant Physiology* 62, 779-780.

Alvim, R., Saunders, P.F. and Barros, R.S. (1979) Abscisic acid and the photoperiodic induction of dormancy in *Salix viminalis* L. *Plant Physiology* 63, 774-777.

Barros, R.S. and Neill, S.J. (1986) Periodicity of response to abscisic acid in lateral buds of willow (*Salix viminalis* L.). *Planta* 168, 530-535.
Bell, A.D. (1991) *Plant Form. An Illustrated Guide to Flowering Plant Morphology.* Oxford University Press, Oxford.
Berridge, M.J. and Irvine, R.F. (1989) Inositol phosphates and cell signalling. *Nature* 341, 197-205.
Bradley, D.J., Kjellbom, P. and Lamb, C.J. (1992) Elicitor- and wound-induced oxidative cross-linking of a proline-rich plant cell wall protein: a novel, rapid defense response. *Cell* 70, 21-30.
Chaloupková, K. and Smart, C.C. (1994) The abscisic acid induction of a novel peroxidase is antagonized by cytokinin in *Spirodela polyrrhiza* L. *Plant Physiology* 105, 497-507.
Cornforth, J.W., Milborrow, B.V., Ryback, G. and Wareing, P.F. (1965) Chemistry and physiology of 'dormins' in sycamore. Identity of sycamore 'dormin' with abscisin II. *Nature* 205, 1269-1270.
Eagles, C.F. and Wareing, P.F. (1964) The role of growth substances in the regulation of bud dormancy. *Physiologia Plantarum* 17, 697-709.
El-Antably, H.M.M., Wareing, P.F. and Hillman, J.R. (1967) Some physiological responses to D,L abscisin (dormin). *Planta* 73, 74-90.
Fry, S.C. (1986) Cross-linking of matrix polymers in the growing cell walls of angiosperms. *Annual Review of Plant Physiology* 37, 165-186.
Hillman, W.S. (1961) The Lemnaceae, or duckweeds. A review of the descriptive and experimental literature. *Botanical Review* 27, 221-287.
Hocking, J.J. and Hillman, J.R. (1975) Studies on the role of abscisic acid in the initiation of bud dormancy in *Alnus glutinosa* and *Betula pubescens*. *Planta* 125, 235-242.
Jacobs, D.L. (1947) An ecological life-history of *Spirodela polyrrhiza* (greater duckweed) with emphasis on the turion phase. *Ecological Monographs* 17, 437-469.
Johansen, L.G., Oden, P.-C. and Junttila, O. (1986) Abscisic acid and cessation of apical growth in *Salix pentandra*. *Physiologia Plantarum* 66, 409-412.
Landolt, E. (1986) *The Family of Lemnaceae - a Monographic Study. Vol. 1. Biosystematic Investigations in the Family of Duckweeds (Lemnaceae) (Vol. 2).* Veröffentlichungen des Geobotanischen Institutes der ETH, Stiftung Rübel, Zürich.
Lenton, J.R., Perry, V.M. and Saunders, P.F. (1972) Endogenous abscisic acid in relation to photoperiodically induced bud dormancy. *Planta* 106, 13-22.
Loewus, F.A. (1990) Inositol biosynthesis. In: Morré, D.J., Boss, W.F. and Loewus, F.A. (eds) *Inositol Metabolism in Plants*. Wiley-Liss, New York, pp. 13-19.
Loewus, F.A., Everard, J.D. and Young, K.A. (1990) Inositol metabolism precursor role and breakdown. In: Morré, D.J., Boss, W.F. and Loewus, F.A. (eds) *Inositol Metabolism in Plants*. Wiley-Liss, New York, pp. 21-44.
Perry, T.O. (1968) Dormancy, turion formation, and germination by different clones of *Spirodela polyrrhiza*. *Plant Physiology* 43, 1866-1869.
Perry, T.O. and Byrne, O.R. (1969) Turion induction in *Spirodela polyrrhiza* by abscisic acid. *Plant Physiology* 44, 784-785.
Perry, T.O. and Hellmers, H. (1973) Effects of abscisic acid on growth and dormancy of two races of red maple. *Botanical Gazette* 134, 283-289.

Phillips, I.D.J., Miners, J. and Roddick, J.G. (1980) Effects of light and photoperiodic conditions in abscisic acid in leaves and roots of *Acer pseudoplatanus* L. *Planta* 149, 118-122.

Powell, L.E. (1976) Effect of photoperiod on endogenous abscisic acid in *Malus* and *Betula*. *HortScience* 11, 489-499.

Powell, L.E. (1988) The hormonal control of bud and seed dormancy in woody plants. In: Davies, P.J. (ed.) *Plant Hormones and Their Role in Plant Growth and Development*. Kluwer, Dordrecht, pp. 539-552.

Qamaruddin, M., Dormling, I., Ekberg, I., Eriksson, G. and Tillberg, E. (1993) Abscisic acid content at defined levels of bud dormancy and frost tolerance in two contrasting populations of *Picea abies* grown in a phytotron. *Physiologia Plantarum* 87, 203-210.

Rinne, P., Saarelainen, A. and Junttila, O. (1994) Growth cessation and bud dormancy in relation to ABA level in seedlings and coppice shoots of *Betula pubescens* as affected by a short photoperiod, water stress and chilling. *Physiologia Plantarum* 90, 451-458.

Robinson, P.M. and Wareing, P.F. (1964) Chemical nature and biological properties of the inhibitor varying with photoperiod in sycamore (*Acer pseudoplatanus*). *Physiologia Plantarum* 17, 314-323.

Saks, Y., Negbi, M. and Ilan, I. (1980) Involvement of native abscisic acid in the regulation of onset of dormancy in *Spirodela polyrrhiza*. *Australian Journal of Plant Physiology* 7, 73-79.

Saunders, P.F., Harrison, M.A. and Alvim, R. (1974) Abscisic acid and tree growth. In: *Plant Growth Substances 1973*. Hirokawa, Tokyo, pp. 871-881.

Silverthorne, J. and Tobin, E.M. (1990) Post-transcriptional regulation of organ-specific expression of individual *rbc*S mRNAs in *Lemna gibba*. *Plant Cell* 2, 1181-1190.

Smart, C.C. and Fleming, A.J. (1993) A plant gene with homology to D-*myo*-inositol-3-phosphate synthase is rapidly and spatially up-regulated during an abscisic acid-induced morphogenic response in *Spirodela polyrrhiza*. *Plant Journal* 4, 279-293.

Smart, C.C. and Trewavas, A.J. (1983a) Abscisic-acid-induced turion formation in *Spirodela polyrrhiza* L. I. Production and development of the turion. *Plant Cell and Environment* 6, 507-514.

Smart, C.C. and Trewavas, A.J. (1983b) Abscisic-acid-induced turion formation in *Spirodela polyrrhiza* L. II. Ultrastructure of the turion; a stereological analysis. *Plant Cell and Environment* 6, 515-522.

Smart, C.C. and Trewavas, A.J. (1984) Abscisic-acid-induced turion formation in *Spirodela polyrrhiza* L. III. Specific changes in protein synthesis and translatable RNA during turion development. *Plant Cell and Environment* 7, 121-132.

Smart, C., Longland, J. and Trewavas, A. (1987) The turion: a biological probe for the molecular action of abscisic acid. In: Fox, J.E. and Jacobs, M. (eds) *Molecular Biology of Plant Growth Control*. Alan Liss, New York, pp. 345-359.

Stewart, G.R. (1969) Abscisic acid and morphogenesis in *Lemna polyrrhiza* L. *Nature* 221, 61-62.

Trewavas, A.J. and Jones, H.G. (1991) An assessment of the role of ABA in plant development. In: Davies, W.J. and Jones, H.G. (eds) *Abscisic Acid: Physiology and Biochemistry*. Bios Scientific Publishers, Oxford, pp. 169-188.

Wareing, P.F. (1956) Photoperiodism in woody plants. *Annual Review of Plant Physiology* 7, 191-214.

Wareing, P.F. and Phillips, I.D.J. (1983) Abscisic acid in bud dormancy and apical dominance. In: Addicott, F.T. (ed.) *Abscisic Acid*. Praeger, New York, pp. 301-329.

20 Characterization of Genes Expressed When Dormant Seeds of Cereals and Wild Grasses are Hydrated and Remain Growth-Arrested

M.K. WALKER-SIMMONS[1] AND PETER J. GOLDMARK[2]
[1]USDA-ARS, Washington State University, Pullman, WA 99164-6420, USA; [2]DJR Research, Star Route 69, Okanogan, WA 98840, USA

Introduction

This chapter describes recent progress in identifying and characterizing genes expressed when dormant seeds of cereals and wild grasses are hydrated and remain growth arrested. The molecular and biochemical constraints that impose dormancy in hydrated seeds are not yet known. During the first hours after imbibition both dormant and non-dormant seeds display the same physiological responses including similar rates of water uptake and new protein synthesis (Bewley and Black, 1985). After further hydration non-dormant seeds exhibit cell expansion and seed germination, while dormant seeds remain growth arrested. By characterizing the genes that are expressed in hydrated dormant seeds, and not in germinating seeds, it is our intent to advance understanding of the mechanisms that impose seed dormancy in mature cereal and wild grass seeds.

Regulation of seed dormancy affects critical agricultural problems in cereal crop production. Field grown wheat and barley seeds with low dormancy levels are vulnerable to preharvest sprouting during rainy weather. Sprouting damage reduces grain quality and the economic value of the grain. Sprouting damage could be reduced by selection for varieties that maintain dormancy through harvest, yet selection markers associated consistently with sprouting resistance are not yet available. Wheat and barley production is also hindered by grassy weed pests. Persistent dormancy of wild grass seeds results in sporadic germination of seeds in the seed bank and the germinated plants compete with cereal plants for light, nutrients and water. Once the wild grass plants are established, it is difficult to control them without harming crop plants. New strategies are thus needed to develop better selection markers for sprouting resistance in cereals and to overcome persistent weed seed dormancy.

Development of new strategies requires improved understanding of the molecular processes that regulate seed dormancy in cereals and wild grasses. Our approach is to determine the molecular and biochemical processes that occur when dormant cereal and grassy weed seeds are hydrated but remain growth-arrested. Here, we review our recent progress.

Gene Expression in Hydrated Dormant Seeds of Wheat

New protein synthesis in hydrated dormant seeds

Comparison of new protein synthesis when dormant and non-dormant wheat seed embryonic axes are imbibed shows similar patterns of new protein synthesis occur during the first 4-6 h of imbibition (Ried and Walker-Simmons, 1990). From 8 h onward dormant seed axes exhibit prolonged synthesis of many heat-soluble proteins. In contrast, non-dormant seed axes initially synthesize a few of the heat-soluble proteins, but after 12 h synthesis declines in these axes.

A comparison of the wheat cultivar Kitakei, which produces dormant seeds, and a mutant line, which is ABA-insensitive and produces non-dormant seeds, has shown that the mRNA for one specific polypeptide, called polypeptide *e*, is not expressed in the mutant embryos (Kawakami and Noda, 1993). When dormant Kitakei seeds are hydrated, polypeptide *e* mRNA and protein levels are maintained during imbibition.

Many proteins synthesized in hydrated dormant seeds are ABA-responsive

Many of the proteins synthesized in dormant seed axes are ABA-responsive. This is expected because in mature wheat seeds embryo sensitivity to ABA as a germination inhibitor correlates with intensity of seed dormancy (Walker-Simmons, 1987). Low (0.05-0.5 µM) ABA concentrations inhibit germination of embryonic axes isolated from dormant grain and prolong synthesis of many ABA-regulated proteins. These same low ABA concentrations have little effect on embryonic axes from non-dormant grain. Many of the ABA-responsive proteins synthesized in hydrated dormant embryos are very hydrophilic and have other unique biochemical properties that make them heat soluble (Jacobsen and Shaw, 1989).

In wheat, we have exploited the enhanced ABA responsiveness of embryonic axes from dormant compared to non-dormant seeds to differentially screen for mRNAs preferentially expressed in hydrated dormant seeds (Morris *et al.*, 1991). Using these methods we obtained cDNA clones for a number of genes expressed preferentially in dormant seeds. Four of the cloned mRNAs correspond to late embryogenesis abundant (LEA) proteins including clones with sequence homology to *Em* (group 1 LEA), dehydrin (*rab*, group 2 LEA), group 3 LEA(I) and group 3 LEA(II) (reviewed in Curry and Walker-Simmons, 1993).

The LEA cDNA clones have been used to measure LEA mRNA levels in imbibed dormant and non-dormant seeds (Fig. 20.1). At the indicated times embryos were isolated from the seeds, RNA extracted and transcript levels measured. Results (Fig. 20.1) reveal that there is prolonged LEA gene expression in hydrated dormant seed embryos. LEA mRNA levels decline and disappear in germinating seeds. Antibodies to the wheat group 2 and group 3

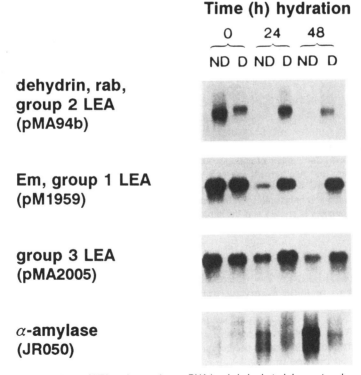

Fig. 20.1. Comparison of LEA and α-amylase mRNA levels in hydrated dormant and non-dormant wheat seed embryos (modified from Morris *et al.*, 1990). Dormant and non-dormant seeds were imbibed in water at time 0. At the times indicated the embryos were isolated from the whole seeds, RNA was extracted and the levels of LEA and α-amylase mRNA determined by Northern blot analysis.

LEA(I) (Ried and Walker-Simmons, 1990, 1993) have also been used to demonstrate that the corresponding LEA proteins are maintained at elevated levels in hydrated dormant seeds. Upon hydration of non-dormant seeds (Fig. 20.1), LEA mRNA levels decline and disappear, but as would be expected in germinating seeds, there is an increase in α-amylase mRNA.

Hydrated dormant seeds of barley also exhibit prolonged expression of the barley group 3 LEA(I), HVA1 (Hong et al., 1992). Levels of the barley LEA decline when non-dormant barley seeds are imbibed. While the functional role of LEA gene expression in dormant seeds of wheat and barley is not known, it is possible that the unique biochemical properties of the LEAs contribute to the maintenance of dormancy in hydrated seeds. Perhaps, LEAs serve as desiccation protectants during repeated drying and hydration of dormant seeds under field conditions (Lane, 1991).

Role of protein phosphorylation/dephosphorylation in dormant seeds

We are investigating the role of protein phosphorylation/ dephosphorylation by protein kinases and phosphatases in seed dormancy. Protein kinases often act in the transduction of external signals and could have a role in the effects of environmental conditions on expression of dormancy. We have cloned a protein kinase mRNA (PKABA1) that accumulates in mature wheat seed embryos and that is responsive to applied ABA (Anderberg and Walker-Simmons, 1991; Holappa and Walker-Simmons, 1995). When dormant seeds are imbibed, embryonic PKABA1 mRNA levels remain high for as long as the seeds are dormant. The kinase mRNA declines and disappears in embryos of germinating seeds (R.J. Anderberg and M.K. Walker-Simmons, unpublished data). Current research focuses on the role of this kinase in dormant seeds including characterization of the kinase activity and its substrate(s).

A potential role of phosphorylation-dependent responses in maintenance of seed dormancy is also supported by recent characterization of the *abi1* mutant of *Arabidopsis*. This ABA-insensitive mutant affects seed dormancy and the *ABI1* gene encodes a protein with similarity to protein phosphatases (Leung et al., 1994; Meyer et al., 1994).

Gene Expression in Hydrated Dormant Seeds of Wild Grasses

Protein synthesis and differential gene expression in dormant seeds

Differences in gene expression in dormant and non-dormant caryopses of *Avena fatua* (wild oats) have been determined (Dyer, 1993; Li and Foley,

1994, 1995; Johnson et al., 1995). By 3 h during imbibition, differences in *in vitro* translation products of embryonic mRNA populations during imbibition of dormant and non-dormant *Avena fatua* (wild oats) caryopses have been detected (Dyer, 1993). The cDNA clones for genes expressed differentially in dormant and non-dormant caryopses are being characterized (Li and Foley, 1994, 1995; Johnson et al., 1995). Recent research focuses on the effects of after-ripening on mRNA stability (Li and Foley, 1995).

Bromus secalinus – *a model system for wild grass seed dormancy*

We have chosen *Bromus secalinus* (cheat or chess) as a model system for wild grass seed dormancy because mature seeds possess high levels of dormancy at maturity but dormancy can be dissipated relatively rapidly by after-ripening (Goldmark et al., 1992). Unlike wheat and barley, the *B. secalinus* seed embryos exhibit true embryonic dormancy in that embryos isolated from dormant seeds remain growth-arrested when hydrated.

pBS128 mRNA levels increase in hydrated dormant seed embryos

Using cDNA cloning and differential screening of RNA isolated from dormant and non-dormant seed embryos, we have cloned mRNAs that are preferentially expressed in the hydrated dormant embryos. One cDNA clone (pBS128) is remarkable in that the clone hybridizes to a mRNA that increases considerably when dormant seed embryos are imbibed (Goldmark et al., 1992). A comparison of pBS128 transcript levels in dormant and non-dormant *B. secalinus* seed embryos upon hydration is shown in Fig. 20.2. The pBS128 mRNA was present in both mature dormant and non-dormant embryos before imbibition at time 0. By 6 h post-imbibition, pBS128 mRNA levels increased significantly. Over a fourfold increase in pBS128 mRNA levels has been measured after 48 h imbibition of dormant seed embryos (Goldmark et al., 1992). When non-dormant seed embryos were imbibed (Fig. 20.2) pBS128 mRNA levels declined with imbibition time and were not detectable by 6 h post-imbibition. The pBS128 mRNA is ABA-responsive and down-regulated by gibberellic acid. Sequence analysis of the pBS128 cDNA indicated a unique sequence not previously reported and that the mRNA corresponds to a basic polypeptide. No similarities to other ABA-responsive genes, including the LEA proteins, were found (Goldmark et al., 1992).

pBS128 transcripts are up-regulated in hydrated dormant seeds from cereals and other wild grass species

The pBS128 transcript is expressed in a wide range of wild grass species. Transcripts that hybridize to the pBS128 cDNA clone have been detected in

Time (h) hydration

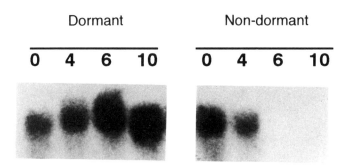

Fig. 20.2. Comparison of embryonic pBS128 mRNA levels in imbibed dormant and non-dormant *B. secalinus* seeds. Seeds were imbibed in water. At the indicated times, embryos were isolated, RNA extracted and the levels of pBS128 mRNA determined by Northern blot analysis.

mature dormant seed embryos of *Bromus tectorum* (downy brome), *Aegilops cylindrica* (jointed goatgrass), *Secale cereale* (common rye), *Avena fatua* (wild oats), and *Oryza sativa* (red rice) (Goldmark *et al.*, 1993, and unpublished data). A comparison of dormant and non-dormant seeds of downy brome after 24 h imbibition shows that the pBS128 transcript is expressed in the hydrated dormant seeds but by that time is not detected in the non-dormant seeds (Fig. 20.3). *Bromus secalinus* seeds exhibit true embryonic dormancy, which is not the case for wheat and barley seed embryos. Embryos isolated from dormant wheat and barley seeds will germinate readily. Thus, it was of interest to determine if the pBS128 gene is expressed in these cereal seed embryos lacking true embryonic dormancy. As shown in Fig. 20.3, the clone does hybridize to transcripts that are expressed in hydrated dormant seeds of both wheat and barley. At the post-imbibition times sampled, the pBS128 transcript was not detectable in the non-dormant seed embryos. Reduced pBS128-related expression in embryos of wheat and barley is thus not an explanation for the lack of true embryonic dormancy in these cereal species.

A barley clone, B15C, corresponding to transcripts expressed in both the aleurone layer and the embryo of developing seeds, has been identified with sequence homology (95% identify) to pBS128 (Aalen *et al.*, 1994). B15C encodes a protein that is identical to the pBS128 encoded protein except for five amino acid substitutions (Aalen *et al.*, 1994).

Results demonstrating that pBS128-related transcripts are expressed in dormant seeds of a wide variety of cereal and wild grass species, and the conservation of the pBS128 (B15C) sequence in *B. secalinus* and barley, suggest that pBS128 expression is associated with maintenance of seed dormancy in hydrated seeds of monocots. Recently, pBS 128 has been shown to have sequence homology with an antioxidant enzyme from mammalian brain (Chae *et al.*, 1994).

Concluding Remarks

Molecular and biochemical differences between dormant and non-dormant seeds are just beginning to be identified. Studies in a variety of cereals and wild grasses indicate that similar genes are often expressed specifically in hydrated dormant seeds. Few genes with known functions have been identified, though the identification of protein kinases and phosphatases associated with seed dormancy is a promising research area. The future challenge is to

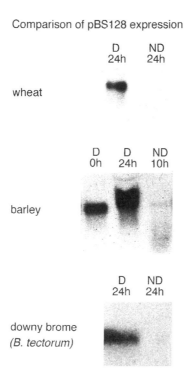

Fig. 20.3. Comparison of embryonic pBS128 mRNA levels in imbibed dormant and non-dormant wheat, barley and downy brome seeds. In wheat, dormant and non-dormant seeds were imbibed for 24 h. In barley, dormant seeds were imbibed for 0 and 24 h, and non-dormant for 10 h. In downy brome, dormant and non-dormant seeds were imbibed for 24 h. pBS128 mRNA levels were determined by Northern blot analysis of RNA extracted from isolated embryos.

determine the critical regulatory genes responsible for expression of seed dormancy in cereals and wild grasses.

References

Aalen, R.B., Opsahi-Ferstad, H.-G., Linnestad, C. and Olsen, O.-A. (1994) Transcripts encoding an oleosin and a dormancy-related protein are present in both the aleurone layer and the embryo of developing barley (*Hordeum vulgare* L.) seeds. *Plant Journal* 5, 385-396.

Anderberg, R.J. and Walker-Simmons, M.K. (1991) Isolation of a wheat cDNA clone for an abscisic acid-inducible transcript with homology to protein kinases. *Proceedings of the National Academy of Sciences, USA* 89, 10183-10187.

Bewley, J.D. and Black, M. (1985) *Seeds: Physiology of Development and Germination*. Plenum Press, New York.

Chae, H.Z., Robison, K., Poole, L.B., Church, G., Storz, G. and Rhee, S.G. (1994) Cloning and sequencing of thiol-specific antioxidant from mammalian brain: alkyl hydroperoxide reductase and thiol-specific antioxidant define a large family of antioxidant enzymes. *Proceedings of the National Academy of Sciences, USA* 91, 7017-7021.

Curry, J. and Walker-Simmons, M.K. (1993) Sequence analysis of wheat cDNAs for abscisic acid-responsive genes expressed in dehydrated wheat seedlings and the cyanobacterium, *Anabaena*. In: Close, T.J. and Bray, E.A. (eds) *Plant Responses to Cellular Dehydration During Environmental Stress*. American Society of Plant Physiologists, Rockville, pp. 128-136.

Dyer, W.E. (1993) Differential gene expression during imbibition of dormant and non-dormant *Avena fatua* embryos. In: Walker-Simmons, M.K. and Ried, J.L. (eds) *Pre-Harvest Sprouting in Cereals 1992*. American Association of Cereal Chemists, St Paul, pp. 303-311.

Goldmark, P.J., Curry, J., Morris, C.F. and Walker-Simmons, M.K. (1992) Cloning and expression of an embryo-specific mRNA up-regulated in hydrated dormant seeds. *Plant Molecular Biology* 19, 433-441.

Goldmark, P.J., Dykes, J. and Walker-Simmons, M.K. (1993) Expression of a *Bromus secalinus* transcript associated with seed dormancy in *Avena fatua* and other grass weeds. In: Walker-Simmons, M.K. and Ried, J.L. (eds) *Pre-Harvest Sprouting in Cereals 1992*. American Association of Cereal Chemists, St Paul, pp. 312-316.

Holappa, L.D. and Walker-Simmons, M.K. (1995) The wheat abscisic acid-responsive protein kinase mRNA, PBAKA1, is up-regulated by dehydration, cold temperature and osmotic stress. *Plant Physiology* 108, 1203-1210.

Hong B., Barg, R. and Ho, T.-H. D. (1992) Developmental and organ-specific expression of an ABA- and stress-induced protein in barley. *Plant Molecular Biology* 18, 663-674.

Jacobsen, J.V. and Shaw, D.C. (1989) Heat-stable proteins and abscisic acid action in barley aleurone cells. *Plant Physiology* 91, 1520-1526.

Johnson, R.R., Cranston, H.J., Chaverra, M.E. and Dyer, W.E. (1995) Characterization of cDNA clones for differently expressed genes in embryos of dormant and non-dormant *Avena fatua* L. caryopses. *Plant Molecular Biology* 28, 113-122.

Kawakami, N. and Noda, K. (1993) Embryonic mRNAs of non-dormant mutants of wheat. In: Walker-Simmons, M.K. and Ried, J.L. (eds) *Pre-Harvest Sprouting in Cereals* 1992. American Association of Cereal Chemists, St Paul, pp. 195-199.

Lane, B.G. (1991) Cellular desiccation and dehydration: developmentally regulated proteins, and the maturation and germination of seed embryos. *FASEB Journal* 5, 2893-2901.

Leung, J., Bouvier-Durand, M., Morris, P.-C., Guerrier, D., Chefdor, F. and Giraudat, J. (1994) Arabidopsis ABA response gene *ABI1*: features of a calcium-modulated protein phosphatase. *Science* 264, 1448-1452.

Li, B. and Foley, M.E. (1994) Differential polypeptide patterns in imbibed dormant and after-ripened *Avena fatua* embryos. *Journal of Experimental Botany* 45, 275-279.

Li, B. and Foley, M.E. (1995) Cloning and characterization of differentially expressed genes in imbibed dormant and after-ripened *Avena fatua* embryos. *Plant Molecular Biology* 29, 823-831.

Meyer, K., Leube, M.P. and Grill, E. (1994) A protein phosphatase 2C involved in ABA signal transduction in *Arabidopsis thaliana*. *Science* 264, 1452-1455.

Morris, C.F., Anderberg, R.J., Goldmark, P.J. and Walker-Simmons, M.K. (1991) Molecular cloning and expression of abscisic acid-responsive genes in embryos of dormant wheat seeds. *Plant Physiology* 95, 814-821.

Ried, J.L. and Walker-Simmons, M.K. (1990) Synthesis of abscisic acid-responsive, heat-stable proteins in embryonic axes of dormant wheat grain. *Plant Physiology* 93, 662-667.

Ried, J.L. and Walker-Simmons, M.K. (1993) Group 3 late embryogenesis abundant proteins in desiccation-tolerant seedlings of wheat (*Triticum aestivum* L.). *Plant Physiology* 102, 125-131.

Walker-Simmons, M. (1987) ABA levels and sensitivity in developing wheat embryos of sprouting resistant and susceptible cultivars. *Plant Physiology* 84, 61-66.

21 Analysis of cDNA Clones for Differentially Expressed Genes in Dormant and Non-dormant Wild Oat (*Avena fatua* L.) Embryos

Russell R. Johnson, Harwood J. Cranston, Martha E. Chaverra and William E. Dyer
Department of Plant, Soil, and Environmental Sciences, Montana State University, Bozeman, MT 59717, USA

Introduction

The control of seed dormancy and its release presumably involves the action of specific genes, which in turn respond to particular environmental and hormonal stimuli. The wild oat, *Avena fatua*, is a good model species for the study of seed dormancy in grasses since its persistence as a problem weed has prompted extensive studies of the physiology, biochemistry, and genetics of dormancy (Simpson, 1990). Previous work (Dyer, 1993) has indicated that while most mRNAs and proteins are equally abundant in dormant and non-dormant embryos during early imbibition, a few mRNA species are preferentially abundant in either dormant or non-dormant embryos. These results imply that there are a number of genes whose levels of expression are different in the embryos of dormant vs. non-dormant wild oat seeds. It is reasonable to hypothesize that some of these genes may be involved in the regulation of dormancy. To gain a better understanding of these differentially expressed genes and their possible roles in the maintenance or release of dormancy, we have begun to clone and analyse cDNAs for such genes.

Materials and Methods

The *A. fatua* line AN265 (Naylor and Jana, 1976) was grown in the greenhouse and seeds were collected as described (Dyer, 1993). Dormant seeds were stored at −20°C for 1 year after harvest while non-dormant seeds were obtained by dry after-ripening for 1 year at 23±4°C. Embryos were excised from hand-peeled seeds (caryopses) that had been imbibed for 0, 3 or 6 h as described (Dyer, 1993) and total RNA was prepared by a standard protocol (Ausubel et al., 1994). Differential display (DD) and cloning of cDNA fragments were performed essentially as described (Liang and Pardee, 1992; Liang et al., 1993). Labelling of cDNA clones with [α-^{32}P]dCTP and northern hybridizations were done using standard techniques (Sambrook et al., 1989; Kroczek and Siebert, 1990). After hybridization, membranes were exposed to X-ray film, and the signal intensities were quantified using a Molecular Dynamics Model 400E Image Analyzer with ImageQuant 3.3 software. Sequencing was performed using the Sequenase Kit (United States Biochemical, Cleveland, Ohio). Database searches were made using the BLAST algorithm (Altschul et al., 1990) (National Center of Biotechnology Information, NIH, Bethesda, Maryland).

Results and Discussion

A variety of gene expression patterns were observed in dormant and non-dormant caryopses at 0 to 6 h of imbibition (Fig. 21.1). Most mRNA species were equally abundant in dormant and non-dormant embryos at 0, 3 and 6 h of imbibition. In addition, some cDNA bands were found to correspond to messages which increased or decreased during the course of imbibition but did not differ in abundance between dormant and non-dormant embryos. In contrast, other cDNA bands represented mRNAs which were differentially abundant in dormant vs. non-dormant embryos. The bands present on a DD gel like those shown (Fig. 21.1) represent 200 to 500 bp of the 3' end of the cDNA. A number of such bands observed to be differential in the dormant vs. non-dormant embryos were excised from the gel and cloned (Johnson et al., 1995).

To date, DD has been performed using 2 of the 12 possible 3' anchored primers, each in combination with 20 arbitrary 5' decamer primers (Liang and Pardee, 1992). Thus, approximately 17% of the mRNA species present in wild oat embryos have been screened. cDNA fragments that appeared to be differential have been utilized as probes for northern hybridizations and five of these clones have been confirmed to correspond to dormancy-associated genes. A much greater number of non-dormancy-associated cDNAs have been identified and analysis has been restricted to only a few of these clones.

The time course of mRNA accumulation corresponding to several differentially expressed genes is shown in Fig. 21.2. The onset of visible germination, as judged by radicle emergence, generally occurs between 24 and 36 h under these conditions. Clone AFD1 (*Avena fatua* dormant) hybridized to a

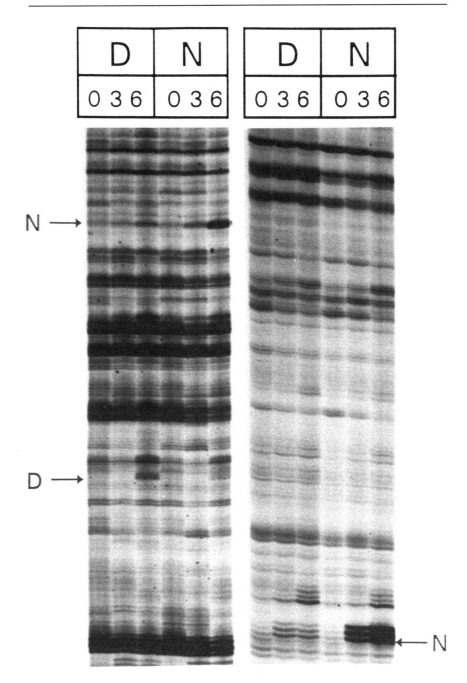

Fig. 21.1. Differential display of cDNA bands corresponding to *A. fatua* embryonic mRNAs. Total RNA from dormant (D) and non-dormant (N) embryos imbibed in water for 0, 3 or 6 h was used for reverse transcription-PCR. Bands corresponding to mRNA species more abundant in dormant or non-dormant embryos are indicated by arrows.

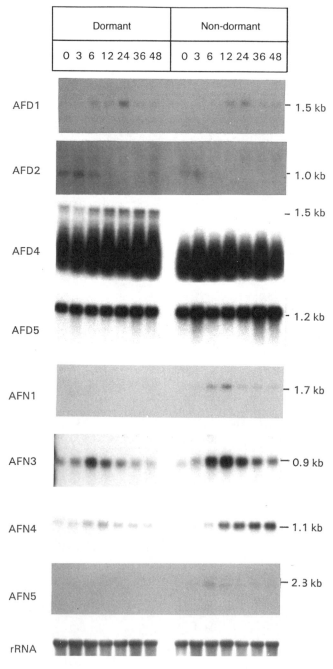

Fig. 21.2. Northern hybridization of cloned cDNAs AFD1, AFD2, AFD4, AFD5, AFN1, AFN3, AFN4 and AFN5 with total RNA from dormant and non-dormant *A. fatua* embryos imbibed for 0 to 48 h in water. The bottom panel shows hybridization to a 26S rRNA probe.

1.5 kb mRNA that was more abundant in dormant embryos than in non-dormant embryos at 6 to 24 h of imbibition. Two mRNA species were recognized by AFD2, a 1.0 kb message that was present in higher amounts in dormant embryos during the first few hours of imbibition and a 1.4 kb message which was constitutively present in both dormant and non-dormant embryos (Johnson et al., 1995).

Clone AFD4 hybridized with an intact 1.5 kb message as well as a large amount of lower molecular weight material which appeared to be partially degraded remnants of this mRNA. While this degraded mRNA, which is not likely to be translationally competent, was present in both dormant and non-dormant embryos, the intact message was only observed in dormant embryos. It is interesting to note that both AFD2 and AFD4 mRNAs were more abundant in the embryos of dry dormant than in dry non-dormant seeds (0 h of imbibition). Since all seeds used in these experiments came from the same batch and the only difference between dormant and non-dormant seeds was the postharvest storage temperature, it appears that these transcripts have disappeared in non-dormant seeds during the period of dry after-ripening (R. Johnson and W. Dyer, unpublished). The phenomenon of differential mRNA degradation during after-ripening may prove to be a mechanism for seed dormancy regulation. Clone AFD5, which initially appeared to be differential on the DD gels, actually hybridized with an abundant constitutive mRNA in northern hybridizations.

Clone AFN1 (*Avena fatua* non-dormant) hybridized with a 1.7 kb mRNA that reached peak levels at 12 h of imbibition and was greater than 10 times more abundant in non-dormant than in dormant embryos. Clone AFN3 hybridized with a 0.9 kb mRNA abundantly expressed in non-dormant embryos during the later stages of imbibition and during the start of germination (Johnson et al., 1995). The 1.1 kb mRNA recognized by AFN4 continued to accumulate in non-dormant embryos up to 48 h. Clone AFN5 hybridized to a 2.3 kb mRNA whose abundance peaked at 6 h of imbibition in non-dormant embryos and declined at later times.

DNA sequences of the cDNA clones described above were compared to sequence databases to determine if they were similar to any known genes. AFD4 appears to code for a wild oat homologue of the barley protein Z (75% identity), a seed protein of unknown function which has sequence identity to mammalian serine proteinase inhibitors (Rasmussen, 1993). AFD5, which is constitutively expressed in wild oat embryos during imbibition, has 84% sequence identity to the *cyc07* gene of *Catharanthus roseus*, which is specifically expressed during the S-phase of the cell cycle in synchronous cultures (Ito et al., 1991). The sequence of AFN3 is similar (75-85% identity) to stress-associated cDNAs from other plants that are thought to encode glutathione peroxidases (Criqui et al., 1992; Holland et al., 1993). AFN4 has 80% sequence identity to the *Arabidopsis thaliana sar1* GTP binding protein (d'Enfert, 1992) which is involved in membrane vesicle transport. Preferential expression of this mRNA in non-dormant embryos may indicate a key role for the gene product during imbibition and germination.

The hormone gibberellic acid (GA_3) is known to break dormancy of many seeds including those of wild oat (Simpson, 1990) and may have an influence on the level of expression of genes involved in maintaining dormancy or in initiating germination. The effect of imbibing seeds in 1.0 mM GA_3 (which causes germination of dormant seeds) on the accumulation of mRNAs recognized by several of the above clones was examined (Fig. 21.3). Both dormant and non-dormant embryos accumulated higher amounts of AFN4 mRNA at 36 and 48 h when seeds were imbibed in GA_3 than when imbibed in water (H.

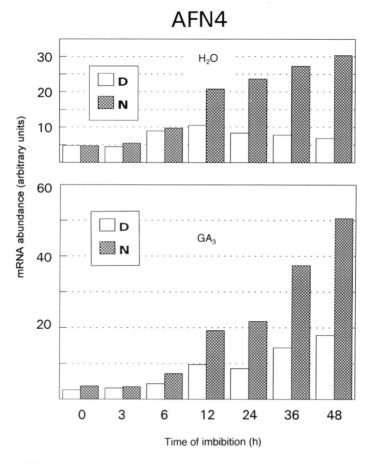

Fig. 21.3. Effect of GA_3 on AFN4 mRNA accumulation during imbibition. Dormant (D) and non-dormant (N) caryopses were imbibed in water or 1 mM GA_3 for 0 to 48 h and embryonic RNA was analysed by northern blot hybridization. Hybridization signals were quantified using a Molecular Dynamics Model 400E Image Analyzer with ImageQuant 3.3 software.

Cranston, R. Johnson, M. Chaverra and W. Dyer, unpublished). Accumulation of AFN3 mRNA in non-dormant embryos also increased when seeds were imbibed in GA_3, to about double the level seen when imbibed in water, although the general pattern of expression was not affected (Johnson et al., 1995). Accumulation of AFD2 mRNA was decreased in both dormant and non-dormant embryos imbibed in GA_3, while expression patterns of the genes corresponding to AFD4, AFD5, AFN1 and AFN5 were unaffected by this hormone. Thus, even though this treatment caused the dormant seeds to germinate, expression patterns of several mRNAs were not identical to those from non-dormant seeds imbibed in water, suggesting that GA_3-induced germination may not be physiologically identical to 'normal' germination.

Conclusions

Partial cDNAs for several genes that are expressed differentially in dormant and non-dormant wild oat embryos have been cloned. The expression patterns of some of these genes are modulated by GA_3. The products of such differentially expressed genes may prove to be important regulators of seed dormancy or germination initiation. For example, the protein encoded by AFD4 (protein Z homologue) could be a repressor of germination that maintains the seeds in a dormant state. A clone such as AFN5, which is expressed very early during imbibition, could code for a protein that is needed to signal or initiate the early steps in germination. Further analysis of these and other differentially expressed genes should help to further our understanding of the molecular and biochemical nature of seed dormancy.

References

Altschul, S., Gish, W., Miller, W., Myers, E. and Lipman, D. (1990) Basic local alignment search tool. *Journal of Molecular Biology* 215, 403-410.

Ausubel, F., Brent, R., Kingston, R.E., Moore, D.D., Seidman, J.G., Smith, J.A. and Struhl, K. (1994) *Current Protocols in Molecular Biology*. John Wiley & Sons, New York.

Criqui, M., Jamet, E., Parmentier, Y., Marbach, J., Durr, A. and Fleck, J. (1992) Isolation and characterization of a plant cDNA showing homology to animal glutathione peroxidases. *Plant Molecular Biology* 18, 623-627.

d'Enfert, C., Gensse, M. and Gaillardin, C. (1991) Fission yeast and a plant have functional homologues of the *sar1* and *sec12* proteins involved in ER to Golgi traffic in budding yeast. *EMBO Journal* 11, 4205-4211.

Dyer, W. (1993) Dormancy associated embryonic mRNAs and proteins in imbibing *Avena fatua* L. caryopses. *Physiologia Plantarum* 88, 201-211.

Holland, D., Ben-Hayyim, G., Faltin, Z., Camoin, L., Strosberg, A. and Eshdat, Y. (1993) Molecular characterization of salt-stress-associated protein in citrus: protein and cDNA sequence homology to mammalian glutathione peroxidases. *Plant Molecular Biology* 21, 923-927.

Ito, M., Kodama, K. and Komamine, A. (1991) Identification of a novel S-phase-specific gene during the cell cycle in synchronous cultures of *Catharanthus roseus* cells. *Plant Journal* 1, 141-148.

Johnson, R.R., Cranston, H.J, Chaverra, M.E. and Dyer, W.E. (1995) Characterization of cDNA clones for differentially expressed genes in embryos of dormant and non-dormant *Avena fatua* L. caryopses. *Plant Molecular Biology* 28, 113-122.

Kroczek, R. and Siebert, E. (1990) Optimization of northern analysis by vacuum blotting, RNA-transfer visualization, and ultraviolet fixation. *Analytical Biochemistry* 184, 90-95.

Liang, P. and Pardee, A. (1992) Differential display of eukaryotic messenger RNA by means of the polymerase chain reaction. *Science* 257, 967-971.

Liang, P., Auerboukh, L. and Pardee, A. (1993) Distribution and cloning of eukaryotic mRNAs by means of differential display: refinements and optimization. *Nucleic Acids Research* 21, 3269-3275.

Naylor, J. and Jana, S. (1976) Genetic adaptation for seed dormancy in *Avena fatua*. *Canadian Journal of Botany* 54, 306-312.

Rasmussen, S. (1993) A gene coding for a new plant serpin. *Biochimica et Biophysica Acta* 1172, 151-154.

Sambrook, J., Fritsch, E. and Maniatis, T. (1989) *Molecular Cloning: A Laboratory Manual*. Cold Spring Harbor Press, Cold Spring Harbor.

Simpson, G. (1990) *Seed Dormancy in Grasses*. Cambridge University Press, Cambridge.

22 Photoperiod-Associated Gene Expression during Dormancy in Woody Perennials

GARY D. COLEMAN[1] AND TONY H.H. CHEN[2]
[1]*Department of Horticulture, University of Maryland, College Park, MD 20742-5611, USA;* [2]*Department of Horticulture, Oregon State University, Corvallis, OR 97331, USA*

Introduction

Photoperiod influences both vegetative and reproductive growth of plants. Many temperate woody perennial species have evolved physiological and developmental responses to changes in photoperiod. These responses range from the expression of specific genes to complex physiological processes. Among the vegetative growth processes influenced by photoperiod are seed germination, stem elongation, leaf growth, root formation, senescence, abscission, cold hardiness, and dormancy (reviewed in Vince-Prue, 1975; Salisbury, 1981).

Woody perennial bud dormancy is a complex physiological state that involves the transition from active growth to the dormant condition. This transition can be divided into a number of individual physiological events, the study of which may assist in elucidation of the complex state. These physiological processes include, but are not limited to, growth cessation, cessation of leaf initiation, bud scale formation, development of cold hardiness and desiccation tolerance, leaf senescence, and carbon and nitrogen storage (Weiser, 1970; Nooden and Weber, 1978; Coleman *et al.*, 1993). The study of seasonal nitrogen cycling may serve as a model for the analysis of seasonal physiological processes associated with bud dormancy.

This chapter provides an overview of photoperiodic responses during dormancy in woody plants and examines the physiological and molecular aspects of photoperiod-associated nitrogen mobilization and storage in temperate deciduous trees.

Photoperiodism and Dormancy

The development of vegetative dormancy in woody perennials is an adaptive mechanism which allows plants to survive seasonal fluctuations in environmental factors such as temperature and water. Early studies established that long days (LD) tend to promote vegetative growth in woody perennials whereas short days (SD) tend to hasten the onset of bud formation and dormancy (Wareing, 1947; van der Veen, 1951; Downs and Borthwick, 1957; Nitsch, 1957). Undoubtedly, in many woody plant species photoperiod is an important environmental signal for the initiation of dormancy. The adaptive value of using photoperiod as a timekeeping mechanism for synchronizing growth transitions is evident since photoperiod is the one environmental cue that does not vary from year to year.

Classically, the effect of photoperiod on the induction of dormancy has been divided into four categories (Nitsch, 1957; reviewed in Vince-Prue, 1975). In the first category, dormancy is induced by SD and prevented by LD. *Populus* is an example of a genus that behaves in this manner. The second category consists of species whereby dormancy is accelerated by SD and delayed, but not prevented, by LD, as is the case for many *Acer* species. The third category includes plants which grow in flushes and the duration between flushes is decreased by LD; examples include many species of *Quercus*. The last category consists of plants whereby dormancy induction is not influenced by photoperiod (i.e. *Malus*).

Vegetative (Bark) Storage Proteins and Photoperiod

Temperate deciduous tree species translocate 50-90% of total leaf nitrogen from senescing leaves to storage sites during autumn (Kang and Titus, 1980; Luxmoore *et al.*, 1981; Ostman and Weaver 1982; Chapin and Kedrowski, 1983; Boerner 1984) and remobilize stored nitrogen to support new growth in spring (Taylor and May, 1967; Ryan and Bormann, 1982). The majority of translocated leaf nitrogen appears to be stored as protein (Taylor and May, 1967; Chapin and Kedrowski, 1983; Oaks *et al.*, 1991). One of the primary sites of nitrogen storage during overwintering is the shoot bark (O'Kennedy and Titus, 1979; Kang and Titus, 1980; Sauter *et al.*, 1989; Wetzel *et al.*, 1989; Coleman *et al.*, 1991). New growth is correlated highly with the level of stored nitrogen (Taylor and May, 1967), which can account for 30-75% of the nitrogen transported to leaves for reuse during spring growth (Nambiar, 1984; Deng *et al.*, 1989; Millard and Neilsen, 1989; Millard and Proe, 1991). Storage protein degradation and nitrogen mobilization during regrowth is controlled by the sink strength of the new growth (Coleman *et al.*, 1993). Presumably this remobilization involves the action of specific proteases and phloem transport of amino acids derived from storage protein breakdown. In addition to the significant contribution of seasonal cycling to annual nitrogen requirements, accumulation of nitrogen for subsequent re-use provides trees with some independence from soil nutrient supplies and therefore reduces annual variations in growth (Ryan and Bormann, 1982; van den Driessche, 1984). Stored nutrients may also reduce potential losses from mineralization and subsequent leaching (Ryan and Bormann, 1982).

Vegetative storage proteins (VSPs) that appear to be important in seasonal nitrogen cycling have been reported in a number of tree species including the angiosperms *Sophora japonica* (Baba *et al.*, 1991), *Sambucus nigra* and *Robinia pseudoacacia* (Nsimba-Lubaki and Peumans, 1986), *Acer saccharum* and *Salix* × *smithiana* (Wetzel *et al.*, 1989), *Populus deltoides* (Wetzel *et al.*, 1989; Coleman *et al.*, 1991), *Populus* × *euramericana* (Van Cleve *et al.*, 1988; Stepien and Martin, 1992), *Malus* (O'Kennedy and Titus, 1979), *Prunus persica* (Arora *et al.*, 1992), *Fagus sylvatica, Fraxinus americana, Tilia americana, Alnus glutinosa, Betula papyrifera, Salix microstachya* and *Quercus rubra* (Wetzel and Greenwood, 1991). VSPs have also been reported for a number of coniferous species as well, including *Larix decidua, Pinus strobus,* and *P. sylvestris* (Wetzel and Greenwood, 1989), *Pseudotsuga menziesii* and *Picea* spp. (Roberts *et al.*, 1991), *Taxodium distichum* and *Metasequoia glyptostorbioides* (Harms and Sauter, 1991) and *Sequoiadendron giganteum* (Harms and Sauter, 1991). Because of the wide distribution of vegetative storage proteins in perennial woody plants, it would seem that a common mechanism for seasonal nitrogen storage and conservation has evolved in trees.

Poplar (*Populus* spp.) provides an excellent model for investigating the physiology and molecular biology of vegetative storage proteins and nitrogen cycling. Attributes of poplars which contribute to the use of this genus as a model system include: (i) a relatively small genome which facilitates genomic library construction and screening; (ii) easy propagation of clonal material, including propagation and regeneration by *in vitro* culture; (iii) the ability for genetic transformation by *Agrobacterium tumefaciens*; (iv) fast growth to rapidly assess introduced traits; (v) the ability to manipulate the transition from active growth to dormancy by SD treatments; and (vi) the ability to break dormancy with low temperature treatment.

A number of vegetative storage proteins have been reported to accumulate in the bark of poplar during the autumn and winter, including a 32 kDa protein (van Cleve *et al.*, 1988; Wetzel *et al.*, 1989; Coleman *et al.*, 1991), 32 and 36 kDa proteins (Langheinrich and Tischner, 1991), and 32, 36 and 38 kDa proteins (Stepien and Martin, 1992). These proteins are glycoproteins that occur as isoforms; however, both deglycosylation analysis (Langheinrich and Tischner, 1991) and peptide mapping (Stepien and Martin, 1992) indicate that these isoforms are simply a result of different degrees of glycosolation of the 32 kDa protein. These proteins are localized in the phloem parenchyma of the inner bark and in the xylem ray cells (van Cleve *et al.*,

1988; Wetzel *et al.*, 1989) and have been termed bark storage proteins (BSPs). Complementary DNA clones have been isolated and sequenced for the 32 kDa bark storage protein (Clausen and Apel, 1991; Coleman *et al.*, 1992). The processed poplar BSP has a calculated molecular weight of 351,000 and a calculated isoelectric point of 7.2, which agrees closely with the behaviour of the poplar 32 kDa BSP in SDS-PAGE and 2D-PAGE (Coleman *et al.*, 1992). The 32 kDa BSP is rich in Ser, Leu, Phe and Lys and contains one glycosylation site. It is encoded by a small multigene family of three genes, one of which (*bspa*) has been isolated and sequenced (Coleman and Chen, 1993).

When soluble bark proteins are extracted and analysed by SDS-PAGE and protein gel blots, digital image analysis reveals that the relative amount of BSP accumulation in field grown plants increases from about 10% of soluble bark proteins at the end of July to about 50% of soluble bark protein by the end of August (Coleman *et al.*, 1991). Accumulation continues throughout autumn and high BSP levels are detected during winter. The accumulation of the 32 kDa BSP is correlated with an increase in the steady-state levels of BSP mRNA (Coleman *et al.*, 1992). However, unlike the accumulation of protein (which remains at high levels throughout overwintering), the levels of BSP mRNA decline to undetectable levels during winter.

The accumulation of the 32 kDa poplar BSP in the autumn suggests that photoperiod could be an environmental signal that triggers gene expression and nitrogen storage. This hypothesis was tested by exposing poplars to either LD (16 h of light and 8 h of darkness) or SD (8 h of light and 16 h of darkness) treatments in environmental growth chambers. Within 7 days of SD treatment, an increase in the 32 kDa BSP mRNA was detected in bark tissues using RNA gel blot analysis (Coleman *et al.*, 1992). The levels of poplar BSP mRNA continued to increase with further exposure to SD and high levels were detected by 28 days of treatment. The accumulation of BSP mRNA in the bark of SD-treated poplars was correlated with an increase in the abundance of the 32 kDa BSP. Accumulation of BSP mRNA was also bark-specific, with no detection in young or mature leaves (Davis *et al.*, 1993).

Photoperiod responses often vary in different ecotypes of woody perennials that are distributed over a broad range of latitudes (Salisbury and Ross, 1992). Pauley and Perry (1954) showed that stem growth cessation in *Populus deltoides* and *P. trichocarpa* varied by ecotype and that this variation was inversely correlated with the latitude of ecotype origin. They also showed that the photoperiod response of hybrids between two contrasting ecotypes was intermediate. Similar results were also reported by Vaartaja (1960) for *Populus deltoides* and *P. tremuloides*. A role for photoperiod in poplar BSP gene expression is established further by the observation that maximum BSP mRNA accumulation occurred in early September for an ecotype native to Minnesota, while maximum mRNA levels were not detected until mid-October for an ecotype native to Texas (G.D. Coleman, unpublished data). Maximum BSP mRNA accumulation of ecotypes native to regions between Minnesota and Texas tended to be intermediate. The date of maximum BSP mRNA accumulation was inversely correlated with the latitude of ecotype origin, accounting for approximately 70% of the variation.

Nitrogen Availability and BSP Gene Expression

Since BSP is thought to be involved with nitrogen storage, it is likely that nitrogen availability may influence both LD and SD gene expression. In LD treated plants, poplar BSP levels were correlated with the amount of NH_4NO_3 provided to the plant (Coleman et al., 1994), suggesting a role in temporary nitrogen storage during growth. In soybean, a VSP accumulated in leaf tissue that appeared to function in temporary nitrogen storage during periods of excess nitrogen availability (Staswick et al., 1991). Similarly, poplar BSP appears to be involved in nitrogen storage when availability exceeds demand. Bañados (1992) also observed that leaf protein levels did not vary among young or mature leaves on plants that received different amounts of nitrogen, whereas bark protein levels increased with increased nitrogen availability. Thus, the 32 kDa BSP appears to have roles in both seasonal nitrogen storage and temporary storage during shoot growth.

In nitrogen deficient plants exposed to SD photoperiods, the accumulation of 32 kDa BSP was reduced, although BSP mRNA levels were similar to plants that were supplied with adequate levels of nitrogen (Coleman et al., 1994). In addition, when LD poplars were treated with high levels of NH_4NO_3, the 32 kDa BSP accumulated to high levels after 4 weeks of treatment while BSP mRNA levels declined. However, if these plants were then exposed to SD conditions, the level of the 32 kDa BSP did not increase beyond LD levels, whereas BSP mRNA accumulated to high levels. Therefore, in LD high nitrogen plants that have accumulated the 32 kDa BSP, SD treatment induced mRNA accumulation without a corresponding increase in the levels of BSP. In SD nitrogen deficient plants, SD treatment induced *bsp* gene expression but BSP levels were reduced. This suggests that although nitrogen availability can influence the level of BSP synthesis, BSP mRNA levels appear to be somewhat independent of nitrogen levels and more closely associated with photoperiod. Clearly, nitrogen availability is important in BSP gene expression under SD conditions; however, the relationship between photoperiod and nitrogen availability has yet to be established.

The influence of nitrogen availability on poplar BSP gene expression could be tied to nitrogen source-sink relations. In soybean, source-sink relations are known to influence the expression of VSP, and strong nitrogen sinks contain elevated levels of VSP (Staswick, 1989). Similar mechanisms may be involved in the induction of BSP accumulation by both SD and nitrogen treatment. Short-day accumulation may involve the induction of leaf senescence and subsequent nitrogen mobilization, which increases nitrogen availability in a manner similar to that resulting from LD nitrogen treatment. Therefore, increased nitrogen availability, whether from leaf senescence or nitrogen treatment, would result in *bsp* expression. However, it is also possible that SD treatment could influence *bsp* gene expression directly, and BSP synthesis would establish a nitrogen sink in bark that may act as the driving force for nitrogen mobilization from foliage.

Conclusions

Seasonal nitrogen storage is undoubtedly influenced by photoperiod. This physiological phenomenon is just one of many responses that precedes the onset of vegetative dormancy. In comparison to dormancy, nitrogen storage represents a less complex process that may serve as a component in the analysis of how photoperiod influences physiological processes associated with dormancy in woody perennial plants.

Acknowledgements

We thank Dr Karen van Zee for helpful comments on previous versions of this manuscript. This research was supported by a grant from the USDA/National Research Initiative Competitive Grants Program (92-37100-7672) to G.D.C. and T.H.H.C.

References

Arora, R., Wisniewski, M.E., and Scorza, R. (1992) Cold acclimation in genetically related (sibling) deciduous and evergreen peach (*Prunus persica* [L.] Batsch). *Plant Physiology* 99, 1562-1568.

Baba, K., Ogawa, M., Nagado, A., Kuroda, H. and Sumiya, K. (1991) Developmental changes in the bark lectin of *Sophora japonica* L. *Planta* 183, 462-470.

Bañados, M.P. (1992) Nitrogen and environmental factors affect bark storage protein gene expression in poplars. MS thesis, Oregon State University, Corvallis, Oregon.

Boerner, R.E. (1984) Foliar nutrient dynamics and nutrient use efficiency of four deciduous tree species in relation to site fertility. *Journal of Applied Ecology* 21, 1029-1040.

Chapin, F.S. and Kedrowski, R.A. (1983) Seasonal changes in nitrogen and phosphorus fractions and autumn retranslocation in evergreen and deciduous taiga trees. *Ecology* 64, 376-391.

Clausen, S. and Apel, K. (1991) Seasonal changes in the concentration of the major storage protein and its mRNA in xylem ray cells of poplar trees. *Plant Molecular Biology* 17, 669-678.

Coleman, G.D. and Chen, T.H.H. (1993) Sequence of a poplar bark storage protein gene. *Plant Physiology* 102, 1347-1348.
Coleman, G.D., Chen, T.H.H., Ernst, S.G. and Fuchigami, L.H. (1991) Photoperiod control of poplar bark storage protein accumulation. *Plant Physiology* 96, 686-692.
Coleman, G.D., Chen, T.H.H. and Fuchigami, L.H. (1992) Complementary DNA cloning of poplar bark storage protein and control of its expression by photoperiod. *Plant Physiology* 98, 687-693.
Coleman, G.D., Englert, J.M., Chen, T.H.H. and Fuchigami, L.H. (1993) Physiological and environmental requirements for poplar (*Populus deltoides*) bark storage protein degradation. *Plant Physiology* 102, 53-59.
Coleman, G.D., Bañados, M.P. and Chen, T.H.H. (1994) Poplar bark storage protein and a related wound-induced gene are differentially induced by nitrogen. *Plant Physiology* 106, 211-215.
Davis, J.M., Egelkrout, E.E., Coleman, G.D., Chen, T.H.H., Haissig, B.E., Riemenschneider, D.E. and Gordon, M.P. (1993) A family of wound-induced genes in *Populus* shares common features with genes encoding vegetative storage proteins. *Plant Molecular Biology* 23, 135-143.
Deng, X., Weinbaum, S.A. and DeJong, T.M. (1989) Use of labeled nitrogen to monitor the transition in nitrogen dependence from storage to current-year uptake in mature walnut trees. *Trees* 3, 11-16.
Downs, R.J. and Borthwick, H.A. (1957) Effects of photoperiod on growth of trees. *Botanical Gazette* 117, 310-326.
Harms, U. and Sauter, J.J. (1991) Storage proteins in the wood of Taxodiaceae and of Taxus. *Journal of Plant Physiology* 138, 497-499.
Kang, S.-M. and Titus, J.S. (1980) Qualitative and quantitative changes in nitrogenous compounds in senescing leaf and bark tissues of the apple. *Physiologia Plantarum* 50, 285-290.
Langheinrich, U. and Tischner, R. (1991) Vegetative storage proteins in poplar: induction and characterization of a 32- and a 36-kilodalton polypeptide. *Plant Physiology* 97, 1017-1025.
Luxmoore, R.J., Grizzard, T. and Strand, R.H. (1981) Nutrient translocation in the outer canopy and understory of an eastern deciduous forest. *Forest Science* 27, 505-518.
Millard, P. and Neilsen, G.H. (1989) The influence of nitrogen supply on the uptake and remobilization of stored N for the seasonal growth of apple trees. *Annals of Botany* 63, 301-309.
Millard, P. and Proe, M.F. (1991) Leaf demography and seasonal internal cycling of nitrogen in sycamore (*Acer pseudoplatanus* L.) seedlings in relation to nitrogen supply. *New Phytologist* 117, 587-596.
Nambiar, E.K.S. (1984) Plantation forests: Their scope and a perspective on plantation nutrition. In: G.D. Bowen and E.K.S. Nambiar (eds) *Nutrition of Plantation Forests*. Academic Press, London, pp. 1-16.
Nitsch, J.P. (1957) Photoperiodism in woody plants. *Journal of the American Society for Horticultural Science* 70, 526-544.
Nooden, L.D. and Weber, J.A. (1978) Environmental and hormonal control of dormancy in terminal buds of plants. In: M.E. Clutter (ed.) *Dormancy and Developmental Arrest*. Academic Press, New York, pp. 221-268.

Nsimba-Lubaki, M. and Peumans, W.J. (1986) Seasonal fluctuations of lectins in barks of elderberry (*Sambucus nigra*) and black locust (*Robinia pseudoacacia*). *Plant Physiology* 80, 747-751.

O'Kennedy, B.T. and Titus, J.S. (1979) Isolation and mobilization of storage proteins from apple shoot bark. *Physiology Plantarum* 45, 419-424.

Oaks, A., Clark, C.J. and Greenwood, J.S. (1991) Nitrogen assimilation of higher plants: strategies for annual and perennial plant species. In: *International Symposium on Nitrogen in Grapes and Wine*, 1991. American Society for Enology and Viticulture, pp. 43-51.

Ostman, N.L. and Weaver, G.T. (1982) Autumnal nutrient transfers by retanslocation, leaching, and litter fall in a chestnut oak forest in southern Illinois. *Canadian Journal of Forestry Research* 12, 40-51.

Pauley, S.S. and Perry, T.O. (1954) Ecotypic variation of the photoperiodic response in *Populus*. *Journal of the Arnold Arboretum* 35, 167-188.

Roberts, L.S., Toivonen, P. and McInnis, S.M. (1991) Discrete proteins associated with overwintering of interior spruce and douglas-fir seedlings. *Canadian Journal of Botany* 69, 437-441.

Ryan, D.F. and Bormann, F.H. (1982) Nutrient resorption in northern hardwood forests. *BioScience* 32, 29-32.

Salisbury, F.B. (1981) Responses to photoperiod. In: Lange, O.L., Nobel, P.S., Osmond, C.B. and Ziegler, H. (eds) *Physiological Plant Ecology I*, Springer-Verlag, Berlin, pp. 135-167.

Salisbury, F.B. and Ross, C.W. (1992) *Plant Physiology*, 4th edn. Wadsworth, Belmont, California.

Sauter, J.J., van Cleve, B. and Wellenkamp, S. (1989) Ultrastructural and biochemical results on the localization and distribution of storage proteins in a poplar tree and in twigs of other tree species. *Holzforschung* 43, 1-6.

Staswick, P.E. (1989) Developmental regulation and the influence of plant sinks on vegetative storage protein gene expression in soybean leaves. *Plant Physiology* 89, 309-315.

Stepien, V. and Martin, F. (1992) Purification, characterization and localization of bark proteins of poplar. *Plant Physiology and Biochemistry* 30, 399-407.

Taylor, B.K. and May, L.H. (1967) The nitrogen nutrition of the peach tree. II. Storage and mobilization of nitrogen in young trees. *Australian Journal of Biological Science* 20, 389-411.

Vaartaja, O. (1960) Ecotypic variation of photoperiodic response in trees especially in two *Populus* species. *Forest Science* 6, 200-206.

van den Driessche, R. (1984) Nutrient storage, retranslocation and relationship of stress to nutrition. In: Bowen, G.D. and Nambiar, E.K.S. (eds) *Nutrition of Plantation Forests*. Academic Press, London, pp. 181-208.

van der Veen, R. (1951) Influence of daylength on the dormancy of some species of the genus *Populus*. *Physiologia Plantarum* 4, 35-40.

van Cleve, B., Clausen, S. and Sauter, J.J. (1988) Immunochemical localization of a storage protein in poplar wood. *Journal of Plant Physiology* 133, 3146-3153.

Vince-Prue, D. (1975) *Photoperiodism in Plants*. McGraw Hill, London.

Wareing, P.F. (1947) Photoperiodism in woody species. *Forestry* 22, 211-222.

Weiser, C.J. (1970) Cold resistance and injury in woody plants. *Science* 169, 1269-1278.

Wetzel, S. and Greenwood, J.S. (1989) Proteins as a potential nitrogen storage compound in bark and leaves of several softwoods. *Trees* 3, 149-153.

Wetzel, S. and Greenwood, J.S. (1991) The 32-kilodalton vegetative storage protein of *Salix microstachya* Turz: characterization and immunolocalization. *Plant Physiology* 97, 771-777.

Wetzel, S., Demmers, C. and Greenwood, J.S. (1989) Seasonally fluctuating bark proteins are a potential form of nitrogen storage in three temperate hardwoods. *Planta* 178, 275-281.

VI Dormancy Modelling

23 Population-Based Models Describing Seed Dormancy Behaviour: Implications for Experimental Design and Interpretation

Kent J. Bradford
Department of Vegetable Crops, University of California, Davis, CA 95616-8631, USA

Introduction

It is a fundamental paradox of plant dormancy that it reveals itself only via the absence of growth or development. That is, we know dormancy is present only when a portion of a population of seeds or buds fails to initiate growth under conditions that would normally be permissive of growth. In essentially all of the plant systems in which it is present, dormancy is characterized by observing the percentage of seeds or buds that are no longer dormant, i.e. those that either germinate or grow. It is then inferred that the remaining members of the population are dormant. This approach has been successful in phenomenologically characterizing dormancy and the influence of various physical and physiological factors on it. However, the fact that dormacy is scored as a quantal, or all-or-none, event for an individual seed or bud is often in conflict with the methods used to explore the physiology, biochemistry and molecular biology of dormancy. For example, assume that some fraction of a given sample population of seeds or buds will be revealed to be dormant after being given an opportunity to germinate or grow. Clearly, studies of the mechanisms controlling the maintenance or release of dormancy must focus on events occurring either in the dormant fraction of the population, or in the non-dormant fraction prior to the initiation of growth. But, before any individuals have germinated or grown, we do not know which are dormant and which are not. Thus, samples of this population for biochemical analysis will be composed of both dormant and non-dormant individuals, and the

© CAB INTERNATIONAL 1996. *Plant Dormancy* (ed. G.A. Lang)

assays will inevitably present an *average* picture of their biochemistry, which may not accurately reflect either subpopulation.

The same complication is present in studies in which a dormant seed or bud population is exposed to conditions conducive to breaking dormancy, and a correlation is drawn between the percentage of remaining dormant individuals and some biochemical or physiological parameter measured in a bulk sample. While the parameter measured *may* be regulatory for dormancy, it may also be associated with the growth that follows the actual breaking of dormancy. In a population where the proportion of dormant individuals is changing due to a given treatment, bulk samples may give excellent correlations with virtually any characteristic associated with the growing state simply because the fraction of individuals in that state increases with the effectiveness of the treatment. However, biochemical events associated with growth *per se* are by definition postdormancy events. The key challenge for dormancy research, therefore, is to identify what prevents the initiation of growth in dormant seeds or buds and what allows the transition to the growing state.

Based upon the pioneering work of E.H. Roberts and R.H. Ellis (Roberts, 1961; Ellis and Roberts, 1981), seed biologists have come to appreciate the power of quantitative models based upon population statistics for analysing seed behaviour. More recently, these methods have been used to develop physiologically based population models that are highly robust in describing seed behaviour with respect to germination timing, water relations, dormancy, hormonal regulation and deterioration (Gummerson, 1986; Bradford, 1990, 1995; Ni and Bradford, 1992, 1993; Bradford *et al.*, 1993; Finch-Savage and Phelps, 1993; Bradford and Somasco, 1994; Dutta and Bradford, 1994). While the presentation of the models and demonstration of their application in dormancy research will be based exclusively upon seed systems, there is no reason why they should not be generalized to other types of plant dormancy. In fact, it has been argued recently that they can be generalized at a cellular level to encompass many aspects of plant growth and development (Bradford and Trewavas, 1994). This chapter will focus on the conceptual basis of the population models, how they can be applied to analyse seed dormancy and the implications of this approach for experimental design and interpretation. Details of the mathematical development of the models and their application are in the original publications cited.

A Population-Based Model of Seed Dormancy

Response threshold variation and dormancy states

A central concept in the population-based model is that individuals vary in their response thresholds for regulatory factors. This is demonstrated by data

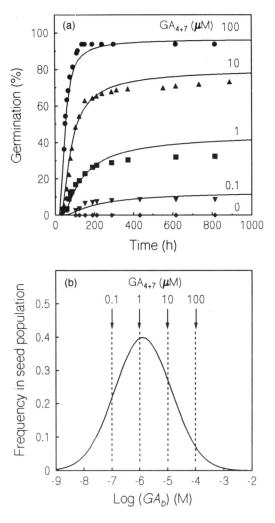

Fig. 23.1. (a) Germination time courses of gibberellin-deficient *gib-1* mutant tomato seeds that have imbibed in 0, 0.1, 1, 10 or 100 μM GA_{4+7} concentrations. The solid curves are the time courses predicted by the GA-time model (Eqn 3), based upon the sensitivity distribution shown in (b). (b) The distribution of threshold sensitivities to GA_{4+7} (GA_b, or the minimum GA_{4+7} concentration required to stimulate radicle emergence) among individual seeds in the population. The normal curve indicates the relative frequency in the seed population of a given value of GA_b. The time to germination is inversely proportional to the amount by which the current level of GA_{4+7} exceeds the GA_b value for a given seed. The arrows at the top and vertical dashed lines indicate the GA_{4+7} concentrations applied in (a); only seeds having thresholds less than the applied concentration will germinate, resulting in the time courses shown. (From Bradford and Trewavas, 1994, adapted from Ni and Bradford, 1993; © American Society of Plant Physiologists, 1994.)

of the type presented in Fig. 23.1, which are common throughout the seed dormancy literature. Gibberellin-deficient *gib-1* mutant tomato (*Lycopersicon esculentum* Mill.) seeds are unable to germinate without exogenous gibberellin (GA), or can be considered to be completely dormant (Groot and Karssen, 1987). When *gib-1* seeds were incubated in increasing concentrations of GA_{4+7}, an increasing fraction of the seed population was able to germinate (Fig. 23.1a; Ni and Bradford, 1993). It is important to note that at all GA_{4+7} concentrations, a final maximum germination percentage was reached that increased only slightly or not at all with extended incubation. These data indicate clearly that the threshold GA concentrations required to stimulate germination varied widely among individual seeds. A given concentration exceeded the response threshold of a specific fraction of seeds in the population, while the remaining seeds did not respond (or at least did not respond sufficiently to achieve radicle emergence). The variation in the individual seed sensitivities could be described by a normal distribution which extended over four orders of magnitude in GA concentration (Fig. 23.1b). When the GA concentration was increased, the threshold (or base) GA concentrations required for germination (GA_b) were exceeded for an increasing fraction of the seed population, resulting in the progressive increase in final germination percentage. Only those seeds with GA_b values less than the applied GA concentration were able to germinate at that concentration. The sensitivity of this seed population to GA can be described by a median GA_b value (the concentration allowing the seed with the median GA_b value to germinate) and the variation among individual seeds in their GA_b values (quantified by the standard deviation, which determines the spread of the normal distribution). Assuming that the GA requirement for germination is indicative of the dormancy state of individual seeds, this also implies a wide variation in dormancy states among seeds.

Regulation of seed dormancy by light can also be analysed using a quantal or population approach (e.g. Frankland, 1975; De Petter *et al.*, 1985; Probert *et al.*, 1987). Figure 23.2a illustrates the induction of the very low fluence response (VLFR) to red light in lettuce (*Lactuca sativa* L.) seeds by low temperature incubation (VanDerWoude, 1985). Without chilling, few seeds germinated until the fluence exceeded -4.5 log mol m^{-2}, when essentially all of the seeds responded over the next tenfold increase in fluence. Germination responses over this region have been termed the low fluence response (LFR). With increasing durations of chilling, an increasing fraction of the seed population was capable of germination at fluences between -9 and -7 log mol m^{-2}, the VLFR range. The cumulative curves in Fig. 23.2a have been generated simply by summing two normal distributions, one centred near -8 and one centred at -4 log mol m^{-2} (Fig. 23.2b). Before chilling, only a small fraction of the population responded in the VLFR range, while most of the seeds required fluences in the LFR range for germination. After 24 h of chilling, most of the seeds were sensitive to fluences in the VLFR range, and

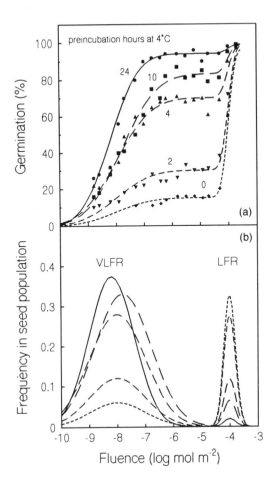

Fig. 23.2. (a) Germination responses of prechilled and unchilled lettuce seeds to 660 nm fluences. Seeds were imbibed in the dark at 4°C for 0, 2, 4, 10 or 24 h, were then moved to 20°C and irradiated. Germination was scored 48 h after irradiation. Prechilling increases the sensitivity to very low fluences in the 10^{-9} to 10^{-7} mol m^{-2} range. Symbols are original data from VanDerWoude (1985). (b) Frequency distributions of sensitivity thresholds for red light in chilled and unchilled lettuce seed populations. Two normal distributions are shown for each curve, one centred near 10^{-8} mol m^{-2} with a standard deviation of about 0.8 log mol m^{-2}, and another centred at 10^{-4} mol m^{-2} with a standard deviation of log 0.2 mol m^{-2}. The maximum frequencies have been adjusted in proportion to the final germination percentages achieved in the very low fluence response (VLFR) and low fluence response (LFR) ranges. During prechilling, the frequency of seeds responding in the VLFR range increases as the frequency of seeds responding only in the LFR range decreases. The curves shown in (a) are the corresponding cumulative normal distributions summed over the two frequency distributions (the widths of the dashes indicate the comparable curves in each panel).

only a few remained which required the higher LFR range of fluences. The progressive decline in the height of the LFR distribution and increase in the height of the VLFR distribution with chilling duration indicate that individual seeds were shifting from one sensitivity state or mode to the other. VanDerWoude (1985) used these data to develop a dimeric model of phytochrome action, which has considerable experimental support. However, the relative proportions of $P_r : P_r$, $P_r : P_{fr}$ and $P_{fr} : P_{fr}$ dimers (P_r inactive and P_{fr} active phytochrome) are assumed to be constant at any given fluence, regardless of the chilling treatment. The response to very low fluences that develops during chilling (or other treatments) must therefore represent a change in sensitivity of individual seeds to a component of the phytochrome signal transduction pathway. Plotting the data as fluence threshold (or sensitivity) distributions (Fig. 23.2b) illustrates that when percentage is the dependent variable and dosage is the independent variable (Fig. 23.2a), the response is quantal in nature and the relationship between dose and response primarily reflects the variation in threshold requirements of individuals making up the sample population (Hewlett and Plackett, 1978).

Variable time scales and germination rates

This discussion of threshold sensitivities and quantal responses is somewhat simplified, as it implies that there is a sharp division between dormant and non-dormant seeds. In fact, it is only with respect to whether or not radicle emergence actually occurs that germination is truly all-or-none, for there is actually a wide range of dormancy and/or vigour states present within a seed population that extends through both the germinable and non-germinable fractions. One way to distinguish 'dormancy' states among germinable seeds is based upon the germination rate, or the inverse of the time to radicle emergence. Gordon (1973) summarized considerable data supporting a close relationship between the final germination percentage of a partially dormant seed lot and the time to radicle emergence of the germinable seeds. As germination percentage increased with a dormancy-breaking treatment, the time to germination was reduced (germination rate increased). He coined the term 'resistance to germination' (the inverse of rate) to draw attention to the fact that, even after all seeds had become germinable, germination rate could continue to increase, indicating residual 'dormancy' was present even in seeds that eventually germinated. By extension, Gordon suggested that dormant seeds also varied in their 'resistance to germination', so that some were more dormant than others and would require more extended or extreme treatments to promote germination.

This concept is illustrated in Fig. 23.1a, which shows that, as the GA concentration increased, the time required to achieve a given germination percentage decreased. This suggests that the more the GA concentration exceeds the GA_b threshold for a given seed (Fig. 23.1b), the more rapidly that seed achieves radicle emergence. Conversely, the more a seed's GA_b threshold exceeds the current GA concentration, the more dormant that seed is. There is therefore a continuum within the seed population, with the most non-dormant seeds being the first to germinate, the just barely non-dormant seeds being the last to germinate, the least dormant seeds requiring only a slight stimulus to germinate and the most dormant seeds requiring the most extreme treatments before completing germination. In this context, the rate of germination (or resistance to germination) is a quantitative expression of the degree of dormancy present on both sides of the quantal threshold of radicle emergence.

Without presenting all of the data and mathematical development here (see Ni and Bradford, 1993), it has been shown that a simple reciprocal model of the relationship between time to germination and the GA concentration can be used to describe the data of Fig. 23.1a. In its simplest form, this model is:

$$(\log [GA] - \log [GA_b]) \times \text{time to germination} = \text{GA-time} \quad \text{(Eqn 1)}$$

where GA is the actual GA concentration and GA-time is a constant in units of log molar-hours or log molar-days. Since the total GA-time to germination is a constant, as the difference between the actual GA concentration and the seed GA threshold increases, the time to germination decreases reciprocally, and vice versa. This is identical in form to the familiar equation used to compare biological rate processes at different temperatures by expressing the time scale in terms of thermal time:

$$(T - T_b) \times \text{time} = \text{thermal time} \quad \text{(Eqn 2)}$$

where T is the current temperature and T_b is the base temperature for the process. The thermal time (degree-days) required for germination of a given seed is constant, but the actual time (clock time) will be much less at a higher temperature than at a lower one (e.g. Garcia-Huidobro *et al.*, 1982; Covell *et al.*, 1986; Dahal *et al.*, 1990). These inverse relationships between temperature or GA concentration and time to germination are illustrated in Fig. 23.3 (Bradford and Trewavas, 1994). Any rectangle formed by the intersection of a given temperature or GA concentration with the curve will have the same area, representing the same total thermal time or GA-time required to achieve radicle emergence (Fig. 23.3). This simple reciprocal model accounts for the more rapid germination of a given seed fraction as the GA concentration is increased (Fig. 23.1a), because higher GA concentrations result in more accumulated GA-time per unit of clock time.

To fully describe the germination time courses of Fig. 23.1a, the variation in GA_b values among individual seeds must be incorporated into Eqn 1. If we define the GA_b value for a given germination fraction or percentage *(g)* as $GA_b(g)$, then Eqn 1 becomes:

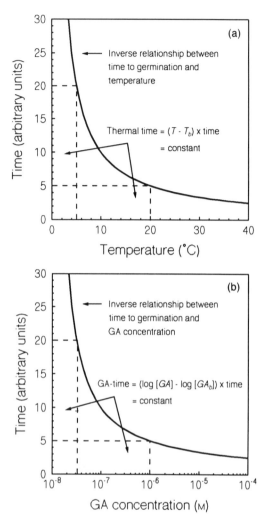

Fig. 23.3. Comparison of the concepts of thermal time and GA-time. (a) The rate of germination is inversely proportional to the degrees (T) in excess of a base, or threshold temperature (T_b). The curve illustrates this relationship for the case where $T_b = 0°C$ and the total thermal time required is 100 degree-days, degree-hours or other appropriate thermal time units. The product of ($T - T_b$) multiplied by the time to a given percentage germination is a constant (i.e. the areas of the dashed rectangles are equal). This allows germination rates at different temperatures to be normalized on a common thermal time scale. (b) Similarly, the GA-time concept postulates that germination rates accelerate or decelerate in proportion to the amount by which the GA concentration exceeds its base or threshold value (GA_b). The log GA concentration (log [GA]) in excess of the seed's sensitivity threshold (log [GA_b], where $GA_b = 10^{-8}$ M in this example) multiplied by the time to germination is a constant (represented by dashed rectangles). As for thermal time, this allows GA-regulated germination rates at different GA concentrations to be normalized on a common GA-time scale (Ni and Bradford, 1993). (Adapted from Bradford and Trewavas, 1994; © American Society of Plant Physiologists, 1994.)

$$(\log [GA] - \log [GA_b(g)]) \times t_g = \text{GA-time} \qquad \text{(Eqn 3)}$$

where $GA_b(g)$ stands for the normal distribution of GA_b thresholds (Fig. 23.1b) and t_g is the time to germination of fraction or percentage g. The difference between the current GA concentration and an individual seed's GA_b threshold will vary depending upon where the seed is in the $GA_b(g)$ distribution. Consequently, the GA-time that each seed accumulates per unit of clock time will also vary in proportion to this difference. Since radicle emergence will occur after the accumulation of a constant total GA-time, the spread in clock time of radicle emergence among individual seeds is due to the differences in their GA_b values. When the GA concentration increases, the times to germination of all seeds are advanced, and an additional fraction of the seed population is able to achieve radicle emergence. The latter will be the last seeds to germinate, as their thresholds will be closest to the current GA concentration, and they will therefore require more clock time to accumulate the required GA-time for radicle emergence. Using this GA-time model (Eqn 3) and allowing $GA_b(g)$ to vary in a normal distribution (as in Fig. 23.1b), the solid curves in Fig. 23.1a can be calculated (Ni and Bradford, 1993). This model matches well to both the time courses and final germination percentages simply by changing the GA concentration in Eqn 3. In addition, and analogous to thermal time analyses, the constant GA-time requirement for germination among all seeds also allows the normalization of time courses on a common GA-time scale (Ni and Bradford, 1993). Increasing the GA concentration in this view can be considered as analogous to increasing the temperature: both shorten the biological time ('biotime') required for a response in proportion to their level in excess of a threshold (Fig. 23.3; Bradford and Trewavas, 1994).

This model demonstrates how both germinable and non-germinable seeds can be considered to be part of the same continuous 'dormancy' distribution. Loss or absence of dormancy is associated with lower thresholds for dormancy-breaking factors, which simultaneously result in more rapid germination (e.g. Derkx and Karssen, 1993). Seeds with higher thresholds near the current stimulus level germinate very slowly, and dormancy (failure to complete germination) is the condition where this delay is extended indefinitely because the stimulus does not exceed the seed's sensitivity threshold. Increasing dormancy is associated with even higher threshold values, which require more extreme dormancy-breaking treatments. While this model has been presented thus far in conceptual terms, it is important to emphasize that it is quantitative, testable and soundly based on the statistics appropriate to quantal systems (Finney, 1971; Hewlett and Plackett, 1978). It therefore provides a straightforward approach to quantifying the dormancy status of a seed population in terms of only two characteristics, the threshold sensitivity distribution (defined by a mean and standard deviation) and a time constant relating the factor level relative to the threshold to the actual time to radicle emergence.

Hydrotime: a physiological basis for dormancy?

While sensitivity threshold distributions and time constants provide a quantitative mathematical description of dormancy states and germination time courses, they do not indicate the mechanism by which dormancy is controlled. We are far from understanding such mechanisms in molecular or biochemical terms, but, physiologically, germination *sensu stricto* is completed by the initiation of embryo growth. The transition from the plateau phase of seed imbibition (which is extended indefinitely in dormant seeds) to the resumption of water uptake associated with growth is a key event distinguishing non-dormant and dormant seeds. The growth potential of seeds, or their capacity for water uptake at a given water potential (ψ), is an indication of their vigour or degree of dormancy. That is, seeds requiring a low (more negative) ψ to prevent embryo growth will generally germinate more rapidly and exhibit greater vigour than seeds with a higher ψ threshold. Obviously, seeds that are unable to germinate even in water ($\psi = 0$ MPa) are effectively dormant. In the same manner as described above for hormonal or phytochrome sensitivity distributions, sensitivity distributions to ψ can also be defined for a seed lot (Gummerson, 1986; Bradford, 1990). The base, or threshold, water potential (ψ_b) for a given seed can be defined as the ψ which just prevents embryo growth (radicle emergence). As for hormonal sensitivity, individual seeds vary in their ψ_b values according to a normal distribution, indicated by $\psi_b(g)$. We can then define a hydrotime constant (θ_H) as

$$\theta_H = (\psi - \psi_b(g)) t_g \quad \text{(Eqn 4)}$$

where $\psi_b(g)$ is the threshold or base ψ that will just prevent germination of percentage g. If θ_H is a constant, then the difference between the ambient ψ and the $\psi_b(g)$ value of a particular percentage g is inversely proportional to the time required for that same percentage to germinate (t_g). Seeds with the lowest ψ_b values will have the highest germination rates (shortest t_g values), while those with higher ψ_b values will have progressively slower germination rates. Since θ_H is a constant, the differences in germination rates among seeds in the population at a given ψ are based entirely upon the variation in their $\psi_b(g)$ values (Bradford, 1990; Dahal and Bradford, 1990, 1994).

An example of the application of this model to seed dormancy is illustrated in Fig. 23.4 (Dutta and Bradford, 1994). Germination of lettuce seeds is inhibited progressively as the temperature is increased above a threshold level ($\sim 31°C$ for the seeds used here). If germination in water only is compared, essentially all seeds were able to germinate at both 25 and 31°C (Fig. 23.4a, b). However, if the seeds were imbibed at reduced ψ at the two temperatures, those at the higher temperature were much more sensitive to inhibition. For example, imbibition at −0.4 MPa had only a slight delaying effect on germination at 25°C, but almost completely blocked germination at 31°C. Using probit analysis to fit the curves across all ψ levels (Bradford, 1990), the $\psi_b(g)$ distributions and θ_H values can be determined (Fig. 23.5). The effect of increasing temperature was to shift the entire $\psi_b(g)$ distribution to higher values. Extrapolating this effect to slightly higher temperatures, part of the distribution would extend above 0 MPa; seeds having nominal ψ_b values in this region will be prevented from germinating in water, or will be dormant. Germination will be prevented completely at a temperature at which the entire $\psi_b(g)$ distribution has shifted above 0 MPa (35°C in this

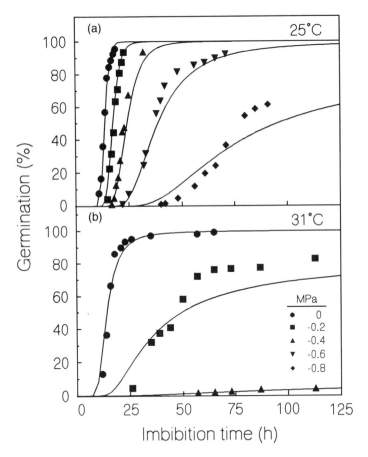

Fig. 23.4. Germination time courses of lettuce seeds that have imbibed in polyethylene glycol (PEG 8000) solutions of 0, −0.2, −0.4, −0.6, and −0.8 MPa at 25°C (a) or 31°C (b). Germination was much more sensitive to reduced ψ at the higher temperature. The solid curves are the time courses predicted by the hydrotime model (Eqn 4) based upon the $\psi_b(g)$ distributions shown in Fig. 23.5. (From Dutta and Bradford, 1994; © CAB International, 1994).

example). The thermodormancy of lettuce seeds can therefore be seen as a consequence of the increase in $\psi_b(g)$ with increasing temperature.

The same situation occurred with GA deficiency in tomato seeds, where GA-deficient seeds had high or positive $\psi_b(g)$ distributions (Fig. 23.6; Ni and Bradford, 1993). Application of GA shifted the $\psi_b(g)$ distribution to lower values and made it more narrow (reduced its standard deviation). This had the effect of allowing all seeds to germinate in water (broke dormancy) and shortened the time to germination, as shown in Fig. 23.1. Essentially the same situation was evident in the wild-type parent line ('Moneymaker'), which, due to postharvest dormancy, had germination rates and percentages and a $\psi_b(g)$ distribution similar to that of *gib-1* seeds in 10 μM GA$_{4+7}$ (Ni and Bradford, 1993). Dormancy of these seeds was also broken by 100 μM GA$_{4+7}$, and the $\psi_b(g)$ distribution shifted to the same position shown in Fig. 23.6. Conversely, abscisic acid (ABA), which is known to be involved in the imposition of dormancy during seed development (Hilhorst and Karssen,

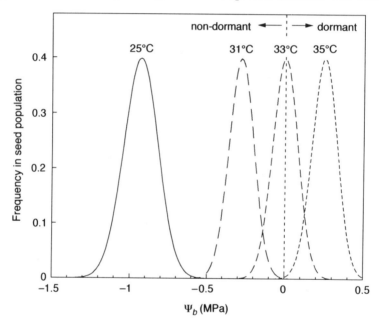

Fig. 23.5. Frequency distributions of ψ_b in lettuce seeds imbibed at 25, 31, 33 or 35°C. The curves for 25 and 31°C are derived from the data in Fig. 23.4, with means of −0.93 (standard deviation = 0.11) and −0.28 MPa (standard deviation = 0.08), respectively (Dutta and Bradford, 1994). Combined with θ_H values of 13 and 4 MPa h at 25 and 31°C, these values can be used in Eqn 4 to predict the germination time courses shown in Fig. 23.4. The curves for 33 and 35°C are extrapolated based upon the decline in final germination percentage as temperature exceeded 31°C. The vertical dashed line at 0 MPa indicates the maximum ψ_b value allowing germination to occur on water. Seeds having predicted ψ_b values in excess of this line represent the dormant fraction of the seed population at that temperature.

1992), causes $\psi_b(g)$ distributions to shift to more positive values (Ni and Bradford, 1992, 1993). The imposition and release of dormancy can therefore be quantified as shifts of the $\psi_b(g)$ distribution to either higher or lower values, with consequent coordinated effects on both germination percentage and germination rate.

It must be acknowledged that ψ_b thresholds greater than 0 MPa cannot be measured directly, but only extrapolated based upon the behaviour of the fraction of the seed population that does germinate. Positive ψ_b values also do not make physical sense, as, by definition, ψ normally does not exceed 0 MPa except under pressure. However, positive ψ_b values do make physiological sense as a measure of the depth of dormancy or the level of a dormancy-

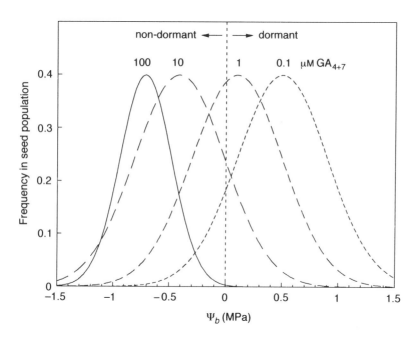

Fig. 23.6. Frequency distributions of ψ_b in gibberellin-deficient *gib-1* mutant tomato seeds imbibed in 0.1, 1, 10 or 100 μM GA_{4+7} concentrations. The distributions for 100 (mean = −0.41 MPa, standard deviation = 0.4 MPa) and 10 (mean = −0.71 MPa, standard deviation = 0.23 MPa) μM GA_{4+7} are from Ni and Bradford (1993). The distributions for 1 and 0.1 μM GA_{4+7} are extrapolated based upon the final germination percentages achieved at these concentrations (Fig. 23.1a). GA has the effect of shifting the distributions to more negative values and narrowing them (reducing the standard deviation), resulting in a higher percentage and more rapid and uniform germination. Using Eqn 4, these distributions can be used to predict germination time courses at different GA concentrations or water potentials. The vertical dashed line at 0 MPa indicates the maximum ψ_b value allowing germination to occur on water. Seeds having predicted ψ_b values in excess of this line represent the dormant fraction of the seed population at that GA concentration.

breaking factor that is required to permit germination. An analogy can be made to the variation in pore apertures among individual stomata (Laisk et al., 1980; Raschke, 1990). A minimum turgor is required to begin to force the guard cells apart, and there is a maximum aperture that will not be exceeded regardless of how high the turgor is. However, we can imagine that individual guard cells may possess turgors either less than the minimum or greater than the maximum required to cause closure or opening of the pore. By using a graded series of opening or closing stimuli, or a dose-response curve, the extremes of the stomatal population can be characterized in terms of how much their turgors differ from the nominal minimum and maximum values revealed by measurements of pore size. This is directly comparable to the situation for seed dormancy, where positive ψ_b values cannot be measured directly, but they can be inferred from the behaviour of the germinable fraction of a seed population in response to a factor that either breaks or imposes dormancy. This ability to extend inferences beyond the observable germinating fraction to characterize the entire seed population is a distinct advantage of the modelling approach.

Experimental data testing the ψ sensitivity of seeds in different dormancy states are not available for all systems. However, with some reasonable assumptions, the same model (Eqn 4) does an excellent job of fitting data obtained from the literature. For example, Perino and Côme (1977) presented data for the escape of apple embryos from thermodormancy at 30°C with prior imbibition time at 15°C. The hydrotime model can match these data very well, simply by assuming that, with increasing time at 15°C, the entire $\psi_b(g)$ distribution shifts to lower values, keeping the θ_H and standard deviation values constant (data not shown). This is directly analogous to the situation for lettuce, where higher temperatures lead to more positive $\psi_b(g)$ values, while lower temperatures are associated with more negative $\psi_b(g)$ values (Fig. 23.5). This relationship appears to hold not only for thermodormancy, but also for low temperature chilling or stratification to break seed dormancy. Simulations have been presented elsewhere (Bradford, 1995) illustrating that the breaking of dormancy in wild rice (*Zizania palustris* var. *interior*) seeds during 26 weeks of imbibition at 5°C (Probert and Longley, 1989) can also be described in a manner essentially identical to that shown in Fig. 23.5, by assuming that the effect of chilling is to gradually shift the entire $\psi_b(g)$ distribution to more negative values. Another example is the breaking of dormancy in apple embryos by chilling (Côme and Thévenot, 1982). With increased chilling duration at 5°C prior to incubation at 20°C, additional embryos germinated and the time to germination was reduced (Fig. 23.7a). These data can be modelled well by Eqn 4 by assuming that the θ_H values decrease first, and then the $\psi_b(g)$ distribution shifts to more negative values without any change in the standard deviation of the population (Fig. 23.7a, b). Thus, changes in germination capacity for the first 3 weeks of chilling apparently were largely due to the reduction in θ_H (decrease in hydrotime constant for germination), while subsequent improvements in germination rate and capacity were due to the shift of $\psi_b(g)$ to more negative values. It is worth noting that the model predicts that, even without chilling, more than 50% of the seeds would eventually germinate, but it would take over 200 days for this to occur due to the low initial θ_H value and high mean ψ_b (simulations not shown). Thus, while we have defined a dormant seed as one having a ψ_b

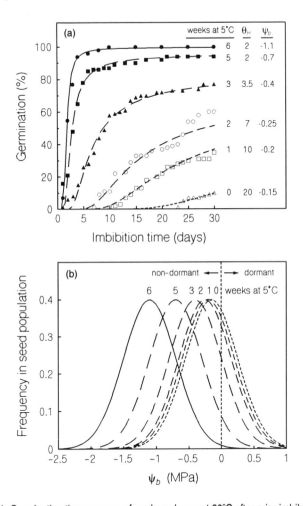

Fig. 23.7. (a). Germination time courses of apple embryos at 20°C after prior imbibition for 0, 1, 2, 3, 5 or 6 weeks at 5°C (Côme and Thévenot, 1982). The curves are the time courses predicted by the hydrotime model (Eqn 4) using the values for θ_H and mean ψ_b shown and a common standard deviation of 0.4 MPa (b). Frequency distributions of ψ_b that were empirically adjusted to match the time courses in (a). In this case, changes in both θ_H and mean ψ_b were required to fit the actual data (tabulated in (a)). The pattern indicated that the θ_H values declined first, followed by a shift in the ψ_b distribution to more negative values with longer chilling duration.

> 0 MPa, a high θ_H value can effectively impose dormancy by delaying germination.

Modelling exercises such as this are certainly not proof that the proposed changes in θ_H or $\psi_b(g)$ actually occurred, but this is experimentally testable. On the other hand, deviations from the model's predictions can be used to identify shifts in physiological parameters that would not have been evident from mere inspection of the germination time course curves (Ni and Bradford, 1992; Dahal and Bradford, 1994). The ability of Eqn 4 to fit so many types of germination data emphasizes that the array of germination time courses generated as dormancy is imposed or released is a direct consequence of physiological changes occurring in concert throughout the seed population while maintaining a wide variability among individual seeds. Only population-based models can readily account for this critical feature of seed dormancy behaviour.

While the hydrotime concept does not in itself identify the mechanism of dormancy, it has a number of advantages for quantifying and analysing the water relations of germination and dormancy. First, it is based upon the simple technique of determining radicle emergence time courses at a range of constant ψ. The model is driven by the timing of germination events that do occur, rather than by growth rates or water uptake after radicle emergence. It therefore recognizes that the initation of growth is a change of state that marks the end of germination *sensu stricto*, and that it is the lag period prior to this event that is critical in understanding the regulation of germination, and particularly the regulation of dormancy. A second advantage of the hydrotime model is that it requires only three parameters to quantify and predict germination behaviour. Once the mean and standard deviation of $\psi_b(g)$ and the hydrotime constant (θ_H) are known for a given seed population, germination time courses at any ψ can be generated simply by varying the value of ψ (Fig. 23.4). Conversely, physiological changes in $\psi_b(g)$ or θ_H can generate a similar pattern of germination time courses at a single ψ (Fig. 23.7). Unlike most models used to fit germination time courses (Brown and Mayer, 1988), changes in the lag period, shape and final asymptote of the curves do not require specific empirical parameters to adjust each characteristic of the curves. Rather, the changing shapes of the time-course curves are generated automatically with changes in ψ due to the dependence of germination rate on the underlying $\psi_b(g)$ distribution. A third feature of the hydrotime model is that it explicitly incorporates the concept that seeds inherently vary in their sensitivity to ψ and that this variation occurs in a normal distribution. There is no logical necessity for this to be the case, but other important aspects of seed biology, such the maximum lifetimes of seeds, also vary in a normal distribution (Ellis and Roberts, 1981; Bradford *et al.*, 1993). Since we are almost always concerned with the behaviour of seed populations, the ability to describe seed characteristics in population terms is critical to a realistic understanding of seed biology.

A final important attribute of the hydrotime model of dormancy might be called its 'ecological rationality'. Regardless of the precise mechanism of dormancy, a decrease in the mean and/or spread of $\psi_b(g)$ can account for the greater percentage, speed and synchrony of germination as dormancy is alleviated. This provides a direct linkage between the depth of dormancy, current water availability, germination rate and the likelihood of successful seedling establishment. In deeply dormant seeds, $\psi_b(g)$ > 0 MPa and germination will not occur at any ψ. Loss of dormancy associated with a lowering of the $\psi_b(g)$ distribution increases the tolerance of germination to low ψ and speeds the rate of approach toward radicle emergence. As the $\psi_b(g)$ distribution shifts to more negative ψ values, some seeds in the population will have relatively low ψ_b values and will be able to germinate rapidly during a period of high ψ, while other seeds may still have high ψ_b values and remain dormant. As the requirements for breaking dormancy are satisfied, the fraction of seeds capable of germination and their rate of germination will increase as $\psi_b(g)$ decreases. However, in a variable rainfall environment, it will still be advantageous to maintain a distribution of ψ_b values and therefore of germination rates. As Meyer and Monsen (1991) have observed, slow germination can be as effective as dormancy in preventing germination under unfavourable conditions. Those seeds with a high ψ_b will have slow rates of progress toward germination and will be less likely to complete radicle emergence during a short window of opportunity. However, if radicle emergence has not occurred, they will survive dehydration and will have another opportunity to complete germination during the next wet cycle (Wilson, 1973; Allen *et al.*, 1993; Finch-Savage and Phelps, 1993). Regardless of the temperature, light or nutrient environment of a seed, water is so critical for the first phase of seedling establishment that seeds have evolved precise mechanisms for gauging their potential water resources before completing germination (Koller and Hadas, 1982). Seasonal or local environmental factors can act via effects on the $\psi_b(g)$ distribution, which would alter the potential for germination at a given water availability. However, the seed's immediate rate of progress toward germination would remain highly dependent upon the current water availability to increase the chances of successful seedling establishment.

Implications of Population Variation for Experimental Design and Interpretation

If one accepts the analysis presented thus far, there are important consequences for experimental design in investigations of dormancy mechanisms. Unfortunately, in many cases these consequences are not widely appreciated or, if appreciated, not uniformly applied. Similarly, certain interpretations are appropriate for quantal data. The intent in this section is to identify some of these implications explicitly and to illustrate how the application of population-based models can alter experimental design and interpretation.

Population variation and experimental design

A key conclusion of the modelling studies is that a seed population contains a wide spectrum of germination and dormancy states distributed among the individual seeds. In most cases, however, we do not know initially which individuals will be more or less dormant. Instead, the dormancy status of a seed is revealed only by its germination rate or its failure to germinate at all. The problem that this poses for scientific investigation is that we would like to know what the early events are that distinguish a dormant from a non-dormant seed, as, once embryo growth has begun, dormancy is over and the seed has passed the relevant physiological state for comparison with dormant seeds (i.e. non-dormant but as yet ungerminated). It is possible to wait until all non-dormant seeds have germinated and then only analyse the remaining dormant seeds, but there are data indicating that dormant seeds return to a state of low activity distinctly different from that occurring soon after imbibition (Ibrahim *et al.*, 1983; Powell *et al.*, 1984). Thus, these seeds may also be in a different physiological state from that when the 'decision' was made to either germinate or not. The usual experimental approach is simply to combine all the seeds together for whatever analysis is being employed, and then monitor changes with time in sequential samples. It is particularly unfortunate that in some cases this is even extended to including in the same sample seeds that have already germinated. This is clearly inappropriate for at least two reasons. First, as is evident from the population models and will be shown experimentally below, this approach blends together both dormant and non-dormant seeds such that any measured parameter will be an average that may be quantitatively representative of only a specific fraction of the population. Second, any characteristic that is expressed only after embryo growth is initiated will almost inevitably show a good correlation with the germination time course and the final germination percentage. As additional seeds enter the growing state and begin to express the particular characteristic, they will contribute to increasing the mean value just as they contribute to increasing the germination percentage. Thus, almost any characteristic associated with growth will show a good correlation with germination percentage if combined samples of all seeds are taken at intervals through the germination time course. While providing information on characteristics that are associated with growth, such studies have relatively limited value for understanding dormancy *per se*, as it is virtually impossible to separate the population effect from quantitative changes that precede the transition from dormancy to non-dormancy.

What can be done about this situation? One can compare completely non-dormant populations with completely dormant populations. Genetic lines differing in innate dormancy, such as wild oat genotypes (Simpson, 1990) or mutants deficient in hormone content (Hilhorst and Karssen, 1992), make this feasible in some cases. However, we are often interested in the action of a specific dormancy-breaking factor, and generally employ dose-response experiments that will result inevitably in populations with mixed dormancy states (e.g. Figs 23.1 and 23.7). As a minimum, biochemical experiments in these situations should separate germinated seeds from ungerminated seeds. In addition, assays that can be applied to single seeds would allow determination of the biochemical variation associated with the variation in dormancy states among seeds. This approach can be criticized as being unfeasible, as it requires a large number of assays on individual seeds to develop a population distribution. However, this is exactly the approach employed in genetic screening programmes, where it is evident that what is often sought is the rare individual with a unique genotype, rather than the average that would be obtained from bulking a segregating population together. This is directly analogous to the 'physiological segregation' that is occurring in a partially dormant seed lot, with individual seeds expressing varying degrees of the property of interest.

We have made some initial efforts in this direction, utilizing partially dormant tomato seed lots. Under some growth conditions, a fraction of 'Moneymaker' tomato seeds is dormant at harvest (approximately 30% in our case; Fig. 23.8F). Considerable evidence suggests that tomato seed germination rates are controlled by the endosperm cap tissue directly opposite and enclosing the radicle tip (Groot and Karssen, 1987, 1992; Dahal and Bradford, 1990; Ni and Bradford, 1993). Since the cell walls of the endosperm cap are composed largely of mannose, mannanase enzyme activity is thought to be involved in cell wall degradation, leading to weakening of the endosperm tissue and subsequent radicle emergence (Groot and Karssen, 1987, 1992; Nonogaki *et al.*, 1992). We have investigated this hypothesis by assaying mannanase activity released from individual tomato endosperm cap tissues ($\sim$ 0.5 mg each) using a sensitive gel diffusion assay (Downie *et al.*, 1994). One of the most striking results from this approach is that the enzyme activity varies over three to four orders of magnitude among individual ungerminated seeds (Figs 23.8A–E; Dahal *et al.*, 1994; D.W. Still and K.J. Bradford, unpublished results). Variation of this magnitude was unexpected, but is in keeping with the logarithmic response to hormonal factors such as GA and ABA (Fig. 23.1; Ni and Bradford, 1992, 1993). It is important to note that this is not experimental error, but rather physiological variation among individual seeds in a genetically homozygous inbred tomato line. With increasing imbibition time, the population separated into two distinct fractions, one with low activity representing approximately 30% of the population, and one with higher activity containing the remaining members of the population

(Figs 23.8B, C). Seeds with higher activity subsequently germinated and disappeared from the remaining seed population and, when only dormant seeds remained, most exhibited low mannanase activity (Fig. 23.8E). This suggests that low mannanase activity is at least associated with dormancy, although it does not prove that high mannanase activity causes germination.

An important conclusion from this experiment is that the distinction between the activities of the dormant and non-dormant seed fractions could only have been made by utilizing single-seed assays. If combined extracts had been assayed, they would have always revealed the average of the total population, rather than the two distinct fractions evident even prior to the initiation of germination (Fig. 23.8B). In addition, due to the logarithmic

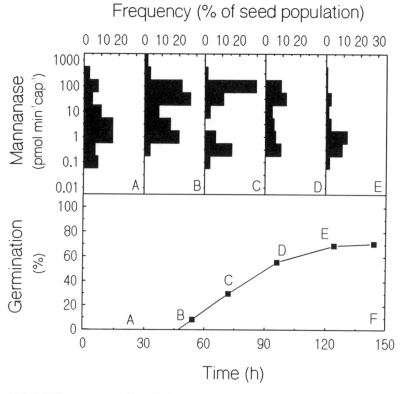

Fig. 23.8. A–E. Mannanase activities of individual endosperm caps of ungerminated tomato seeds at various times after imbibition. Frequencies of mannanase activities (pmol m^{-1} cap $^{-1}$) are shown in categories of 0.5 log units at times corresponding to the labels on the germination time course in F. Note the logarithmic scale for enzyme activity, spanning almost four orders of magnitude. Note also the bimodal distribution in B and C. This 'Moneymaker' seed lot contained 30% dormant seeds, which coincides with the percentage of seeds present in the lower activity fraction in C–E. (D.W. Still and K.J. Bradford, previously unpublished results.)

variation in enzyme activity, a few seeds with very high activity can skew the average, obscuring the activity of the majority of seeds. We have examples of cases where 95% of the total activity in a bulk sample would have been contributed by only 6% of the seed population (D.W. Still and K.J. Bradford, unpublished results). This possibility of a few seeds skewing the result of a composite assay is well recognized in other areas of seed technology, such as in leakage (conductivity) tests (Steere *et al.*, 1981; Reusche, 1987), and has now been demonstrated for biochemical activities as well. Thus, we must consider the possibility that elucidating the biochemical mechanisms underlying dormancy will require alternative experimental designs and methods to directly address the variation inherent in seed populations. In particular, development of non-invasive methods to determine physiological and biochemical states of individual seeds, such as by transformation with reporter genes under the control of dormancy-associated promoters, should be a priority for future research.

Interpretation of quantal data

The collection of quantal data (i.e. percentage of germinated seeds in a population) leads to certain interpretations of data that seem to be generally underappreciated. A plot of percentage germination versus the dosage of a dormancy-breaking factor (e.g. Fig. 23.2a) actually represents the distribution of response thresholds among individual seeds. It provides no information about the rate of response, as only the final all-or-none criterion is reported. None the less, it is common for mathematical models derived from growth analysis or enzyme kinetics to be applied to such curves, implicitly converting the percentage into a rate parameter (Brown and Mayer, 1988; Hilhorst, 1990a,b). Such models can fit germination response curves very well, and useful information can be obtained from them, but, when they are extended to mechanistic explanations, it is important to distinguish between percentages and rates. For example, if a seed population germinates 50% in response to a particular stimulus, this does not mean that it was only half-effective; rather, it was fully effective for half of the seeds and ineffective for the other half. This rather subtle distinction is none the less important: a percentage response curve only documents the percentage of the population that responded at a particular stimulus level. This is quite different from an enzyme-catalysed reaction rate that would occur at the K_m, or half-maximal activity. The latter assumes that the same enzyme is present with a constant substrate affinity and that the rate of conversion to product is dependent upon the substrate concentration. In a quantal response, additional individuals are responding at each increment in dosage level, and the 'maximum rate' is simply the percentage of individuals whose response threshold has been exceeded at a particular dose. Referring again to the curves in Fig. 23.2a, additional seeds were induced to germinate with each increment in light intensity. The individual data are therefore end-points or steady states that reveal the variation in thresholds among seeds in the population. The slope of such a curve is not a rate, but rather reflects the spread in thresholds among individuals; i.e. how much additional stimulus is required to trigger the next most sensitive seed in the population.

Dormancy, as represented in a percentage response plot, is not a continuous variable in the same sense that a reaction rate or a growth rate is. It is a discrete state of an individual seed that can be assessed by determining the dosage of a stimulus needed to exceed its response threshold. On the other hand, a major argument of the models presented here is that there is a continuous (generally normal) distribution of dormancy states among seeds in a population. This variation is more readily revealed by germination rates (timing) than by final percentages, because rates reveal the differences among germinable seeds while final percentages do not. Final percentages and rates are linked by the biotime models presented here, so there is a specific relationship between them. However, much more information about the seed lot can be obtained from applying the models to germination time courses than from final germination percentages alone. We therefore advocate that time-to-response data be collected in dormancy studies and that the biotime models be utilized in the characterization of seed populations and the interpretation of data.

A final aspect that needs to be considered is the application of the biotime concept to dormancy studies. For example, if a seed is slow to germinate, was it initially dormant and then lost dormancy? Was it partially dormant? Was it non-dormant but of low vigour? These are important questions when investigating the mechanisms of dormancy maintenance and release. Although not presented in detail here, the biotime models allow time courses at different dosage levels to be normalized on a common time scale. For example, all of the curves of Fig. 23.1(a) collapse into a single common time course when the time axis is presented in terms of GA-time rather than clock time (Ni and Bradford, 1993). Similarly, the array of time courses at different ψ (Fig. 23.4) can be collapsed to a single time course on a common hydrotime scale (Bradford and Somasco, 1994). This is directly analogous to the use of thermal or degree-day time-scales to normalize biological rate processes across a range of temperatures. A possible application of the biotime concept is to determine whether a particular biochemical process or event is simply correlated with the rate of progress toward germination or is actually differentially regulated or expressed in relation to dormancy status. For example, genes have been identified that are expressed for a longer time after imbibition in dormant embryos than in non-dormant embryos, but, since the embryos are excised, even the dormant ones eventually germinate (Morris *et al.*, 1991). We can ask: is the extended expression of a particular gene involved in the maintenance of dormancy, or is the embryo simply slower to complete a transition to germination that occurs rapidly for the non-dormant embryos? As an analogy, if the two embryo populations were imbibed at two different temperatures, we would find it neither surprising nor mechanistically insightful to determine that germination was delayed at the lower temperature, and so was the disappearance of a particular message or protein. However, if the two populations represent two extremes of dormancy, we do tend to ascribe mechanistic significance to the same observations. According to the biotime model, however, the two cases are somewhat analogous, since the condition of dormancy can be viewed as a stretching of the time axis in much the same way that lowering the temperature stretches the time axis. By using the population models to determine the biotime constants and normalize the time-scales, it should be possible to distinguish between processes that are proceeding in concert with the accumulation of biotime and those that are not. This may be useful in distinguishing between processes that are simply correlated with dormancy status and those that are involved in regulating that status.

Conclusions

A quantitative mathematical model has been presented that can be used to characterize, describe and predict seed germination behaviour with respect to dormancy and many of the factors influencing it. It utilizes statistical methods appropriate to quantal (all-or-none) data, and explicitly recognizes the seed-to-seed variation in sensitivity thresholds that is evident in all seed populations. A biotime concept is introduced to account for the specific relationships between the level of a factor influencing germination and the time to response (e.g. GA-time, hydrotime). Together, these features give the model considerable power in understanding the patterns of germination evident in most seed dormancy studies. Some of the conclusions of this approach seem intuitively obvious ('common sense'), such as the reciprocal relationship between final germination percentage and time to achieve that percentage. Other implications of the model are less intuitive, such as the concept of biotime and its normalization across dormancy states, but they provide unique insights into the nature of dormancy. Regardless of whether the model itself survives further experimental testing, some of the implications of the population-based approach for experimental design and interpretation seem unavoidable if we are to make progress in deciphering the mechanisms underlying dormancy. Emphasis should be placed on individual seed assays and particularly on non-invasive methods of identifying dormancy states. The same approach can be extended to many aspects of plant growth and development, as these can often be analysed in terms of the threshold responses of populations of cells within tissues (Bradford and Trewavas, 1994).

An analogy is appropriate to the current state of dormancy research. Before Newton developed his laws of gravitational motion, it would have been obvious to all that, when an object is dropped, it will fall to the ground. People could have readily described the various ways in which they could drop a myriad of objects and probably even intuitively predict the individual trajectories the objects might take on their way to the ground. Any number of measurements might have been made of how long it took the objects to fall, how much they weighed, where they landed, what they looked like. Similarly, we have an overwhelming wealth of phenomenological data about dormancy and about the factors influencing it in many diverse biological systems. It was Newton's achievement to show how all of those individual trajectories of all the various objects were a direct consequence of a simple underlying relationship. Newton's laws did not mechanistically explain what gravity is, but they quantitatively defined its attributes, allowed the measurement of specific properties and relationships, enabled prediction of future outcomes and provided a foundation for exploration to discover new properties. They also changed the nature of the questions asked and the experimental methods appropriate for answering them. Quantitative models of dormancy behaviour, while certainly not on the same scale of importance as Newton's accomplishments, provide similar advantages for understanding dormancy. They reveal for us the underlying patterns (threshold distributions) and simple relationships (biotime) that result in the germination time courses (biological trajectories) observed under a given condition. By focusing our attention on the patterns and relationships, population models may not define for us what dormancy is, but they will give us a foundation on which to build.

References

Allen, P.S., Debaene, S.B.G. and Meyer, S.E. (1993) Regulation of grass seed germination under fluctuating moisture regimes. In: Côme, D. and Corbineau, F. (eds) *Fourth International Workshop on Seeds: Basic and Applied Aspects of Seed Biology*, Vol. 2. Association pour la Formation Professionelle de l'Interprofession Semences, Paris, pp. 387-392.

Bradford, K.J. (1990) A water relations analysis of seed germination rates. *Plant Physiology* 94, 840-849.

Bradford, K.J. (1995) Water relations in seed germination. In: Kigel, J. and Galili, G. (eds) *Seed Development and Germination*. Marcel Dekker, New York, pp. 351-396

Bradford, K.J. and Somasco, O.A. (1994) Water relations of lettuce seed thermoinhibition. I. Priming and endosperm effects on base water potential. *Seed Science Research* 4, 1-10.

Bradford, K.J. and Trewavas, A.J. (1994) Sensitivity thresholds and variable time scales in plant hormone action. *Plant Physiology* 105, 1029-1036.

Bradford, K.J., Tarquis, A.M. and Durán, J.M. (1993) A population-based threshold model describing the relationship between germination rates and seed deterioration. *Journal of Experimental Botany* 44, 1225-1234.

Brown, R.F. and Mayer, D.G. (1988) Representing cumulative germination. 2. The use of the Weibull function and other empirically derived curves. *Annals of Botany* 61, 127-138.

Côme, D. and Thévenot, C. (1982) Environmental control of embryo dormancy and germination. In: Khan, A.A. (ed.) *The Physiology and Biochemistry of Seed Development, Dormancy and Germination*. Elsevier Biomedical Press, Amsterdam, pp. 271-298.

Covell, S., Ellis, R.H., Roberts, E.H. and Summerfield, R.J. (1986) The influence of temperature on seed germination rate in grain legumes. I. A comparison of chickpea, lentil, soyabean, and cowpea at constant temperatures. *Journal of Experimental Botany* 37, 705-715.

Dahal, P. and Bradford, K.J. (1990) Effects of priming and endosperm integrity on seed germination rates of tomato genotypes. II. Germination at reduced water potential. *Journal of Experimental Botany* 41, 1441-1453.

Dahal, P. and Bradford, K.J. (1994) Hydrothermal time analysis of tomato seed germination at suboptimal temperature and reduced water potential. *Seed Science Research* 4, 171-80.

Dahal, P., Bradford, K.J. and Jones, R.A. (1990) Effects of priming and endosperm integrity on seed germination rates of tomato genotypes. I. Germination at suboptimal temperature. *Journal of Experimental Botany* 41, 1431-1439.

Dahal, P., Still, D.W. and Bradford, K.J. (1994) Mannanase activity in tomato endosperm caps does not correlate with germination timing. *Plant Physiology* 105 (Suppl.), abstract 914.

De Petter E., Van Wiemeersch L., Rethy, R., Dedonder, A., Fredericq, H., De Greef, J., Steyaert, H. and Stevens, H. (1985) Probit analysis of low and very-low fluence-responses of phytochrome-controlled *Kalanchoë blossfeldiana* seed germination. *Photochemistry and Photobiology* 42, 697-703.

Derkx, M.P.M. and Karssen, C.M. (1993) Changing sensitivity to light and nitrate but not to gibberellins regulates seasonal dormancy patterns in *Sisymbrium officinale* seeds. *Plant, Cell and Environment* 16, 469-479.

Downie, B., Hilhorst, H.W.M. and Bewley, J.D. (1994) A new assay for quantifying endo-β-mannanase activity using Congo Red dye. *Phytochemistry* 36, 829-835.

Dutta, S. and Bradford, K.J. (1994) Water relations of lettuce seed thermoinhibition. II. Ethylene and endosperm effects on base water potential. *Seed Science Research* 4, 11-18.

Ellis, R.H. and Roberts, E.H. (1981) The quantification of ageing and survival in orthodox seeds. *Seed Science and Technology* 9, 373-409.

Finch-Savage, W.E. and Phelps K. (1993) Onion (*Allium cepa* L.) seedling emergence patterns can be explained by the influence of soil temperature and water potential on seed germination. *Journal of Experimental Botany* 44, 407-414.

Finney, D.J. (1971) *Probit Analysis*, 3rd edn. Cambridge University Press, Cambridge.

Frankland, B. (1975) Phytochrome control of seed germination in relation to light environment. In: Smith, H. (ed.) *Light and Plant Development*. Butterworths, London, pp. 477-491.

Garcia-Huidobro, J., Monteith, J.L. and Squire, G.R. (1982) Time, temperature and germination of pearl millet (*Pennisetum thyphoides* S. and H.). I. Constant temperatures. *Journal of Experimental Botany* 33, 288–296.

Gordon, A.G. (1973) The rate of germination. In: Heydecker, W. (ed.) *Seed Ecology*. Butterworths, London, pp. 391–409.

Groot, S.P.C. and Karssen, C.M. (1987) Gibberellins regulate seed germination in tomato by endosperm weakening: a study with gibberellin-deficient mutants. *Planta* 171, 525–531.

Groot, S.P.C. and Karssen, C.M. (1992) Dormancy and germination of abscisic acid-deficient tomato seeds: studies with the *sitiens* mutant. *Plant Physiology* 99, 952–958.

Gummerson, R.J. (1986) The effect of constant temperatures and osmotic potential on the germination of sugarbeet. *Journal of Experimental Botany* 37, 729–741.

Hewlett, P.S. and Plackett, R.L. (1978) *An Introduction to the Interpretation of Quantal Responses in Biology*. University Park Press, Baltimore, Maryland.

Hilhorst, H.W.M. (1990a) Dose-response analysis of factors involved in germination and secondary dormancy of seeds of *Sisymbrium officinale*. I. Phytochrome. *Plant Physiology* 94, 1090–1095.

Hilhorst, H.W.M. (1990b) Dose-response analysis of factors involved in germination and secondary dormancy of seeds of *Sisymbrium officinale*. II. Nitrate. *Plant Physiology* 94, 1096–1102.

Hilhorst, H.W.M. and Karssen, C.M. (1992) Seed dormancy and germination: the role of abscisic acid and gibberellins and the importance of hormone mutants. *Plant Growth Regulation* 11, 225–238.

Ibrahim, A.E., Roberts, E.H. and Murdoch, A.J. (1983) Viability of lettuce seeds. II. Survival and oxygen uptake in osmotically controlled storage. *Journal of Experimental Botany* 142, 631–640.

Koller, D. and Hadas, A. (1982) Water relations in the germination of seeds. In: Lange, O.L., Nobel, P.S., Osmond, C.B. and Ziegler, H. (eds) *Physiological Plant Ecology II. Water Relations and Carbon Assimilation*. Encyclopedia of Plant Physiology, New Series, Vol. 12B, Springer-Verlag, Berlin, pp. 401–431.

Laisk, A., Oja, V. and Kull, K. (1980) Statistical distribution of stomatal apertures of *Vicia faba* and *Hordeum vulgare* and the Spannungsphase of stomatal opening. *Journal of Experimental Botany* 31, 49–58.

Meyer, S.E. and Monsen S.B. (1991) Habitat-correlated variation in mountain big sagebrush (*Artemisia tridentata* ssp. *vaseyana*) seed germination patterns. *Ecology* 72, 739–742.

Morris, C.F., Anderberg, R.J., Goldmark, P.J. and Walker-Simmons, M.K. (1991) Molecular cloning and expression of abscisic acid-responsive genes in embryos of dormant wheat seeds. *Plant Physiology* 95, 814–821

Ni, B.-R. and Bradford, K.J. (1992) Quantitative models characterizing seed germination responses to abscisic acid and osmoticum. *Plant Physiology* 98, 1057–1068.

Ni, B.-R. and Bradford, K.J. (1993) Germination and dormancy of abscisic acid- and gibberellin-deficient mutant tomato seeds: sensitivity of germination to abscisic acid, gibberellin, and water potential. *Plant Physiology* 101, 607–617.

Nonogaki, H., Matsushima, H. and Morohashi, Y. (1992) Galactomannan hydrolyzing activity develops during priming in the micropylar endosperm tip of tomato seeds. *Physiologia Plantarum* 85, 167-172.

Perino, C. and Côme, D. (1977) Influence de la température sur les phases de la germination de l'embryon de pommier (*Pirus malus* L.). *Physiologie Végétale* 15, 469-474.

Powell, AD, Dulson, J. and Bewley, J.D. (1984) Changes in germination and respiratory potential of embryos of dormant Grand Rapids lettuce seeds during long-term imbibed storage, and related changes in the endosperm. *Planta* 162, 40-45.

Probert, R.J., and Longley, P.L. (1989) Recalcitrant seed storage physiology in three aquatic grasses (*Zizania palustris, Spartina anglica* and *Porteresia coarctata*). *Annals of Botany* 63, 53-63.

Probert, R.J., Gajjar, K.H. and Haslam, I.K. (1987) The interactive effects of phytochrome, nitrate and thiourea on the germination response to alternating temperatures in seeds of *Ranunculus sceleratus* L.: a quantal approach. *Journal of Experimental Botany* 38, 1012-1025.

Raschke, K. (1990) How abscisic acid causes depressions of the photosynthetic capacity of leaves. In: Pharis, R.P. and Rood, S.B. (eds) *Plant Growth Substances 1988*. Springer-Verlag, Berlin, pp. 383-390.

Reusche, G.A. (1987) Comparison of the AOSA recommended conductivity analysis and the alternative single seed procedure. *Association of Official Seed Analysts Newsletter* 61, 79-97.

Roberts, E.H. (1961) Domancy in rice seed. I. The distribution of dormancy periods. *Journal of Experimental Botany* 12, 319-329.

Simpson, G.M. (1990) *Seed Dormancy in Grasses*. Cambridge University Press, Cambridge.

Steere, W.C., Levengood, W.C. and Bondie, J.M. (1981) An electronic analyzer for evaluating seed germination and vigour. *Seed Science and Technology* 9, 567-576.

VanDerWoude, W.J. (1985) A dimeric mechanism for the action of phytochrome: evidence from photothermal interactions in lettuce seed germination. *Photochemistry and Photobiology* 42, 655-661.

Wilson, A.M. (1973) Responses of crested wheatgrass seeds to environment. *Journal of Range Management* 26, 43-46.

24 An Integrating Model for Seed Dormancy Cycling: Characterization of Reversible Sensitivity

Henk W. M. Hilhorst[1], Maria P. M. Derkx[2] and Cees M. Karssen[1]

[1]*Department of Plant Physiology, Wageningen Agricultural University, Arboretumlaan 4, NL-6703 BD Wageningen, The Netherlands;* [2]*Research Institute for Nursery Stock, Rijneveld 153, 2770 AC Boskoop, The Netherlands*

Introduction

The emergence of many annual weed species is often restricted to certain periods of the year. In temperate zones, emergence often occurs within a limited period in the spring, sometimes followed by additional flushes in summer, e.g. *Sisymbrium officinale* (Bouwmeester and Karssen, 1993). Species that originate from climates with a hot dry summer and a cool humid winter, such as *Arabidopsis thaliana* (Baskin and Baskin, 1983; Derkx and Karssen, 1994), mainly germinate in autumn, surviving the winter as rosette plants. This periodicity of seasonal emergence may be due to seasonal fluctuations in the number of seeds in the soil, particularly for grasses that form transient seed banks in which none of the seeds remain viable in the soil for more than 1 year (Thompson and Grime, 1979). However, dicotyledonous annual species form large persistent seed banks, of which the size outnumbers the annual input of seeds. Survival of seeds in these seed banks may be as long as decades to centuries (Baskin and Baskin, 1977; Ödum, 1978; Priestley and Posthumus, 1982). Emergence from such seed banks is strongly stimulated by disturbance of the soil (e.g. cultivation), but the period of emergence is not significantly affected (Roberts and Feast, 1973; Roberts and Lockett, 1978). The timing of emergence is best explained by changes in seed dormancy (Karssen, 1982; Baskin and Baskin, 1985).

During the past decade, evidence has accumulated that temperature is the main regulatory factor for these annual dormancy cycles (Bouwmeester, 1990; Probert, 1992). Apart from temperature, natural factors that may influence emergence include light and nitrate. It is unlikely that these factors play a role in the regulation of annual dormancy cycles. Light can only penetrate to a limited depth in the soil, whereas soil nitrate levels show a large variability between years and thus are not associated with seasonal timing of seedling emergence (Karssen and Hilhorst, 1992). However, temperature, light and nitrate are cues for the germination of many species. Thus, alleviation of dormancy and germination must be considered separate phenomena (Karssen, 1982; Hilhorst and Karssen, 1989, 1992).

Breaking of dormancy is characterized by a widening of the germination temperature range (temperature 'window') and a reduction in the requirement for other germination promoters (e.g. light and nitrate). On the other hand, induction of dormancy is associated with a narrowing of the germination temperature window and an increase in requirements for environmental factors. It is important to note that temperature has a dual role in seedling emergence: (i) in a seasonal fashion it influences the state of dormancy; and (ii) it is a regulatory element in the germination process. Consequently, optimum temperatures for dormancy release and germination are not necessarily similar.

Apart from water and oxygen, light (through phytochrome) and nitrate may be considered the two most important naturally occurring factors required for seed germination of many species (for reviews, see Roberts and Smith, 1977; Bewley and Black, 1985; Hilhorst and Karssen, 1989; Karssen and Hilhorst, 1992). Interactions between these factors and with temperature have been reported (Vincent and Roberts, 1977; Hilhorst *et al.*, 1986). Evidently, fully dormant seeds do not respond to light and nitrate. Responsiveness increases during dormancy alleviation and decreases during dormancy induction. Thus, the degree of dormancy can be expressed as the (germination) responsiveness to stimulating factors. Studying these changes in responsiveness may provide new insights into the regulation of dormancy. This chapter will review recent research in this field and discuss its relevance for understanding dormancy and its regulation.

Sensitivity to Light, Nitrate and Gibberellins

A certain responsiveness or sensitivity results from the sum total of processes which precede the measured parameter. Thus, sensitivity includes not only the reception of a certain chemical by its receptor, but also all steps of the transduction chain that lead to the final response, e.g. uptake and transport. Figure 24.1 provides a schematic view of elements of the signal-transduction chain that are proposed to be involved in the stimulation of *Sisymbrium officinale* and *Arabidopsis thaliana* germination by light and nitrate (Hilhorst and Karssen, 1988; Derkx and Karssen, 1994). Germination of the former species depends absolutely on the simultaneous presence of light and nitrate; germination of the latter depends on light only, but nitrate may modify light-induced germination to some extent. After penetrating the seed coat (and being modified by it), the red light component will convert inactive phytochrome (P_r) to the active P_{fr} (Fig. 24.1). The uptake of nitrate by the seed is at least partly active (Hilhorst, 1990b). With or without interacting with nitrate, P_{fr} action leads to synthesis of gibberellins (GAs). Circumstantial evidence for this comes from the observation that GAs may replace light (Taylorson and Hendricks, 1976), from studies with synthetic inhibitors of GA biosynthesis (Hilhorst and Karssen, 1988) and from experiments with hormone-deficient mutants (Karssen *et al.*, 1988, 1989; Derkx and Karssen, 1993b). Finally, these synthesized GAs induce a subsequent chain of events that results in germination. All these elements of the signal-transduction chain may contribute to the final responsiveness.

An approach to establishing the relative importance of singular elements of this pathway is to perform dose-response experiments under conditions

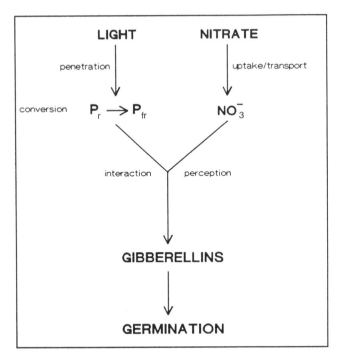

Fig. 24.1. Schematic overview of elements involved in the regulation of germination of *S. officinale* and *A. thaliana* seeds.

that are limiting for one particular element. In *S. officinale*, nitrate is limiting at optimal light and temperature conditions, whereas light (or P_{fr}) is limiting when seeds are incubated in the dark at optimal temperature (Hilhorst, 1990a,b). When seeds are deprived of both factors, GAs are limiting (Derkx and Karssen, 1993a). Alternatively, responsiveness to GA can be studied in GA-deficient *A. thaliana* mutants (Derkx and Karssen, 1994).

Light dose–response studies

Seeds of *A. thaliana* and *S. officinale* gradually develop (secondary) dormancy when incubated in the dark at 15°C (Fig. 24.2). Depending on the seed lot, primary dormancy may be broken prior to reinduction of dormancy. Both species are light-dependent for germination at 24°C. Application of nitrate had no effect on changes in dormancy of *A. thaliana* but delayed the induction of dormancy in *S. officinale*. It has been shown that germination of *S. officinale* in the absence of nitrate depended on the level of endogenous nitrate (Hilhorst, 1990b). Thus, germination of both species appeared to require an increasing dose of light, and, for *S. officinale*, nitrate during dormancy induction. These changes in dormancy have been studied in more detail by means of dose–response experiments for light and nitrate (Hilhorst, 1990a,b; Derkx and Karssen, 1993a, 1994).

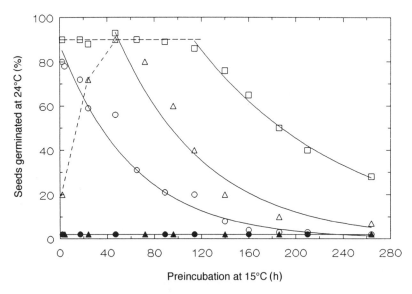

Fig. 24.2. Germination at 24°C of *S. officinale* and *A. thaliana* seeds after various periods of dark preincubation at 15°C. Germination conditions: *S. officinale*: dark, water (●); light, water (○); light, 25 mM KNO$_3$ (□). *A. thaliana*: dark, water (▲); light, water (△). (Redrawn from combined data from Hilhorst, 1990a, and Derkx and Karssen, 1994.)

Irradiation of dark-incubated *S. officinale* seeds (Fig. 24.2) with a range of red light (660 nm) fluences resulted in a set of fluence–response curves that shifted to higher fluence values when dormancy became deeper (Fig. 24.3). Maximum germination levels declined, as could be anticipated from Fig. 24.2. Interestingly, curves were also shifted after incubation periods up to 120 h that hardly affected maximal germination.

To test the ecological relevance of this observation, the experiments were repeated with *S. officinale* seeds buried in the field at a depth of 10 cm. Under these conditions seeds are subject to seasonal, as well as daily, fluctuations of the temperature (Derkx and Karssen, 1993a). Moreover, seeds will also experience other influences, such as changes in soil water potential and chemical components. Germination was tested at 15°C in 25 mM KNO_3 at monthly intervals during a burial period of 2 successive years (Fig. 24.4). Results compared well with those obtained under laboratory conditions (Fig. 24.3). During winter and early spring when dormancy was broken, maximal germination rose and fluence–response curves shifted to the left. During summer and autumn the reverse occurred.

A result that was limited to the field situation was that, during spring, part of the seed population germinated independently of light. As discussed above, this may be due to pre-existing P_{fr} combined with a very high sensitivity to P_{fr}, known as the very low fluence response (Cone and Kendrick, 1985) and observed before in buried *Datura ferox* seeds (Scopel *et al.*,

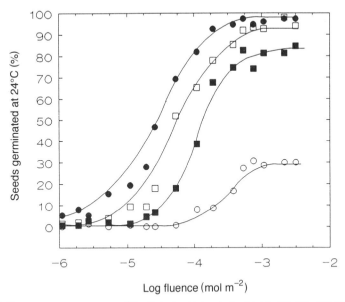

Fig. 24.3. Fluence–response curves of *S. officinale* obtained after 24 (●), 120 (□), 192 (■) and 264 (○) h of dark preincubation at 15°C in 25 mM KNO_3. Seeds were irradiated and germinated at 24°C. Curves were calculated from population parameters. (Redrawn from Hilhorst, 1990a.)

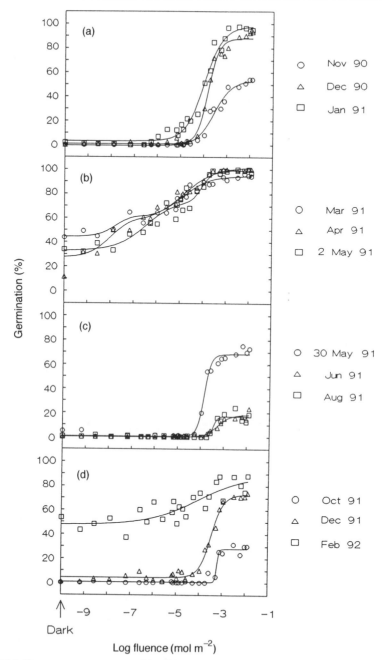

Fig. 24.4. Fluence–response curves of *S. officinale* seeds. After exhumation at the indicated dates, germination capacity was determined at 15°C in 25 mM KNO_3 at a range of fluence values. Germination data were fitted as logistic dose–response curves. The first point of each curve represents dark germination. (From Derkx and Karssen, 1993a.)

1991). One can only speculate why the very low fluence response appears to occur in buried seeds and not in water-imbibing seeds. This trait may be associated with phytochrome transformations and the intermediates involved. It is possible that periods of partial seed dehydration in the soil result in accumulation of P_{fr} or its immediate precursors (Kendrick and Spruit, 1977).

Apart from this very low fluence response, dose-response relations of seeds after burial up to 2 years were remarkably similar to those of the laboratory experiment. In the burial experiment, curve shifts and changes in maximal germination were fully reversible. Similar behaviour was displayed by *A. thaliana* seeds, not buried but water imbibed under outside temperature conditions (Derkx and Karssen, 1994). Interestingly, these seeds did not show the very low fluence response. It can be concluded that temperature is indeed the major regulatory factor in reversibly changing sensitivity to light upon release or induction of dormancy. However, there is a possibility that changes in seed water status caused by changes in soil water potential may affect this sensitivity to some extent.

Nitrate dose–response studies

Analysis of the response to nitrate has been carried out in a similar fashion to that for light, with the nitrate dose as limiting factor under saturating light conditions (Hilhorst, 1990b; Derkx and Karssen, 1993a). At the same intervals as used for studying responses to light (Figs 24.2 and 24.3), nitrate responses were determined during dormancy induction of *S. officinale* seeds under laboratory conditions (Fig. 24.5). Although the shapes of the nitrate-response curves were more complex than those of the fluence-response curves, two observations were similar: (i) the curve shifted to the right during dormancy induction; and (ii) maximal germination level decreased slightly later than the shift. At the start of dormancy induction (24 and 48 h imbibition), germination was partially independent of nitrate. It has been shown unambiguously (Hilhorst, 1990b) that this was the result of pre-existing nitrate, combined with a high sensitivity to nitrate, which is analogous to what we suggested above for the very low fluence response. Furthermore, nitrate-independent germination was associated with a biphasic dose-response curve. It has been suggested that *S. officinale* seeds possess high- and low-affinity nitrate binding sites that are related to the very low and low nitrate responses, respectively (Hilhorst, 1990b).

The results from the nitrate response experiments with *S. officinale* seeds after burial (for periods up to 16 months) were comparable with the fluence-response test results (Fig. 24.6). Again, curves shifted to the left during winter (dormancy release), followed by an apparently nitrate independent germination during spring. During the summer period the pattern

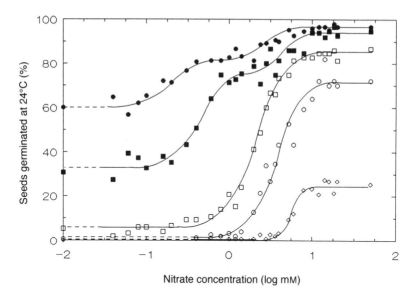

Fig. 24.5. Red light-induced germination at 24°C of *S. officinale* seeds preincubated at 15°C for 24 (●), 48 (■), 120 (□), 192 (○) and 264 (◊) h in a range of nitrate concentrations. Curves were calculated from population parameters. The first point of each curve represents germination in water. (From Hilhorst, 1990b.)

was reversed. In these experiments, biphasic responses were not observed, possibly due to a lack of sufficient detail.

Gibberellin dose–response studies

It is essential to know whether modulations of sensitivity towards primary components of the germination signal-transduction chain, such as light and nitrate, are also reflected in comparable changes in responsiveness to elements further downstream on the signal transduction pathway. If so, changing sensitivity may be due to overriding factors affecting the pathway as a whole. As pointed out before, there are strong indications that environmental signals which induce germination act via synthesis of GAs. The response to exogenous GAs can be studied when the presence of GAs is limiting for germination. This has been done for seeds of *S. officinale* in the absence of light and nitrate (Hilhorst and Karssen, 1988; Derkx and Karssen, 1993a) and *A. thaliana* (Derkx and Karssen, 1994). GA response curves of *S. officinale* seed germination shifted to the left (or the $[GA]_{50}$, the GA concentration required for half-maximal germination, decreased) during winter (dormancy release) (Fig. 24.7). However, during periods of dormancy induction, the curves did not shift back to higher GA concentrations. In fact, during the test

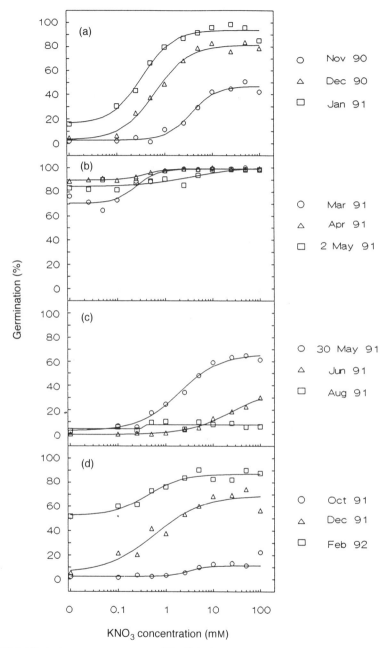

Fig. 24.6. Nitrate dose–response curves of *S. officinale* seeds. After exhumation at the indicated dates, germination capacity was determined at 15°C after a brief exposure to red light. Germination data were fitted as logistic dose–response curves. (From Derkx and Karssen, 1993a.)

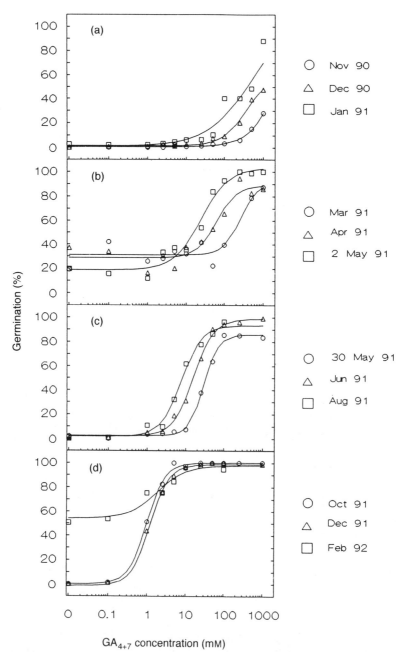

Fig. 24.7. Gibberellin dose–response curves of *S. officinale* seeds. After exhumation at the indicated dates, germination capacity was determined at 15°C in darkness in a range of GA_{4+7} concentrations. The germination data were fitted as logistic dose-response curves. (From Derkx and Karssen, 1993a.)

period, sensitivity to GA apparently increased independently of seasonal influences. Virtually similar results were obtained with *A. thaliana* seeds incubated in water at outdoor temperatures (Fig. 24.8). The sensitivity to GA increased, then decreased during the first release and induction of dormancy, but was not associated with further dormancy cycling. From these results it can be concluded that, contrary to the fluence and nitrate responses, responsiveness to exogenous GAs is not correlated with changes in dormancy.

Calculation and interpretation of the dose–response curves

Many dose-response curves for the action of plant hormones are sigmoidal in shape (Nissen, 1985, 1988). Fluence, nitrate and gibberellin dose-response curves for the stimulation of seed germination of *A. thaliana* and *S. officinale* have also been shown to be sigmoidal (Hilhorst and Karssen, 1988; Hilhorst, 1990a,b; Derkx and Karssen, 1993a). This allows a quantitative approach to the concept of sensitivity (Weyers *et al.*, 1987). Calculation of dose-response curves from germination data is preferably done through probit analysis

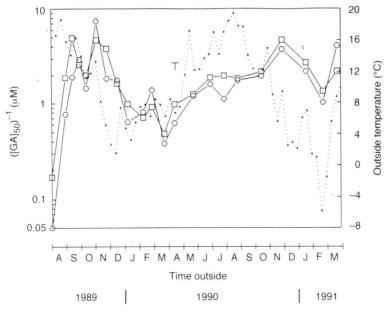

Fig. 24.8. Seasonal variation in germination capacity of wild-type (○) and GA-deficient *ga1-2* (□) *A. thaliana* seeds incubated at outside temperatures. Incubation occurred in Petri dishes that were packed in light-tight plastic boxes. Portions of seeds were transferred to the laboratory at the indicated times and germination was tested at 24°C in darkness in a range of GA_{4+7} concentrations. Reciprocal values of GA concentrations required for half-maximal response were calculated from the fitted GA dose–response curves. The dotted line indicates the air temperature at 1.50 m. (Redrawn from Derkx and Karssen, 1994.)

because of the all-or-none response type (DePetter et al., 1985). However, calculating logistic dose–response curves from the individual data points gives essentially similar results (Derkx and Karssen, 1993a, 1994). Sigmoidal dose–response curves can also be described by an equation derived from the following simple interaction model of a ligand with its receptor, developed by Clark in 1933 (see e.g. Boeynaems and Dumont, 1980):

$$H + Rec \underset{k_{-1}}{\overset{k_1}{\rightleftharpoons}} HRec \overset{k_r}{\rightarrow} Response \qquad (Eqn\ 1)$$

where H = ligand; Rec = receptor; $HRec$ = ligand–receptor complex; k_1 and k_{-1} are kinetic association and dissociation constants; and k_r is a rate constant for the rate-limiting step in the signal-transduction path between the formation of $HRec$ and the final response. Assuming that the receptors for one ligand are equivalent and independent, that the response is proportional to the number of occupied receptors, that the ligand can exist in only two states (free or bound to its receptor) and that the identity or rate of the rate-limiting step does not change during the response, the following equation can be derived from the simple interaction model (Weyers et al., 1987; Hilhorst, 1990a; Derkx and Karssen, 1993a):

$$R = R_{min} + \frac{R_{max} - R_{min}}{1 + \left[\frac{[H]_{50}}{[H]}\right]^p} \qquad (Eqn\ 2)$$

where R = response; R_{min} = minimum response in the absence of exogenous $[H]$; R_{max} = maximum response; $[H]$ = applied dose concentration; $[H]_{50}$ = dose concentration required for half-maximum response; and p = Hill or cooperativity coefficient.

This relationship is equivalent to Michaelis–Menten kinetics for a cooperative interaction. Cooperativity is expressed by the Hill coefficient p. It has been shown that modification of Clark's (1933) equation (Eqn 1) by introducing a degree of cooperativity results in better fits (Hilhorst, 1990a). The cooperativity coefficient can be derived from the slope of the curve at the inflection point. If the calculated curve parameters, R_{min}, R_{max}, slope and $[H]_{50}$ are substituted in this equation, very good fits have been obtained that hardly deviate from the original calculated curves (Fitzsimons, 1989; Hilhorst, 1990a,b; Derkx and Karssen, 1993a).

On the basis of the simple bimolecular interaction model, curve parameters may, with caution, be related to binding parameters. The parameter R_{min} represents the base level of response in the absence of added ligand. R_{min} can be a ligand-independent process or the result of endogenous levels of the ligand. The response range R_{max} minus R_{min} may be taken as a measure of the capacity to respond when ligand availability is not limiting. R_{max} is a function of the total level of receptors according to $R_{max} = k_r[\text{Rec}]_T$, with $[\text{Rec}]_T = [\text{Rec}] + [H\text{Rec}]$.

Thus, changes in R_{max} (or response range) may reflect differences in the numbers of receptors. However, since R_{max} also depends on k_r, changes in R_{max} may also reflect changes in the rates of processes following ligand-receptor complex formation. The parameter $[H]_{50}$ may be used to calculate the dissociation constant K_D of the ligand-receptor complex according to $K_D = ([H]_{50})^p$. The parameter p is a measure of the degree of cooperativity of ligand binding. However, it is unsafe to give a biological interpretation to p values, and thus to K_D values, especially when the p value is not an integer (Boeynaems and Dumont, 1980; Weyers et al., 1987).

Despite these limitations and the assumptions that are basic to the simple model (see above), interpretation of curve parameters as binding parameters is still possible. Clark's (1933) equation (Eqn 1) has been applied successfully to a number of drug-receptor interactions and latter proved to be valid for the actual ligand-receptor binding and initiation of the signal-transduction chain (Hollenberg, 1985a). The receptor occupancy theory, to be discussed below, has been used to explain the observed shifts and changes in shape of the fluence, nitrate and GA response curves described above.

The Receptor Occupancy Theory

From Clark's (1933) equation (Eqn 1), it follows that changes in the maximal response can be related to changes in the number of receptors. Likewise, curve shifts may be linked to alteration of the dissociation constant during changing dormancy levels. However, parallel curve shifts can also occur as a result of changing receptor levels, without a necessary change in K_D. The fluence, nitrate and GA response curves described above all showed parallel shifts without significant changes in the slope of the curves. Furthermore, in a study of the binding of phytochrome to its putative receptor X, it was calculated that approximately 40% of the total amount of active receptors was sufficient to maximize the low fluence response of lettuce seeds (VanDerWoude, 1985). Thus, more receptors may be present than are required for a maximal response. This assumption has been used for the interpretation of the changes in fluence response during induction of dormancy in *S. officinale* seeds (Hilhorst, 1990a; Figs 24.2 and 24.3). If the (exponential) *S. officinale* decay curves of Fig. 24.2 are extrapolated to time zero, 'germination' values of 220% and 88% are found for conditions with or without applied nitrate, respectively. From these values, it follows that occupation of approximately 40% of the total amount of receptors that are present (under high nitrate conditions) may result in a maximal response.

Another implication is that nitrate appears to create this receptor reserve. It has been known for a long time that nitrate and light (phytochrome) may interact in the promotion of seed germination. Hilhorst and Karssen (1988) suggested that nitrate could be regarded as a cofactor for phytochrome action. In the receptor occupancy model, nitrate action as a cofactor would be on the level of receptor activation and/or synthesis. If so, changes in nitrate response should be linked closely to changes in the fluence response. Indeed, a strong linear correlation between $[P_{fr}/P_{tot}]_{50}$ and $[NO_3]_{50}$ was found (Hilhorst, 1990b). This role of nitrate is also in agreement with the report that nitrate is not assimilated as a nutrient, via successive reductions to ammonium, but is active as the nitrate ion (Hilhorst and Karssen, 1989). Small ions have been reported to alter receptor properties in animal cells, e.g. by neutralizing surface charges on proteins or altering the membrane potential in the case of membrane-associated receptor complexes (Hollenberg, 1985b).

With the concept of spare receptors, shifts of curves without changes in maximal response can be explained. From Eqn 1 it follows that:

$$[H\text{Rec}] = \frac{[\text{Rec}]_T [H]}{K_D + [H\text{Rec}]} \quad \text{(Eqn 3)}$$

Thus, a decreasing level of $[\text{Rec}]_T$ can result only in a maximum response when the concentration of $[H]$, here P_{fr} or nitrate, is increased to achieve the same level of $[H\text{Rec}]$, provided p remains unchanged (Eqn 3), which was generally the case for the results discussed above. Hence, the fluence, nitrate and gibberellin dose–response curves shift to the right. This shift can occur only when spare receptors are present. When the receptor number is reduced below the level required for the maximal response, R_{max} will be reduced. Conversely, an increasing level of $[\text{Rec}]_T$ (i.e. dormancy release) will require less of the ligand to attain a maximal response. Above the level required for the maximal response, this will result in a left-hand shift. Below this level, R_{max} will increase. However, it should be kept in mind that $R_{max} = k_r[\text{Rec}]_T$ and thus reduction of R_{max} may also be the result of postreceptor defects, expressed in the value of k_r.

Receptor-regulated dormancy cycling: a model

From the dose–response analysis of factors involved in germination and dormancy of *A. thaliana* and *S. officinale*, it can be concluded that dormancy cycling in these species is associated with changes in sensitivity to environmental factors. This opens the possibility that changes in dormancy are regulated at the level of perception of these cues. The most simple approach is to assume that the availability of active receptors determines the final response. As shown above, application of the receptor occupancy theory supports this idea. Earlier studies of the action of phytochrome on *Portulacca oleracea* and *Rumex crispus* seed germination resulted in the hypothesis that the amount of interaction between P_{fr} and its receptor X may be limiting

(Duke *et al.*, 1977; Duke, 1978). Furthermore, studies by VanDerWoude (1985) and DePetter *et al.* (1985) on light-induced germination of *Lactuca sativa* and *Kalanchoë blossfeldiana*, respectively, also suggest an important role for the availability of active phytochrome receptors. However, a phytochrome receptor has not been found thus far. The same is true for a possible nitrate receptor, apart from the well-studied binding to nitrate reductase. Nevertheless, evidence is available that nitrate may act in the unreduced state (Hilhorst and Karssen, 1989; Redinbaugh and Campbell, 1991). Thus, binding of nitrate to a receptor cannot be excluded.

The supply of active receptors may be regulated in several ways. In a one-step model, the level of active receptors depends solely on the net result of synthesis and degradation of the receptor protein or on an activation-inactivation mechanism of a large receptor pool. For a number of species, it is known that Arrhenius plots of germination rate vs. temperature have properties that are also found in protein denaturation kinetics (Bewley and Black, 1982). These observations would fit in a model in which receptor synthesis and denaturation are regulated by a temperature-time relationship. Ideas concerning possible activation of dormancy proteins or receptors have also been reported. Trewavas (1988) proposed the existence of a bistable metabolic switch for the transition between the dormant and non-dormant states. The metabolic bistability could be provided by autophosphorylating protein kinases (Lisman, 1985). The enzyme could exist in the active unphosphorylated form in germination and in the highly autophosphorylated unresponsive form in dormancy. In the unphosphorylated form, the enzyme is capable of phosphorylating and activating other proteins. The switch from the inactive to the active form would be mediated by a phosphatase.

The results discussed above, with support from other studies, have led to a model of receptor-regulated dormancy cycling, incorporating the respective actions of temperature, light and nitrate (Hilhorst, 1990a, b). To this can be added the results from the GA response experiments. As discussed before, GA sensitivity is not readily associated with dormancy cycling (Derkx and Karssen 1993a, 1994). Only the breaking of primary dormancy was accompanied clearly by in increase in sensitivity to GA, which remained fairly constant during subsequent dormancy cycling (Figs 24.7 and 24.8). However, GA sensitivity can be influenced positively by light upon induction of germination (Hilhorst and Karssen, 1988; Derkx and Karssen, 1993b). In our model, this could be envisaged as an effect on synthesis or activation of the GA receptor.

The dormancy model presented here (Fig. 24.9) accounts for the often observed interactions among environmental factors affecting dormancy and germination of many phytochrome-requiring weed species. It also accounts for the dual role of temperature: first on dormancy cycling (T_1), and second on the process of germination *per se* (T_2). The basis of the model is the synthesis and/or activation of the phytochrome receptor X. The level of X is determined by the ratio of the rate constants for synthesis and denaturation of the receptor protein. If a separate activation step is involved, nitrate (Hilhorst and Karssen, 1988; Hilhorst, 1990a) and/or (de)phosphorylation (Trewavas, 1988) may be involved in its regulation. Interestingly, nitrate has been shown to be involved in the light-mediated phosphorylation of phosphoenolpyruvate carboxylase and sucrose phosphate synthase in wheat leaves (VanQuy and Champigny, 1992). Moreover, phytochrome itself may have protein kinase activity (Wong et al., 1989). Thus, phosphorylation/dephosphorylation at the reception site of nitrate and P_{fr} could well be part of the signal transduction chain, as suggested by Trewavas (1988).

Subsequent formation of the $P_{fr}X$ complex initiates a signal–transduction path leading to GA synthesis. That GA synthesis is indeed a later event than P_{fr} and nitrate action has been shown by means of escape experiments with seeds of *S. officinale* (Hilhorst et al., 1986). It was shown that the escape time for the inhibiting effect of far-red irradiation on light induced germination was much shorter (8 h) than the escape time for inhibition by the GA-biosynthesis inhibitor tetcyclacis (16 h).

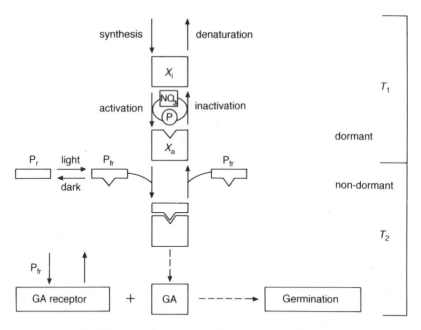

Fig. 24.9. An integrating model for seed dormancy cycling. The inactive phytochrome receptor (X_i) is activated by nitrate and/or phosphorylation, resulting in a receptor that is accessible to P_{fr}. The phytochrome–receptor complex initiates a signal–transduction chain to induce GA synthesis. GA binds to its receptor (possibly activated by P_{fr}). This complex induces a chain of reactions ultimately leading to germination. The dual role of temperature is expressed in T_1 acting on dormancy relief and induction and T_2 regulating germination-related processes.

Presumably, GA binds to a receptor. Conclusive evidence for the existence of a true GA receptor is still lacking but indications for a GA binding site associated with the plasma membrane of *Avena fatua* aleurone protoplasts are very strong (Hooley *et al.*, 1991). The cues involved in the regulation of synthesis and/or activation of GA receptors remain unknown. There is evidence that P_{fr} may positively affect sensitivity to GA in *S. officinale* and *A. thaliana* (Hilhorst *et al.*, 1986). It has been shown that the presence of GA-receptors does not depend on T_1 (Derkx and Karssen, 1993a, 1994).

The signal-transduction route downstream from GA perception remains largely unknown. In endosperm-retaining seeds, GA may induce hydrolytic enzyme activity required to degrade endosperm cell walls to allow radicle protrusion (Groot *et al.*, 1988).

Conclusions

The present model favours a subtle switch between the dormant and non-dormant state, rather than a gradual metabolic shift as proposed in numerous earlier studies (see Bewley and Black, 1985, for extensive review). Shifts in general metabolic activity occur upon induction of germination. As the distinction between dormancy release and germination induction has seldom been made, increased metabolic activity during germination has often been ascribed to the breaking of dormancy. True alleviation of seed dormancy in *Avena fatua* and *Sisymbrium officinale* was not accompanied by an increase in oxygen consumption (Simmonds and Simpson, 1971, as cited in Bewley and Black, 1982; Derkx *et al.*, 1993). From the viewpoint of energy expenditure, dormancy regulation at the level of perception of primary stimulatory factors, such as P_{fr} and nitrate, seems to be most favourable. Seeds in a seed bank may undergo annual dormancy cycling for many years before conditions for emergence are favourable. Only with a very low basic level of respiration and general metabolism could energy reserves be maintained for such long periods.

Future studies of dormancy regulation may be aimed at locating the primary reception sites. Furthermore, the involvement of membranes in regulation must be considered. Several studies have indicated that membrane transitions may be involved as regulatory elements in dormancy and germination (Bewley and Black, 1985; Hilhorst, 1990a, b). However, little is known about interactions among membrane-associated (receptor) proteins and membrane properties in higher plants.

References

Baskin, J.M. and Baskin, C.C. (1977) Role of temperature in the germination ecology of three summer annual weeds. *Oecologia* 30, 377-382.

Baskin, J.M. and Baskin, C.C. (1983) Seasonal changes in the germination responses of buried seeds of *Arabidopsis thaliana* and ecological interpretation. *Botanical Gazette* 144, 540-543.

Baskin, J.M. and Baskin, C.C. (1985) The annual dormancy cycle in buried weed seeds: a continuum. *BioScience* 35, 492-498.

Bewley, J.D. and Black, M. (1982) *Physiology and Biochemistry of Seeds*, Vol. 2. Springer-Verlag, New York, pp. 276-339.

Bewley, J.D. and Black, M. (1985) *Seeds: Physiology of Development and Germination*, 1st edn. Plenum Press, New York, USA.

Boeynaems, J.M. and Dumont, J.E. (1980) *Outlines of Receptor Theory*. Elsevier/North Holland Biomedical Press, Amsterdam.

Bouwmeester, H.J. (1990) The effect of environmental conditions on the seasonal dormancy pattern and germination of weed seeds. Unpublished PhD thesis, Wageningen Agricultural University.

Bouwmeester, H.J. and Karssen, C.M. (1993) Annual changes in dormancy and germination in seeds of *Sisymbrium officinale* (L.) Scop. *New Phytologist* 124, 179-191.

Clark, A.J. (1933) *The Mode of Action of Drugs on Cells*. Williams and Williams, Baltimore.

Cone, J.W. and Kendrick, R.E. (1985) Fluence-response curves and action spectra for promotion and inhibition of seed germination in wild type and long-hypocotyl mutants of *Arabidopsis thaliana* L. *Planta* 163, 43-54.

DePetter, E., VanWiemeersch, L., Rethy, R., DeDonder, A., Fredericq, H., DeGreef, J., Steyaert, H. and Stevens, H. (1985) Probit analysis of low and very-low-fluence-responses of phytochrome-controlled *Kalanchoe blossfeldiana* seed germination. *Photochemistry and Photobiology* 42, 697-703.

Derkx, M.P.M. and Karssen, C.M. (1993a) Changing sensitivity to light and nitrate but not to gibberellins regulates seasonal dormancy patterns in *Sisymbrium officinale* seeds. *Plant, Cell and Environment* 16, 469-479.

Derkx, M.P.M. and Karssen, C.M. (1993b) Light- and temperature-induced changes in gibberellin biosynthesis and -sensitivity influence seed dormancy and germination in *Arabidopsis thaliana*: studies with gibberellin-deficient and -insensitive mutants. *Physiologia Plantarum* 89, 360-368.

Derkx, M.P.M. and Karssen, C.M. (1994) Are seasonal dormancy patterns in *Arabidopsis thaliana* regulated by changes in seed sensitivity to light, nitrate and gibberellin? *Annals of Botany* 73, 129-136.

Derkx, M.P.M., Smidt, W.J., VanDerPlas, L.H.W. and Karssen, C.M. (1993) Changes in dormancy of *Sisymbrium officinale* seeds do not depend on changes in respiratory activity. *Physiologia Plantarum* 89, 707-718.

Duke, S.O. (1978) Significance of fluence-response data in phytochrome-initiated seed germination. *Photochemistry and Photobiology* 28, 383-388.

Duke, S.O., Egley, G.H. and Reger, B.J. (1977) Model for variable light sensitivity in imbibed dark-dormant seeds. *Plant Physiology* 59, 244-249.

Fitzsimons, P.J. (1989) The determination of sensitivity parameters for auxin-induced H^+-efflux from *Avena* coleoptile segments. *Plant, Cell and Environment* 12, 737-746.

Groot, S.P.C., Kieliszewska-Rokicka, B., Vermeer, E. and Karssen, C.M. (1988) Gibberellin-induced hydrolysis of endosperm cell walls in gibberellin-deficient tomato seeds prior to radicle protrusion. *Planta* 174, 500-504.

Hilhorst, H.W.M. (1990a) Dose response analysis of factors involved in germination and secondary dormancy of seeds of *Sisymbrium officinale*. I. Phytochrome. *Plant Physiology* 94, 1090-1095.

Hilhorst, H.W.M. (1990b) Dose response analysis of factors involved in germination and secondary dormancy of seeds of *Sisymbrium officinale*. II. Nitrate. *Plant Physiology* 94, 1096-1102.

Hilhorst, H.W.M. and Karssen, C.M. (1988) Dual effect of light on the gibberellin- and nitrate-stimulated seed germination of *Sisymbrium officinale* and *Arabidopsis thaliana*. *Plant Physiology* 86, 591-597.

Hilhorst, H.W.M. and Karssen, C.M. (1989) The role of light and nitrate in seed germination. In: Taylorson, R.B. (ed.) *Recent Advances in the Development and Germination of Seeds*. NATO ASI Series A 187, Plenum Press, New York, USA, pp. 191-205.

Hilhorst, H.W.M. and Karssen, C.M. (1992) Seed dormancy and germination: the role of abscisic acid and gibberellins and the importance of hormone mutants. *Plant Growth Regulation* 11, 225-238.

Hilhorst, H.W.M., Smitt, A.I. and Karssen, C.M. (1986) Gibberellin-biosynthesis and -sensitivity mediated stimulation of seeds germination of *Sisymbrium officinale* by red light and nitrate. *Physiologia Plantarum* 67, 285-290.

Hollenberg, M.D. (1985a) Receptor models and the action of neurotransmitters and hormones: some new perspectives. In: Hollenberg, M. and Yamamura, H.I. (eds) *Neurotransmitter Receptor Binding*. Raven Press, New York, pp. 1-39.

Hollenberg, M.D. (1985b) Biochemical mechanisms of receptor regulation. *Trends in Pharmacological Sciences* 6, 299-302.

Hooley, R., Beale, M.H. and Smith, S.J. (1991) Gibberellin perception at the plasma membrane of *Avena fatua* aleurone protoplasts. *Planta* 183, 274-280.

Karssen, C.M. (1982) Seasonal patterns of dormancy in weed seeds. In: Khan, A.A. (ed.) *The Physiology and Biochemistry of Seed Development, Dormancy and Germination*. Elsevier Biomedical Press, Amsterdam, The Netherlands, pp. 243-270.

Karssen, C.M. and Hilhorst, H.W.M. (1992) Effect of chemical environment on seed germination. In: Fenner, M. (ed.) *Seeds: The Ecology of Regeneration in Plant Communities*. CAB International, Wallingford, UK, pp. 327-348.

Karssen, C.M., Derkx, M.P.M. and Post, B.J. (1988) Study of seasonal variation in dormancy of *Spergula arvensis* L. seeds in a condensed annual temperature cycle. *Weed Research* 28, 449-457.

Karssen, C.M., Zagórski, S., Kepczynski, J. and Groot, S.P.C. (1989) Key role for endogenous gibberellins in the control of seed germination. *Annals of Botany* 63, 71-80.

Kendrick, R.E. and Spruit, C.J.P. (1977) Phototransformations of phytochrome. *Photochemistry and Photobiology* 26, 201-204.

Lisman, J.E. (1985) A mechanism for memory storage insensitive to molecular turnover: a bistable autophosphorylating kinase. *Proceedings of the National Academy of Sciences USA* 82, 3055-3057.

Nissen, P. (1985) Dose responses of auxins. *Physiologia Plantarum* 65, 357-374.

Nissen, P. (1988) Dose responses of gibberellins. *Physiologia Plantarum* 72, 197-203.
Ödum, S. (1978) *Dormant Seeds in Danish Ruderal Soils*. Report of the Royal Veterinarian and Agricultural University, Horsholm, Denmark.
Priestley, D.A. and Posthumus, M.A. (1982) Extreme longevity of lotus seeds from Pulantien. *Nature* 299, 148-149.
Probert, R.J. (1992) The role of temperature in germination ecophysiology. In: Fenner, M. (ed.) *Seeds: The Ecology of Regeneration in Plant Communities*. CAB International, Wallingford, UK, pp. 285-325.
Redinbaugh, M.C. and Campbell, W.H. (1991) Higher plant responses to environmental nitrate. *Physiologia Plantarum* 82, 640-650.
Roberts, E.H. and Smith, R.D. (1977) Dormancy and the pentose phosphate pathway. In: Khan, A.A. (ed.) *The Physiology and Biochemistry of Seed Development, Dormancy and Germination*. Elsevier Biomedical Press, Amsterdam, The Netherlands, pp. 385-411.
Roberts, H.A. and Feast, P.M. (1973) Emergence and longevity of annual weeds in cultivated and undisturbed soil. *Journal of Applied Ecology* 10, 133-143.
Roberts, H.A. and Lockett, P.M. (1978) Seed dormancy and field emergence in *Solanum nigrum* L. *Weed Research* 18, 231-241.
Scopel, A.L., Ballaré, C.L. and Sánchez, R.A. (1991) Induction of extreme light sensitivity in buried weed seeds and its role in the perception of soil cultivations. *Plant, Cell and Environment* 14, 501-508.
Simmonds, J.A. and Simpson, G.M. (1971) Increased participation of pentose phosphate pathway in response to afterripening and gibberellic acid treatment in caryopses of *Avena fatua*. *Canadian Journal of Botany* 49, 1833-1840.
Taylorson, R.B. and Hendricks, S.B. (1976) Interactions of phytochrome and exogenous gibberellic acid on germination of *Lamium amplexicaule* L. seeds. *Planta* 132, 65-70.
Thompson, K. and Grime, J.P. (1979) Seasonal variation in the seed banks of herbaceous species in ten contrasting habitats. *Journal of Ecology* 67, 893-921.
Trewavas, A.J. (1988) Timing and memory processes in seed embryo dormancy - a conceptual paradigm for plant development questions. *Bioessays* 6, 87-92.
VanDerWoude, W.J. (1985) A dimeric mechanism for the activation of phytochrome: evidence from photothermal interactions in lettuce seed germination. *Photochemistry and Photobiology* 42, 655-661.
VanQuy, L. and Champigny, M.-L. (1992) NO_3 enhances the kinase activity for phosphorylation of phosphoenolpyruvate carboxylase and sucrose phosphate synthase protein in wheat leaves: evidence from the effects of mannose and okadaic acid. *Plant Physiology* 99, 344-347.
Vincent, E.M. and Roberts, E.H. (1977) The interaction of light, nitrate and alternating temperature in promoting the germination of dormant seeds of common weed species. *Seed Science and Technology* 5, 659-670.
Weyers, J.D.B., Paterson, N.W. and A'Brook, R. (1987) Towards a quantitative definition of plant hormone sensitivity. *Plant, Cell and Environment* 10, 1-10.
Wong, Y.-S., McMichael, Jr, R.W. and Lagarias, J.C. (1989) Properties of a polycation-stimulated protein kinase associated with purified *Avena* phytochrome. *Plant Physiology* 91, 709-718.

25 Modelling Climatic Regulation of Bud Dormancy

SCHUYLER D. SEELEY
Plants, Soils, and Biometeorology, Utah State University, Logan, UT 84322-4820, USA

Introduction

The lifelong helix of a perennial plant can be divided into many arcs (Fig. 25.1); in fact, during the period of embryoid and fruit growth, it becomes a double helix. Arcs of the helix (e.g. endodormancy, full bloom, vegetative growth, leaf fall, to identify but a few) flow smoothly into adjacent arcs. Each is responsive to previous and current conditions and is dependent on, and modified by, circumstances that affect the preceding arcs. Some arcs have been studied extensively, others only superficially. Vegetative growth, with its driving photosynthetic and respiratory functions, has been characterized in depth. Endodormancy inception, which occurs during the annual senescence of perennials, has been the subject of few endeavours.

Each complete arc can be subdivided into component arcs as more information is acquired. Thus far, endodormancy can be divided into inception, transition and release (Fig. 25.1); these terms will be used to subdivide endodormancy in this chapter. Our knowledge of plant growth and development can be measured by our ability to subdivide the helix. As more is learned about dormancy inception, for instance, characterizing component arcs will become evident.

To understand the perennial helix thoroughly, all arcs and their modifying influence on subsequent arcs need to be known and understood. In the absence of knowledge of the biochemical or biophysical mechanisms which have their effects on the transition of a plant through each arc, we are limited to descriptive analysis of an arc with empirical or phenological models. Even though they are quite simple (Einstein said 'Everything should be made as simple as possible, but not simpler.'), these mathematical descriptions may be very useful in efforts to eventually elucidate the basic endodormancy mechanism(s).

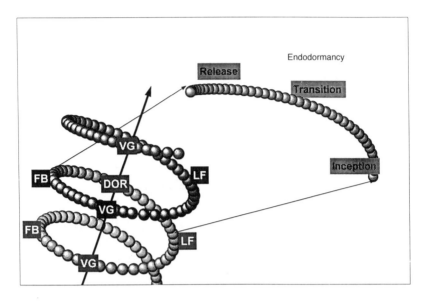

Fig. 25.1. The perennial plant life cycle represented as a helix, with endodormancy subdivided into the subarcs Inception, Transition, and Release. The bold arrow represents time. FB = full bloom, DOR = endodormancy, VG = vegetative growth, LF = leaf fall.

Chill Units and the Utah Model

The evolution of the chill unit (CU) began about 100 years ago with Jost, who initiated the technical study of plant cold temperature responses (Samish, 1945). In 1932, Hutchins proposed 7°C as the threshold for chilling phenomena (Weinberger, 1950). Weldon (1934) observed that insufficient chilling, which resulted in delayed foliation, occurred when the average temperatures for December and January were above 9.4°C. Based on these and other observations, Weinberger (1950) correlated the performance of peach cultivars to the accumulated hours at temperatures below 7°C, which he called chill hours. He also noted that warmer temperatures were not only unfavourable for endodormancy development, but that they negated the effect of previously accumulated chilling temperatures. Da Mota (1957) developed an equation to calculate chill hours under the conditions of Rio Grande do Sul in order to evaluate chilling in mild winters. Munoz Santa Maria (1969), in a review of chill hour systems, concluded that the values obtained by Da Mota's equation were the most suitable for Central Mexico.

During the mid-1950s, temperature response curves for vernalization of rye and endodormancy transition of apple seeds were published by Hansel (1953) and Schander (1955). The surprisingly similar curves (Fig. 25.2) suggest a chemical or physiochemical process with a temperature optimum

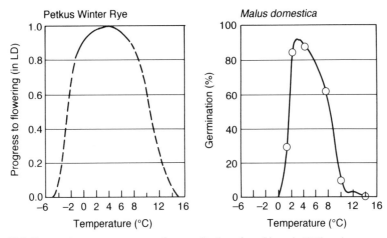

Fig. 25.2. Temperature response curves for vernalization of rye (Hansel, 1953) and endodormancy transition of apple seeds (Schander, 1955).

in the mid single digits, a minimum around the freezing point of plant tissuewater, and a maximum near 15°C. In 1971, Erez and Lavee published a vague temperature response curve for peach bud chilling. They reported that chilling at 3 and 10°C was about half as effective as the optimum 6°C. In 1972, Gonzalez-Cepeda introduced a concept of adjusted chill hours in which the final value for the day was calculated by summing the hours below 7°C and subtracting from this sum twice the hours above 18°C.

The Utah Chill Unit (CU) model for temperate zone endodormancy release of peach flower buds was introduced in 1974 (Richardson *et al.*, 1974). It has evolved through several interim versions and is still being modified as it gradually replaces chill hours. Currently, the CU is defined as 1.0 h at the optimum chilling temperature at the optimum chilling time. The chilling response curve for peaches (Fig. 25.3) was submitted in 1974, but not published. Instead, a very rough step function approximation of the response curves for 'Redhaven' and 'Elberta' flower buds was published, i.e. 0 CU below 1.4°C, 0.5 CU between 1.5 and 2.4°C, 1.0 CU between 2.5 and 9.1°C, 0.5 CU between 9.2 and 12.4°C, 0 CU between 12.5 and 15.9°C, −0.5 CU between 16.0 and 18.0°C, and −1.0 CU above 18°C. A more advanced curve (Fig. 25.4) illustrates the sigmoidal response of the curve from 0 CU at the freezing point of water in plant tissue (−2.0°C) to a maximum (1.0 CU) at the optimum temperature; a reverse sigmoidal curve from the optimum to the compensation point temperature (i.e. 0 CU at 14°C); and a negatively increasing sigmoidal curve for chilling negation temperatures (i.e. ⩾ 16°C).

The transition from endodormancy to growth in response to chilling temperatures is not linear over time. Stratification of seeds of various deciduous fruit tree species at −4 to 16°C for up to 120 days (2880 CU) yielded

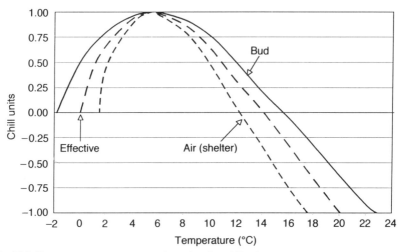

Fig. 25.3. Temperature response curve for estimating chill unit accumulation in peaches (as published in Anderson *et al.*, 1986, from the original manuscript of Richardson *et al.*, 1974).

differential increments of increased germination in response to 10-day intervals of increased stratification (e.g. Fig. 25.5 for peach). In general, the rate of response to any temperature was sigmoidal, with slow initial germination, a long phase of rapid germination and a terminal phase of slowed response again (Fig. 25.6). The initial rate extended too long to be a simple hydration effect and, in subsequent studies using a short hydration period at the compensation temperature, similar stratification did not change the results

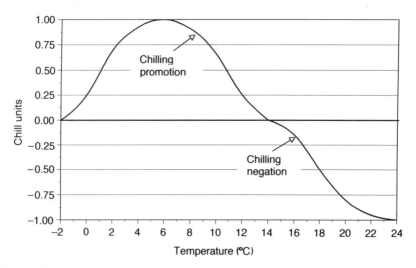

Fig. 25.4. Temperature response curve for estimating chill units and chill unit negation.

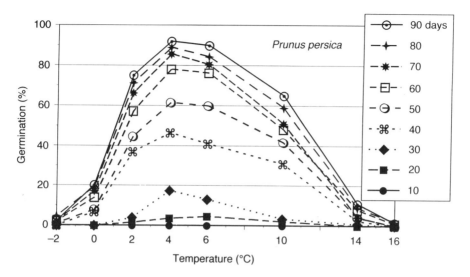

Fig. 25.5. Ten day succeeding stratification temperature-response curves for peach seed germination.

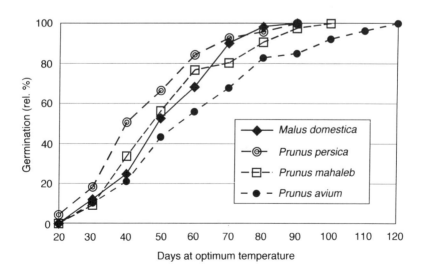

Fig. 25.6. Stratification curves for peach, cherry (*avium* and *mahaleb*) and apple.

significantly. Thus, temperature effects change according to the physiological time of application and length of exposure; altered chilling temperatures during the grand phase of the sigmoid response curve have larger positive and negative effects on endodormancy transition than those that occur at other times.

Since negative CUs accumulate above 16°C and average late summer temperatures exceed this threshold, a plot of CU accumulation in late summer gives an increasingly negative curve until average temperatures drop below negation levels and positive CU accumulation begins (Fig. 25.7). This maximum negative accumulation can be used as the starting point for physiological CU accumulation. Such a 'point event' works well in temperate zone climates with definite seasons.

Problematic Aspects and Evolution of the Utah Model

All good models require experimental testing and modification as appropriate. Experimental work with temperatures is fraught with difficulty and requires discriminating attention. An obvious problem in CU research occurs when temperatures other than the compensation point for imbibition or forcing of plant material are used. At certain times during endodormancy, if forcing is done at a temperature higher than the compensation point, negative CUs accumulate; if forcing is done at temperatures below the

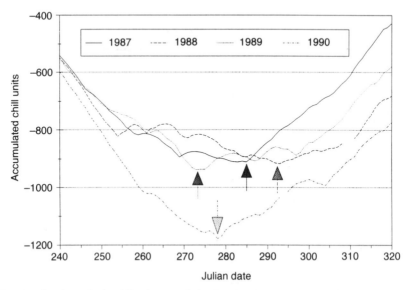

Fig. 25.7. Starting point for chill unit accumulation in cold climates. The point of maximum negative accumulation of chill units occurs around the beginning of autumn in Utah.

compensation point, positive CUs accumulate. Both scenarios can modify the forced seed or bud response. Furthermore, the release of endodormancy is not abrupt. A common criterion for the end of endodormancy (e.g. 2 weeks to a certain bloom stage) is useful for measurement purposes and experimental comparisons, but it is not intended to indicate that there is a precipitous end to the process.

The calculation of hourly temperatures from maximum and minimum temperatures, as in the original Utah model (Richardson *et al.*, 1974), has been an area of potential difficulty. While this 'straight line' method worked fairly well in Utah, it certainly did not work in coastal areas with onshore afternoon breezes (Aron, 1975). A sine-exponential model for synthesizing hourly temperatures from maximum/minimum has subsequently been developed that more readily fits any location in question (Fig. 25.8).

Likewise, in warmer climates where temperatures may vary for considerable lengths of time above and below the compensation point, determining when to begin accumulating CUs is problematic. One potential alternative to the use of maximum negative CU accumulation (as described above) is the use of vegetative maturity (Fuchigami *et al.*, 1982), the absence of regrowth after defoliation or decapitation. However, this physiological benchmark is labour and plant material intensive, time consuming, and dependent on climatic factors

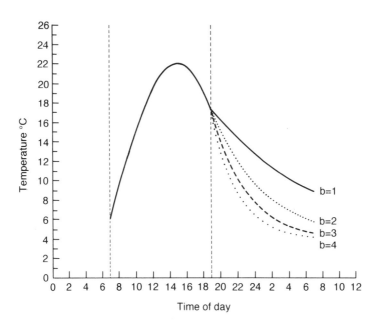

Fig. 25.8. Sine-exponential model for synthesizing hourly temperatures from maximum and minimum temperatures. The *beta* parameter is shown here. Three other parameters and all combinations were tested.

that occur after the point event. Another alternative is natural defoliation, which has been suggested as a phenological benchmark for beginning CU accumulation (Walser et al., 1981). However, leaves may senesce over varying lengths of time, become inactive due to stress or other stomatal closure conditions while still attached and/or be inactivated by cold temperatures yet remain on the tree for extended periods. At present, the maximum negative accumulation is best for distinctly seasonal climates while the search remains for other methods better suited to more variable climates.

Temperature is definitely the primary factor in the endodormancy transition. For buried seeds and for seedlings in shade or close to the ground, temperature is all important. For buds exposed to radiation, fog, mist, dew, rain, snow and sleet, most results can be explained through temperature modification. Indeed, bud cooling by evaporation or shade accelerated the endodormancy transition (Buchanan et al., 1977). However, mist and low light intensity have been shown to accelerate bud break even under controlled temperatures in growth chambers (Freeman and Martin, 1981). Mist increased peach floral bud break about 25%; high light had a small negative effect at 6°C, but not at 10°C. Similarly, Westwood and Bjornstad (1968) found that winter rainfall accelerated the endodormancy transition although they did not segregate the bud temperature effect. Factors such as these have yet to be incorporated into the CU model.

As the CU model is based on optimum temperatures at the optimum time during the chilling process, it requires calibration for each species or cultivar. The cardinal temperatures, the temperature response curve, and the changing ability of the organism to react to temperature over time need to be delineated. For instance, conditions which affect the plant at the inception of endodormancy may modify endodormancy depth and length (Walser et al., 1981). The development of a complete phenological model for a crop will bring other possibilities to light.

When it was developed, the temperature response curve was thought to be constant throughout the course of endodormancy. Indeed, subsequent experiments with interruption of chilling at higher and lower temperatures have indicated that the CU temperature response *optimum* does not change during the course of endodormancy (S.D. Seeley, unpublished data). However, as Askenasy noted in 1877 (Vegis, 1973) in his study of dormant cherry flower buds, *plant responses* to temperature change during dormancy. Endodormancy development was studied in apple seeds stratified at 4°C for 10 weeks with sequential weekly interruptions at 0–24°C (Fig. 25.9). After treatment, all seed groups had received 9 weeks at 4°C (1512 CUs) plus 1 week at one of the other temperatures, except for controls (10 weeks at 4°C). After stratification, seed germination and seedling growth revealed very significant differences during weeks 6 to 8 at interruption temperatures in the negation range. The ability of temperatures above 16°C to negate chilling gradually decreased. This suggests that a change occurs in the physiology of

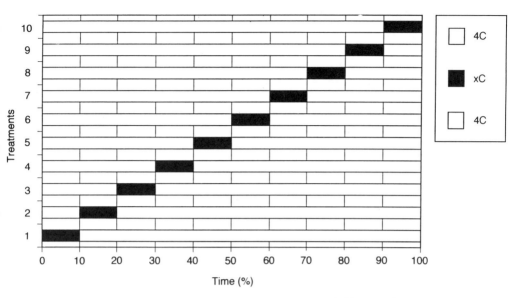

Fig. 25.9. Plan of a temperature interruption study with apple seeds. The control was a constant 4°C; xC temperatures ranged from 2 to 24°C.

the plant such that temperatures above a threshold around 4°C also begin to contribute to thermal unit accumulation for anthesis (Fig. 25.10). Results from other experiments with interruptions up to 25% of the total stratification time were about the same. Thus, as the CU accumulation nears culmination, the same temperatures may negate chill accumulation and promote bud development, with a net promotion of anthesis.

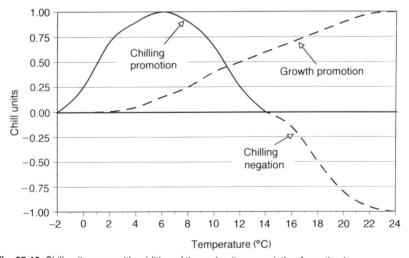

Fig. 25.10. Chill unit curves with addition of thermal unit accumulation for anthesis.

The studies described above were used to generate a revised Utah model describing endodormancy development in regard to temperature, physiological time, and CUs which can be described in three ways: a value matrix, a topographic map, and a three dimensional response surface (Fig. 25.11). This revision was tested with ten years (1972-1984) of phenological data from northern Mexico, a climate for which the original Utah Chill Unit Model gave a coefficient of determination of 66%; the revised model gave a coefficient of determination of 74% (del Real-Laborde, 1989). The revised model gives greater accuracy in colder climates as well, especially for years in which warm temperatures occur during the chilling period.

Dormancy/Anthesis Transition Models

Couvillon and Erez (1985) showed that chilling beyond that amount required for bud break reduced the specific heat requirement for bloom development. In warm climates where significant chilling negation occurs, the chilling requirement often remains unfulfilled and high temperatures prevail during anthesis. Under such conditions, the time of bloom probably is reflected mainly through the chilling requirement and not through the energy required for bloom. Hanninen (1990) synthesized a 'growth competence' model that postulated: (i) an absolute chilling requirement before which no growth competence occurred; (ii) an increasing growth competence with increased chilling beyond the absolute threshold; and (iii) an increased growth competence with higher forcing temperatures near the end of endodormancy. Both sets of authors provide equations that may be used in modelling the endodormancy-anthesis transition. In many temperate climates, these are not needed since the chilling required for complete endodormancy release is almost always ($\sim$ 90% of the time in Utah) reached long before temperatures high enough to drive anthesis are encountered. However, for warm climates where chilling may be marginal, they improve results when included.

In modelling temperature response to chilling, the CU is a tool for measuring a phenological process. This concept can be used also for the phenological modelling of bloom development. Anthesis Units (AUs) have advantages over the concept of growing degree hours, which is cumbersome and does not parallel CUs. As with the CU model, AUs represent a process that is temperature dependent. Temperature response curves for anthesis, with cardinal minimum, maximum and optimum temperatures, have been established for several tree fruit cultivars. The AU is defined as 1 h at the optimum temperature for blossom bud development, with higher and lower temperatures giving fractional values. Anthesis units accumulate according to how temperature drives each phenotypical flower stage, e.g. 'Montmorency' tart cherry (*Prunus cerasus*) phenotypic stages and their corresponding AU accumulations are given in Table 25.1.

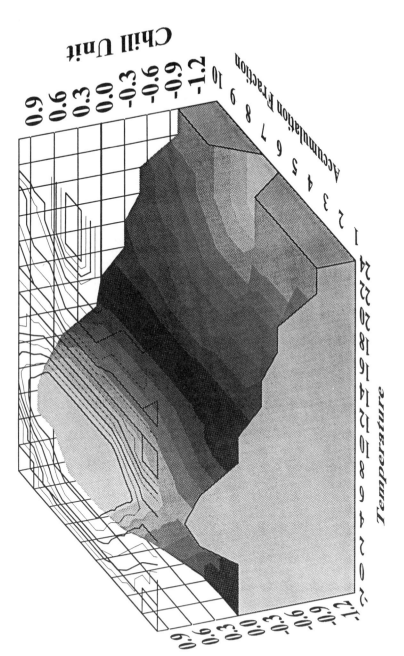

Fig. 25.11. Chill unit model for apple in 1992.

Table 25.1. Anthesis unit accumulation values for Montmorency tart cherry flower bud stages.[1]

Bud stage	Anthesis unit accumulation
Bud swell	128
Green tip	182
Swollen bud	258
Flower separation	330
First white	390
First bloom	470
Full bloom	566

[1] Cardinal temperatures: minimum = 5°C, optimum = 25°C, maximum = 35°C.

From such data, a predictive endodormancy/anthesis computer program has been developed which uses maximum and minimum temperature inputs and calculates the CU and AU accumulations for the season. After the seasonal temperatures to date have been analysed, the program switches to mean historic temperatures to complete calculations through full bloom and thus provide tentative dates of endodormancy release and bloom. This program has made possible the use of evaporative cooling to delay apple anthesis and produce bloom on a given target date in Utah orchards.

Growing Season Models

Models exist, then, for endodormancy transition/release and for anthesis. No phenological models for endodormancy inception currently exist; thus, a complete endodormancy model has yet to be synthesized. This is a deserving study area. Perhaps the endodormancy inception model can be approached best by development of a post bloom growth and development model which would lead into a model of endodormancy inception and leaf senescence.

A phenological model for spring and summer growth and development has been based on studies of whole tree net photosynthesis (Pn) (Heinicke and Childers, 1937) and the leaf area index concept (e.g. see Fig. 25.12 for apple). Analysis of data from these studies suggests that net photosynthate productivity of fruit trees in the late spring season can be expressed by the equation:

Spring Pn (g CO_2day^{-1}) = $-1080 + 7.90x$(Day of Year) $- 1.113x$(Min Temp of Day) $- 2.18x$(Max Temp of Day) $+ 0.0443x$(Langleys day^{-1})

This equation explains about 90% of the variation in Pn during this time period. It is predominantly a foliation index. In summer, radiation becomes the most important factor, expressed by the equation:

Summer Pn (g CO_2day^{-1}) = $58.23 + 0.2468x$(Langleys day^{-1}) $- 1.19x$(Max Temp of Day)

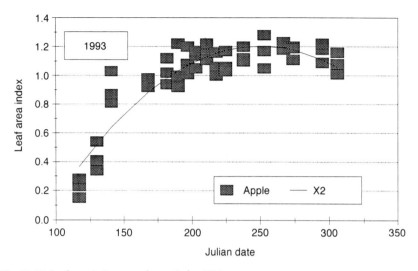

Fig. 25.12. Leaf area index curve for apple for 1993.

This equation explains about 67% of the variation in summer Pn, with temperature comprising only 1.6% of the variation.

These Pn values can be normalized to the greatest potential value per day (Fig. 25.13). This includes a temperature factor which is not very important in photosynthate accumulation. A temperature term (heat units, Fig. 25.14) was derived from an optimum of 1.0 at the optimum temperature for development and was also normalized to the best day in the season. The daily

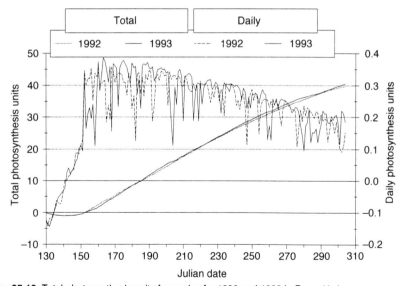

Fig. 25.13. Total photosynthesis units for apples for 1992 and 1993 in Perry, Utah.

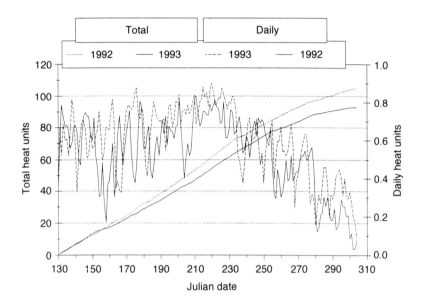

Fig. 25.14. Total heat units for apples for 1992 and 1993 in Perry, Utah.

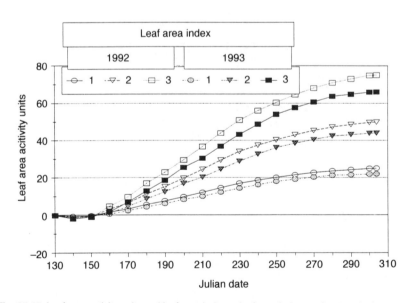

Fig. 25.15. Leaf area activity units and leaf area index at leaf area indexes of 1, 2 and 3 for 1992 and 1993 in Perry, Utah.

photosynthate term was multiplied by the heat unit term to get a leaf area activity (LAA) accumulation value for each succeeding day. Summation of LAA units over time gives a leaf area activity index (LAAI). Typical LAAI curves for leaf area indices of 1, 2 and 3 are shown in Fig. 25.15. The growth unit is a measure of photosynthate production by an orchard canopy.

As this model develops further, component arcs will become apparent and the effects of differing growth seasons on subsequent endodormancy development may be discovered. Then, attention can be focused on the late stages of the summer model to begin the development of a dormancy inception model, which may use vegetative maturity, leaf senescence, or some other phenological marker for its initialization.

References

Anderson, J.L., Richardson, E.A. and Kesner, C.D. (1986) Validation of chill unit and flower bud phenology models for 'Montmorency' sour cherry. *Acta Horticulturae* 184, 71- 77.

Aron, R.H. (1975) Letters. *HortScience* 10, 559-560.

Buchanan, D.W., Bartholic, J.F. and Biggs, R.H. (1977) Manipulation of bloom and ripening dates of three Florida grown peach and nectarine cultivars through sprinkling and shade. *Journal of the American Society for Horticultural Science* 102, 466-470.

Couvillon, G.A. and Erez, A. (1985) Influence of prolonged exposure to chilling temperatures on bud break and heat requirement for bloom of several fruit species. *Journal of the American Society for Horticultural Science* 110, 47- 50.

Da Mota, F.S. (1957) Os invernos de Pelotas, R. S. en relacao as exigencias dos arbores fruitiferas de folhas caducas. *Rio Grande do Sul. Biological and Technical Institute for Agriculture do Sul* 8 (as quoted in Munoz Santa Maria, 1969).

del Real-Laborde, J.I. (1989) An apple rest model for mild winter conditions. PhD thesis. Utah State University, USA.

Erez, A. and Lavee, S. (1971) The effect of climatic conditions on dormancy development of peach buds. I. Temperature. *Journal of the American Society for Horticultural Science* 96, 711-714.

Freeman, M.W. and Martin, G.C. (1981) Peach floral bud break and abscisic acid content as affected by mist, light and temperature treatments during rest. *Journal of the American Society for Horticultural Science* 106, 333-336.

Fuchigami, L.H., Weiser, C.J., Kobayashi, K., Timmis, R. and Gusta, L.V. (1982) A degree growth stage (°GS) model and cold acclimation in temperate woody plants. In: Li, P.H. and Sakai, A. (eds) *Plant Cold Hardiness and Freezing Stress, Mechanisms and Crop Implications*, CRC Press, Boca Raton; pp. 93-116.

Gonzalez-Cepeda, I.A. (1972) Control de los efectos de inviernos benignos en manzano (*Malus sylvestris* Mill) para la region de Navidad, Nuevo Leon. BS Thesis. ITESM, Mexico. 100 pp.

Hanninen, H. (1990) Modelling bud dormancy release in trees from cool and temperate regions. *Acta Forestalia Fennica* 213, 1–47.

Hansel, H. (1953) Vernalisation of winter rye by negative temperatures and the influence of vernalisation upon the lamina length of the first and second leaf in winter rye, spring barley, and winter barley. *Annals of Botany* 67, 418-431.

Heinicke, A. and Childers, N. (1937) The daily rate of photosynthesis during the growing season of 1935 of a young apple tree of bearing age. *Cornell University Agricultural Experiment Station Memoir* 201, 1-52.

Munoz Santa Maria, G. (1969) Evaluacion de dormulas para el calculo de horas frio en algunas zonas fruiticolas de Mexico. *Proceedings of the Tropical Region of the American Society for Horticultural Science* 13, 345-366.

Richardson, E.A., Seeley, S.D. and Walker, D.R. (1974) A model for estimating the completion of rest for 'Redhaven' and 'Elberta' peach trees. *HortScience* 9, 331-332.

Samish, R.M. (1945) The use of dinitrocresol mineral oil sprays for the control of prolonged rest in apple orchards. *Journal of Pomology and Horticultural Science* 21, 164-179.

Schander, H. (1955) Keimungsphysiologische studien an kernobst II. Untersuchungen uber die allgemeinen temperaturanspruche der kernobstsamen wahrend der keimung. *Zeitschrift Pflanzenzucht* 34, 421-440.

Vegis, A. (1973) Effect of temperature on growth and development. In: Precht, H., Christophersen, J., Hensel, H. and Larcher, W. (eds) *Temperature and Life*. Springer-Verlag, New York, pp. 145-170.

Walser, R.D., Walker, D.R. and Seeley, S.D. (1981) Effect of temperature, fall defoliation, and gibberellic acid on the rest period of peach leaf buds. *Journal of the American Society for Horticultural Science* 106, 91-94.

Weinberger, J.H. (1950) Chilling requirements of peach varieties. *Proceedings of the American Society for Horticultural Science* 56, 122-128.

Weldon, G.P. (1934) Fifteen years study of delayed foliation of deciduous fruit trees in Southern California. *California Department of Agriculture, Monthly Bulletin* 23, 7-9;160-181.

Westwood, M.N. and Bjornstad, H.O. (1968) Chilling requirements of dormant seeds of 14 pear species as related to their climatic adaptation. *Proceedings of the American Society for Horticultural Science* 92, 141-149.

Index

Aardvarks 13
ABA (Abscisic acid) 3, 17, 19-20, 22-24, 30, 33-36, 60, 119-121, 123, 127-128, 137, 139-140, 158, 185, 213-225, 227-230, 259, 271, 273-278, 284-287, 324, 331
    analogues 214-219
    enantiomers 215-219
    receptors 213-220
    structural formula 214-216
Abies alba 176, 185
Abies magnifica var. *shastensi* 185
Abies procera 185
Abscisic acid *see* ABA
Acacia 13
ACC (1-Amino-cyclopropane-1-carboxylic acid) 38, 119
Acer 32, 302
Acer pseudoplatanus 18, 273
Acer rubrum 50-51
Acer saccharinum 51
Acer saccharum 50-51, 303
Acrotony 85, 88, 105
Adenosine triphosphate *see* ATP
Aegilops cylindrica 288
Aesculus hippocastanum 18
After-ripening 245-254, 260, 287, 294, 297
Agrobacterium tumefaciens 303

Alfalfa (*Medicago sativa*) 19-24, 41
Almonds 52
Alnus glutinosa 303
Althaea rosea 148
Amaranthus retroflexus 164
Ambrosia 5, 7
Ambrosia artemisiifolia 31, 39
1-Amino-cyclopropane-1-carboxylic acid *see* ACC
Anaesthetic action hypothesis 258
Anaesthetics 31
Ancymidol 33, 41
Anthesis 369-370, 372
Anthesis units 370, 372
Apical meristems 61-78, 85, 152, 194-195, 197-198
Apium graveolens see Celery
Apples (*Malus domestica*) 52-53, 88-89, 96, 98-99, 124-128, 158, 203, 326-327, 362-363, 365, 368-369, 371-374
Arabidopsis 5, 7, 9, 11, 53, 121, 127, 166, 221-222, 228-229, 276, 286
Arabidopsis thaliana 34, 150, 153, 236, 297, 341, 343-344, 347-348, 351, 354, 357
Arachidonic acid 135
Ash *see Fraxinus*
Astericus pygmaeus 8

ATP (Adenosine triphosphate) 94-95, 100-101, 103, 129, 235-236, 238, 240-241
ATPase 159, 233-243
Auxins 85-86, 116, 119, 128, 137-139
Avena fatua 8, 31, 39, 41, 245-254, 286-288, 293-300, 331, 357
5-Azacytidine 153

Baeria 11
BAP *see* 6-Benzylaminopurine
Barley 216, 221, 228, 283, 286-289, 297
Basitony 86, 88, 105
Batatasins 139-140
Beet *see Beta vulgaris*
Begonia 5-7
6-Benzylaminopurine (BAP) 93, 119, 124
Bermuda grass 4, 6
Beta vulgaris 148, 150
Betula 47, 273
Betula papyrifera 164, 303
Betula pendula 246
Bimolecular interaction model 352
Biotime models 334-336
Birch *see Betula*
Birds 13-14
Black henbane *see Hyoscyamus niger*
Black spruce *see Picea mariana*
Blackcurrants 273
Blue light 118
Blueberries *see Vaccinium*
Brassica oleracea see Brussels sprouts
Broadleaf plantain *see Plantago major*
Bromegrass *see Bromus*
Bromoethane 135
Bromus 219
Bromus secalinus 287-288
Bromus tectorum 288-289
Brussels sprouts (*Brassica oleracea*) 148
Bud break 49-53, 85-93, 96-98, 100-103, 171-174, 177-178, 184, 186, 193-199, 202-204, 208, 239, 270, 272, 368, 370
biochemistry 202

chilling requirement 49-50
conifers 171-174, 177-178, 184, 186, 193-199
environmental factors 368, 370
near-lethal stress 202-204, 208
Spirodela polyrrhiza 270, 272
woody plants 52-53, 85-93
Buds, definitions 85
dormancy 47-131, 136, 139, 148, 154, 171-199, 201-210, 233, 236, 241, 269-281, 301-309, 361-376
biochemistry 96-98
climatic regulation 361-376
conifers 171-199
genetics 51-53
models 361-376
physiology 47-56, 61-63
woody plants 49-53, 83-113, 201-210, 301-309
formation 85, 172-173
supercooled 205-206
Bulbils 49-50, 139-140
Bulblets 115-124, 128
Bulbs xix, 99, 115-116, 121, 128
dormancy 115

Cactus 13
Calcium 157-163, 165-166, 259
Calcium cyanamide 93
Calmodulin 157-166
Canarygrass *see Phalaris arundinacea*
Canola *see* Rape
Carbohydrates, determination 248
metabolism 245-256
Carrots (*Daucus carota*) 32-33, 37, 148
Cassava (*Manihot esculenta*) 134
Castor beans 18-21, 24, 222
Cataphylls 83, 85
Catharanthus roseus 70-71, 297
Cattle 13
Cedars 13
Celery (*Apium graveolens*) 33, 35, 38, 148
Cell signalling networks 67-68
Cereals 29, 31, 133, 259, 283-291

Chaparral shrubs 9
Chara 71
Cheat *see Bromus secalinus*
Chemical stress 203-204
Chemosensing 11-12, 14
Chenopodiaceae 8
Chenopodium bonus-henricus 31
Cherries (*Prunus avium, P. cerasus* and *P. mahaleb*) 13, 93, 158, 365, 368, 370, 372
Chess *see Bromus secalinus*
Chill unit dormancy models 204, 207, 362-372
Chilling 49-52, 90-93, 147-155, 157-169, 174-175, 179, 182, 194-197, 199, 206-208, 234-235, 238-240, 270, 272, 303, 316-318, 326-327, 362-372
2-Chloroethylphosphonic acid 119
Chrysanthemums (*Dendranthema grandiflora*) 148
Cicer arietinum 158
Cichorium endivia see Endives
Citrus 235
Citrus sinensis see Oranges
Climax timothy *see Phleum pratense*
Cocklebur *see Xanthium strumarium*
Cold hardiness 171-192, 201, 203-208, 246, 301
Colocasia esculenta 134
Colorado spruce *see Picea pungens*
Common ragweed *see Ambrosia artimisiifolia*
Compositae 8, 11
Cones 9
Conifers 171-199, 203
Corms 115, 134
 dormancy 115
Cornus sericea 202, 204
Corylus 53
Corylus avellana 51, 53, 157-166
Cress 216-217
Crimson clover *see Trifolium incarnatum*
Cucumis 13
Curly dock *see Rumex crispus*
Cyclins 136

Cytokinins 12, 30, 33, 36, 38-39, 93, 100, 119, 128, 135, 137-139, 274, 278

Datura ferox 345
Daucus carota see Carrots
Degree growth stage model 172-176, 178-181, 184, 186, 202-204, 206
Dehydrins 207, 285
Dendranthema grandiflora see Chrysanthemums
Desiccation 18-25, 53, 177, 205-207, 222, 253, 286, 301
 perception 20-25
 tolerance 253, 301
Diapause 4
Dieback 201, 204-205, 208
Digitalis purpurea 148
Dinitro-o-cresol 93
Dioscorea see also Yams
Dioscorea alata 50
Dioscorea bulbifera f. *spontanea* 50
Dioscorea izuensis 50
Dioscorea japonica 50
Dioscorea nipponica 50
Dioscorea pentaphylla 50
Dioscorea quinqueloba 50
Dioscorea septemloba 50
Dioscorea tenuipes 50
Dioscorea tokoro 50
DNA 14, 135-136, 152-153, 275-277, 285, 287, 293-300, 304
 clone analysis 293-300
 methylation 152-153
Dormancy breaking chemicals 257-265
Dormancy release index 174
Dormancy
 biochemistry 206-208
 cell biology 135-137
 classification 48, 60-64
 control 76-77, 184-186, 194
 definitions xix, 17, 59-60, 63, 77, 193
 genetics 51, 53, 60-61, 70, 134, 218-219, 254, 334
 hormonal regulation 137-140

inhibitors 10-11, 48-49, 157
mechanisms 62, 254
models 62-63, 311-376
molecular genetics 136, 275-278, 283-291, 293-309
partial 39, 49, 318
primary 30-32, 39, 254, 344, 355
roots 177-180
secondary 30, 32, 39, 48-49, 157, 254, 344
sequential 8
transient 234-241
woody plants 49-53, 83-113, 201-210, 301-309
Dormancy/anthesis transition models 370, 372
Dose-response studies 343-353
Douglas fir *see Pseudotsuga menziesii*
Downy brome *see Bromus tectorum*
Drought 172, 174, 178, 183, 246

Earthworms 13
Easter lily *see Lilium longifolium*
Eastern red cedar *see Juniperus virginiana*
Ecodormancy 48, 60, 62-63, 75, 77-78, 83, 87, 90-92, 94, 96-97, 106, 201-203, 206, 270
Electrolyte leakage 202-204, 333
Elephants 13
Endives (*Cichorium endivia*) 35
Endodormancy 48, 62-63, 74-75, 77-78, 83-87, 89-92, 94, 96-98, 100, 102-103, 105-106, 201-206, 208, 269, 361-363, 366-368, 370, 372, 375
Endoplasmic reticulum 68-70, 72, 74-77
Engelmann spruce *see Picea engelmannii*
Episodic growth 98, 102-106, 178
Ethanol 234-236, 239-241, 259-260
Ethephon 33, 38-39, 185
Ethrel 138
Ethylene 3, 30, 33, 36, 38-39, 93, 119, 128, 135, 137-139, 202, 204
Ethylene chlorohydrin 135-136

Eucalyptus 9
Euphorbia 10

Fagus 51
Fagus sylvatica 50-51, 303
Ferns 165
Festuca rubra 41
Field pennycress *see Thlaspi arvense*
Filbert *see Corylus avellana*
Fire 9-10, 12-14
First International Symposium on Plant Dormancy xix-xx
Flower initiation 70-71, 147, 149-154, 157, 203, 363, 368, 370
Fluence response curves 316-318, 345-347, 351, 353-354
Fluorescent dyes 72-73
Fluridone 23-24, 119-121, 127, 139, 227-229
Flushing 11, 84-85, 89, 98, 100-102, 105, 172, 302
Foxglove *see Digitalis purpurea*
Fraxinus americana 303
Fraxinus excelsior 90, 96-98, 102-103
Fraxinus ornus 49
Free radicals 202, 205, 208
Freezing temperatures *see* Frost
Frost 172, 175, 179, 182-186, 205-206, 269, 272
  tolerance 172, 175, 179, 182, 184-186, 272
Fruit trees 363, 370
Fruits 13, 17, 48-49, 202
Fungi 11
β-Furfuryl-β-glucoside 76
Fusicoccin 33, 235, 238

GA *see* Gibberellins
GA-time model 315, 319-321, 335
Gas exchange 9-10, 48
Germination response curves 333-334
Germination time courses 323, 328, 330, 334, 336
Geum urbanum 148
Gibberellic acid *see* Gibberellins
Gibberellins 29-41, 85, 93, 96, 100, 121, 127-128, 135, 137-140,

157-158, 165, 221-230, 245-249, 253, 274, 287, 298-299, 315-316, 318-321, 324-325, 331, 334, 342-344, 348, 350-351, 353-357
   inhibitors 33-35, 40-41
Gleditsia triacanthos 91-92
Glucuronic acid 276-277
Grasses 29, 31, 41, 283-291, 341
Growing season models 372-375
Growth competence model 370
Growth regulators 33, 47, 91-94, 119, 175

Hairy vetch *see Vicia villosa*
Hard seeds 8, 10, 13, 32-33, 36, 39, 48, 53, 159
Hardening off 177
Hazelnuts *see Corylus avellana*
Heat shock proteins 202, 207
Heat stress 202-204, 206
Heat tolerance 38-39, 176-177, 201, 203
Helianthus annuus see Sunflowers
Helianthus tuberosus see Jerusalem artichokes
Herbicides 40, 177
Heteroblasty 83, 85, 100
Hevea 102
Hollyhocks *see Althaea rosea*
Honesty *see Lunaria annua*
Hormones 3, 30, 39, 53, 64, 67-68, 74, 93-94, 107, 125-128, 137-140, 158, 174, 221-231, 322
   sensitivity 221-231
Horse chestnuts *see Aesculus hippocastanum*
Horses 13
Hydrogen cyanamide 202, 208
Hydrotime model 322-329, 334-335
Hyoscyamus niger 148

IAA (Indoleacetic acid) 119, 138, 277
IBA (Indolebutyric acid) 125-126
Ice nucleation 205-207
Indoleacetic acid *see* IAA

Indolebutyric acid *see* IBA
Inositol 276-277
   metabolism 277
Insects 13
Ipomoea batatas see Sweet potatoes
Irradiation 31-32, 35-36
Isocitrate lyase 21
Isoetes 13
Isopentenyladenosine 138

Japanese yew *see Taxus cuspidata*
Jerusalem artichokes (*Helianthus tuberosus*) 84, 94-95, 233-239, 259
Jointed goatgrass *see Aegilops cylindrica*
Juglans regia 92
Juniperus chinensis 'Pfitzerana' 180
Juniperus virginiana 175

Kalanchoë blossfeldiana 355
Kinases 135-136, 286, 289, 355-356
Kinetin 33, 38, 274, 278

Lactuca sativa see Lettuces
Larix decidua 303
Late embryogenesis abundant proteins 285-287
Leaf area activity units 374-375
Leaf area index 372-375
Leguminosae 10-11, 41
Lemnaceae 270, 272
Lettuces (*Lactuca sativa*) 30-35, 37-38, 40, 164-165, 216, 316-318, 322-324, 326, 353-355
Light effects 9-12, 30-31, 33-35, 37-38, 90, 123-124, 159, 161-164, 166, 233, 259, 316-318, 334, 342-345, 347-348, 354-356, 368
Lilies *see Lilium*
Lilium 115-124, 127-128, 150
Lilium auratum 119
Lilium longiflorum 116-118, 120
Lilium longifolium 150

Lilium speciosum 116-121, 123-124
Linear hormonal hypothesis 84
Lizards 13
Loblolly pine *see Pinus taeda*
Lolium perenne 41
Lotus nelumbo 8
Lucifer Yellow dye 72-76
Lunaria annua 150
Lycopersicon 13
Lycopersicon esculentum see Tomatoes

Maize (*Zea mays*) 11, 41, 228, 235
Malus 32, 124, 302-303
Malus domestica see Apples
Mangifera indica see Mangoes
Mangoes (*Mangifera indica*) 99, 102, 104
Manihot esculenta see Cassava
Mannanase 331-332
Matriconditioning 35, 37, 39
Maturity induction point 172-173
Medicago sativa see Alfalfa
Mesembrianthemum 5
Mesquite plants 13
Metasequoia glyptostorbioides 303
Michaelis-Menton kinetics 352
Mitotic index 173-174, 194-195, 198
Mixed cropping 41
Models, experimental design 330-333
Moisture 194-198, 253, 283-291, 294-297, 322-329
 stress 194-198
Molecular modelling 214
Monkeys 13
Monterey pine *see Pinus radiata*
Mung beans 235

NAA (1-Naphthaleneacetic acid) 119, 138
NAD 158, 162, 166
NAD kinase 158-163, 165
NADH dehydrogenase 235-236
NADP 158-160, 162-163, 166
1-Naphthaleneacetic acid *see* NAA
Nitrate, activity, germination 342-349, 351, 353-357

Nitrogen cycle 301-309
Nitrogen dioxide 259
Non-adenylic nucleotides *see* NTP
Norway spruce *see Picea abies*
NTP (Non-adenylic nucleotides) 94-98, 100-103

Oaks *see Quercus*
Oats 40, 235
Olea europaea see Olives
Olives (*Olea europaea*) 49, 51
Onions 37
Onoclea sensibilis 71
Oranges (*Citrus sinensis*) 49, 51
Orchids 11
Orobanche 11
Oryza sativa 257, 261-263, 288
Osmoconditioning 35, 37-38
Osmotic effects 11, 19, 24, 33, 35, 37-38, 118, 128, 276

Paclobutrazol 33, 39, 41, 121
Panax ginseng 32
Paradormancy 48, 62-63, 77-78, 83, 87, 89-92, 94, 96-98, 102-103, 106, 201-202, 233, 269
PCR *see* Polymerase chain reaction
Peaches (*Prunus persica*) 49, 51-52, 96, 203, 205, 207-208, 234, 238-239, 303, 362-365, 368
Pears 122
Peas 165, 221, 228
Pectis 10
Pentose phosphate pathway 158, 163, 205, 245
Peppers 32-34, 37-38
Perennial goosefoot *see Chenopodium bonus-henricus*
Perennial ryegrass *see Lolium perenne*
Periodic growth 98, 100-102, 105-106
Periwinkle *see Catharanthus roseus*
Peroxidases 277-278
Pfitzer juniper *see Juniperus chinensis* 'Pfitzerana'
pH effects 258-261
Phalaris arundinacea 41
Phaseic acid 218

Phaseolus 19
Phaseolus vulgaris 158
Phleum pratense 41
Phosphatases 286, 289, 355
Photoperiodism 4-7, 14, 47, 50-51,
    91-92, 118, 126-128, 147-148,
    150-151, 172-173, 175,
    179-180, 185-186, 194-199,
    273-274, 301-309
Phytic acid 276
Phytoalexins 135
Phytochrome 3, 6, 29, 90, 157, 159,
    161, 163-166, 318, 322,
    342-345, 347, 353-357
Phytophthora infestans 135
Picea 172, 174, 303
Picea abies 176, 185-186
Picea engelmannii 176, 179
Picea glauca 176, 178-180, 185-186
Picea mariana 176, 180, 185-186
Picea pungens 175, 185
Picea sitchensis 179
Pigweed *see Spergula*
Pines *see Pinus*
Pinitol 276-277
Pinus 53, 172, 273
Pinus banksiana 9
Pinus cembra 176
Pinus elliottii 185
Pinus lambertiana 246
Pinus ponderosa 179
Pinus radiata 185
Pinus strobus 303
Pinus sylvestris 180, 185, 303
Pinus taeda 185, 246
Plantago major 39
Plasma membrane 233-243, 357
Plasmalemma 233-243, 259
Plasmodesmata 67-72, 74-77
    primary 71-72
    secondary 70-72
    ultrastructure 68-70
Pollen 166
Pollution 205
Polyethylene glycol 31, 323
Polygonum pensylvanicum 39
Polymerase chain reaction (PCR) 136,
    295

Ponderosa pine *see Pinus ponderosa*
Poplars *see Populus*
Populus 203, 207, 302-305
Populus deltoides 303-304
Populus nigra 203
Populus tremuloides 304
Populus trichocarpa 304
Populus × *canadensis* 203
Populus × *euramericana* 303
Portulacca oleracea 354
Potatoes (*Solanum tuberosum*) 49, 64,
    66, 68, 70, 72, 74-75, 133-138,
    140
    postharvest losses 133
Presoaking 38-39
Prolepsis 106
Proteins, stress induced 206-208
Prunus 53
Prunus avium see Cherries
Prunus cerasus see Cherries
Prunus mahaleb see Cherries
Prunus persica see Peaches
Pseudotsuga menziesii 174, 176-177,
    179, 181, 185-186, 303
Pseudotsuga menziesii var. *glauca* 174
Pseudotsuga menziesii var. *menziesii*
    194
Pyrus 51-52
Pyrus calleryana 51

Quantal data 316, 329, 333-335
    interpretation 333-335
Quercus 96, 98-102, 104, 302
Quercus mirbeckii 104
Quercus pedunculata 104
Quercus robur 99-101
Quercus rubra 303
Quercus suber 104
Quiescence 31, 48, 135, 173-174,
    179-180

Radishes 158

Ragweed *see Ambrosia*
Rape 217, 221
Receptor occupancy theory 353-354
Receptor regulated dormancy cycling model 354-357
Reciprocal model 319-320
Red clover *see Trifolium pratense*
Red fescue *see Festuca rubra*
Red light 17, 47, 118, 159, 161-162, 164-165, 316-318, 343, 345, 348-349
Red rice *see Oryza sativa*
Red-osier dogwood *see Cornus sericea*
Relative stress resistance 177
Reptiles 13
Resistance to germination 318-319
Response thresholds 314-325, 333-335
Rhamnus frangula 88
Rhododendron 205
Rhus laurina 9
Rice 4-5, 257
Ricin D 21
RNA 21, 135, 275-277, 284-289, 293-299, 304-305
Robinia pseudoacacia 51, 303
Root growth potential 178-179, 183
Roots 134, 152, 171-172, 174, 177-186, 216, 301
   dormancy 177-180
Rosa 127
Rumex crispus 31-32, 39, 354
Rye (*Secale cereale*) 288, 362-363

Salix microstachya 303
Salix × *smithiana* 303
Sambucus nigra 303
Scots pine *see Pinus sylvestris*
Sea urchins 259
Seasonal synchrony 4-6, 39, 302, 341-342
Secale cereale *see* Rye
Seed water potential 322-329, 334, 347
Seeds, dimorphism 7-8
   dispersal 4, 12-14
     dormancy 1-56, 105, 115, 121-123, 127, 157, 160, 163, 213, 221-231, 245-265, 283-291, 293-300, 313-339, 341-360
   annual cycle 39, 341-360
   Avena fatua 293-300
   cereals 283-291
   control 29-45
   grasses 283-291
   mechanisms 283-291
   models 313-339, 341-360
     population-based 313-339
   molecular genetics 293-300
   natural history 3-16
   physiology 47-56
   reversible sensitivity 341-360
   woody plants 49-53
  embryo growth potential 31-33
  production 147, 150
  survival 6-9
Seed germination
  deferment 8
  failure 17-27
  inhibitors 21-22, 24
  molecular genetics 19-20, 24
  regulation 343
Sequoiadendron giganteum 303
Sesquiterpenes 135
Setaria lutescens 246
Sine-exponential model 367
Sisymbrium officinale 341, 343-351, 353-354, 356-357
Sitka spruce *see Picea sitchensis*
Smartweed *see Polygonum pensylvanicum*
Solanaceae 13
Solanum tuberosum *see* Potatoes
Sophora japonica 303
Sorghum 11
Source-sink relations 95-97, 100-101, 233, 236, 239-241, 302, 305-306
Soybeans 221, 305
Spergula 8
Spirodela polyrrhiza 246, 269-281
  turion formation 269-281
Spruce *see Picea*
Storage proteins 302-306
Stress resistance 171-192
Stress, near lethal 201-210

Striga 11
Subterranean clover *see Trifolium subterraneum*
Sugar pine *see Pinus lambertiana*
Sulphydryl compounds 135
Sunflowers (*Helianthus annuus*) 221-231
Sweet potatoes (*Ipomoea batatas*) 134
Sycamores *see Acer pseudoplatanus*
Syllepsis 106
Symplasmic networking 59-81, 95-96

Taro *see Colocasia esculenta*
Taxodium distichum 303
Taxus cuspidata 180
Temperature effects 4-6, 32-33, 36-39, 47, 49-50, 90-91, 99, 102-103, 115, 118, 121-124, 128, 134, 147-154, 157-169, 171-186, 201-205, 272-277, 319-324, 344-345, 355-356, 362-373
Temperature resistance 174-177, 180-184
Temperature response curves 362-365, 368, 370
Terminalia 102
Tetcyclacis 31, 33-35, 37-41, 356
Thermodormancy 38-39, 326
Thidiazuron 96, 119
Thiourea 12
Thlaspi arvense 148-151, 153
Thrushes 14
Tilia americana 303
Tillage 9-10, 14, 341
Tissue culture 116, 118, 124-128
Tobacco 166
Tomatoes (*Lycopersicon esculentum*) 32-33, 37, 53, 315-316, 324-325, 331-332
Tortoises 13
Total heat units 374-375
Total photosynthesis units 373
Trifolium incarnatum 41
Trifolium pratense 41
Trifolium repens 41
Trifolium subterraneum 41
Triticum aestivum see Wheat

Tsuga heterophylla 174, 185
Tubers xix, 15, 49, 64, 72, 74-75, 77, 84, 94-96, 99, 133-143, 234-241, 259
  dormancy 15, 133-143
    cell biology 135-137
    hormonal regulation 137-140
  sprouting 49, 75, 77, 94, 133-134, 136-140
  nucleotide metabolism 94
Tulips 128
Tunica-corpus layer model 65-66, 70, 74-75
Turions 269-281

Uniconazol 33, 41
Utah chill unit model 362-372

Vaccinium ashei 207
Vaccinium corymbosum 207
Vernalization 147-155, 362-363
  models 152-154
  molecular genetics 152-154
  physiology 149
Vicia villosa 41
Viruses 69, 72
Vivipary 17, 158

Walnuts *see Juglans regia*
Water availability 91-92
Weak acids 258-262
Weeds 9, 39-40, 257, 283, 341, 355
  control 257
Western hemlock *see Tsuga heterophylla*
Wheat (*Triticum aestivum*) 22, 148, 150, 158, 213-222, 283-289, 356
  embryos 213-220
White clover *see Trifolium repens*
White spruce *see Picea glauca*
Whole tree net photosynthesis 372-373
Wild oats *see Avena fatua*
Wild rice *see Zizania palustris* var. *interior*

Woody plants 49-53, 83-113, 125, 201-210, 269-270, 272, 301-309

Xanthium strumarium 48, 260

Yams (*Dioscorea*) 49, 134-135, 137, 139-140

Yeasts 276

Zea mays see Maize
Zeatin riboside 138
Zizania palustris var. *interior* 326
Zonation model 65-66, 70, 76